FLORA & FAUNA HANDBOOK No. 10

T0199825

NORTH AMERICAN
PSOCOPTERA
(Insecta)

Flora & Fauna Handbooks

This series of handbooks provides for the publication of book length working tools useful to systematists for the identification of specimens, as a source of ecological and life history information, and for information about the classification of plant and animal taxa. Each book is sequentially numbered, starting with Handbook No. 1, as a continuing series. The books are available on a standing order basis, or singly.

Each book treats a single biological group of organisms (*e.g.*, family, subfamily, single genus, etc.) or the ecology of certain organisms or certain regions. Catalogs and checklists of groups not covered in other series are included in this series.

The books are complete by themselves, not a continuation or supplement to an existing work, or requiring another work in order to use this one. The books are comprehensive, and therefore, of general interest.

The books in this series to date are:

Handbook No. 1, 1985. **THE SEDGE MOTHS,** by John B. Heppner

Handbook No. 2, 1992, 2nd printing, slightly revised. **INSECTS AND PLANTS: Parallel Evolution and Adaptation,** by Pierre Jolivet

Handbook No. 3, 1988. **THE POTATO BEETLES,** by Richard L. Jacques, Jr.

Handbook No. 4, 1988. **THE PREDACEOUS MIDGES OF THE WORLD,** by Willis W. Wirth and William L. Grogan, Jr.

Handbook No. 5, 1990. **A CATALOG OF THE NEOTROPICAL COLLEMBOLA,** by José A. Mari Mutt and Peter F. Bellinger

Handbook No. 6, 1990. **THE ENDANGERED ANIMALS OF THAILAND,** by Stephen R. Humphries and James R. Bain

Handbook No. 7, 1990. **A REVIEW OF THE GENERA OF NEW WORLD MYMARIDAE,** by Carl M. Yoshimoto

Handbook No. 8, 1990. **MAYFLIES OF THE WORLD: A Catalog of the Family and Genus Group Taxa,** by Michael D. Hubbard

Handbook No. 9, 1993. **CATALOG OF THE SOFT SCALE INSECTS OF THE WORLD,** by Yair Ben-Dov

Handbook No. 10, 1993. **NORTH AMERICAN PSOCOPTERA,** by Edward L. Mockford

Handbook No. 11, 2nd Edition, 1993. **INSECT AND SPIDER COLLECTIONS OF THE WORLD,** by Ross H. Arnett, Jr. and G. Allan Samuelson

Handbook No. 12, 1993. **SPIDERS OF PANAMA** by W. Nentwig

FLORA & FAUNA HANDBOOK No. 10

NORTH AMERICAN
PSOCOPTERA
(Insecta)

by

Edward L. Mockford

Illinois State University

Normal, Illinois

CRC Press
Taylor & Francis Group
Boca Raton London New York

CRC Press is an imprint of the
Taylor & Francis Group, an **informa** business

Published 1993 by CRC Press, Inc.
Taylor & Francis Group
6000 Broken Sound Parkway NW, Suite 300
Boca Raton, FL 33487-2742

©1993 by Taylor & Francis Group, LLC
CRC Press is an imprint of Taylor & Francis Group, an Informa business

First issued in paperback 2019

No claim to original U.S. Government works

ISBN 13: 978-0-367-45009-0 (pbk)
ISBN 13: 978-1-877743-12-2 (hbk)

Visit the Taylor & Francis Web site at
http://www.taylorandfrancis.com

and the CRC Press Web site at
http://www.crcpress.com

TABLE OF CONTENTS

Preface vii

Acknowledgements x

Glossary xii

Chapter 1. Order Psocoptera 1
   a) General Information 1
   b) Biology 1
   c) Classification 8
   d) Synonymy and Diagnosis 8

Chapter 2. Suborder Trogiomorpha 15

Chapter 3. Suborder Troctomorpha 56

Chapter 4. Suborder Psocomorpha 104

Chapter 5. Distribution Patterns 312

Literature Cited 322

Illustrations 346

Index 448

PREFACE

This book treats the fauna of the insect order Psocoptera for the United States and Canada. It would be more desirable to treat a natural region of the world than a political one, but inclusion of Mexico at this time is impossible. Despite great effort, most of the bountiful Mexican psocid fauna remains undescribed. Where the term "North America" is used in the text that follows, only the U.S. and Canadian portions are meant.

The book contains keys to all of the known (i.e., described) taxa of Psocoptera which have been found in the study area, including three genera named as new. Not only are the native and established species included, but also those which have been taken at ports of entry in human commerce. It contains differential diagnoses of the taxa above species level. For each named species there is an account consisting of synonymy, recognition features, relationships, distribution, and habitat.

The synonymies are, for the most part, taken from the literature. Several new synonymies have been found. Some doubtful synonymies have been re-investigated. Trinomials which have been treated as synonyms are, for the most part, omitted, as it is possible that some of them represent valid races. Over the years, I have attempted to examine primary types whenever I encountered doubt about a name. I have examined several in preparation of this book, but still a few have not been seen.

The sections on recognition features are based entirely on original observations except in a few instances where it was not possible for me to examine representitives of a species. In those instances, recognition features are taken from literature accounts. The recognition features do not give a complete description of a species but include only the characters which one needs in order to establish a sound confirmation of an identification, i.e., primarily characters which place a species in a smaller group within its genus and characters unique to the species.

The section on relationships are brief statements indicating probable phylogenetic alignment of the species with some other species or group of species. Such information is of interest of its own account, may be of biogeographical use, and may be of use in helping to confirm an identification. In some cases it was not possible to make a reasonable guess at relationships.

The sections on distribution and habitat attempt to rough out the region of 'continuous' distribution of a species, i.e., the region where, given the right set of environmental conditions, one is likely to find the species. A number of distribution patterns are repeated many times, and the final chapter in the book is devoted to an examination of the biogeographic data. Unfortunately, in a group as poorly studied as the Psocoptera, the distribution data still mark primarily the distributions of collectors. Thus, such centers of learning as Ithaca, New York, Gainesville, Florida, and Normal, Illinois, and their environs are frequently corners of the continuous distributions. Outlier populations have sometimes been found and are noted in the distributions sections. In cases where a species occurs both within and outside the study area, its total distribution is noted. Western Hemisphere distributions outside the study area are primarily from my own published records, also from papers by Badonnel (1962, 1963, 1968, 1971b, 1971c, 1972, 1978, 1979, 1986c, 1986d), Badonnel and García Aldrete (1980), and the listed works of García Aldrete. Eastern Hemisphere distributions are primarily from Smithers (1967) and Günther (1974a); other sources are usually cited in the text.

The sections on distribution and habitat also include notes on habitat, i.e., the kind of plants and the type of plant material (dead leaves, live leaves, branches, trunks, etc.) or other substrate on which the species has been collected.

It is my hope that enough information has been included in the diagnoses and species accounts to allow the incipient student or investigator to verify or refute identifications arrived at by use of the keys, and also to pursue the study of any particular species. Only adults are treated. This acknowledged bias results from scarcity of information on the immature spages. Only named species are included, except for the brief mention in a few instances of an undescribed species where that is the sole representative of its genus in the study area. The un-named species from the study area constitute a sizeable percent, perhaps as much as 25 percent, of the total psocid fauna of the study area. It would be prohibitive in terms of time and space to try to include descriptions of these new species in the present work.

It is likely that a student or investigator using this book will encounter an undescribed species, hence one not included in the book. Is this book, then, premature? I have thought so in the recent past, but several colleagues have convinced me that a comprehensive work, even at this relatively early stage of our knowledge, is important. The Europeans have long had the advantage of the work by Badonnel (1943) on the French fauna, that by Günther (1974a) on the German fauna, and that by New (1974) on the British fauna. There has been no comprehensive volume comparable to these works available for North America. It has been suggested that this is one of the reasons why students have tended to avoid working on this group in North America.

The three chapters of key and species accounts, corresponding to the three suborders of Psocoptera, are preceded by a chapter about the order in general (chapter 1). Chapter 1 treats the biology and includes a diagnosis of the order, which contains explanations of the terms used in the keys and species accounts. A few special terms are introduced subsequently, where needed.

The keys are in some cases new and untried. This is true of the keys to species groups (combined with keys to species) in the large genera Caecilius and Lachesilla. In some cases one must know both sexes and have them at hand in order to negociate the key correctly. The key to the common booklice, genus *Liposcelis*, is based entirely on females. Males are scarce, and information on them remains too limited to allow their proper systematic treatment.

Illustrations are primarily original for this volume, but some have been used in my previous publications and a few for which I had no specimens, were re-drawn from other (cited) sources. All original drawings were done either with camera lucida or micro-projector. Surface lines are usually solid (dotted where very fine), while sub-surface lines are dashed except where two surfaces are essentially fused, as in the hypandrium and underlying phallosome of some psyllipsocids and some species of *Lachesil-*

la. In these cases, all lines are usually solid. Most figures are oriented so that structures directed forward on the insect point upward while those directed backward on the insect point downward. In a few cases internal structures on slide preparations, the orientation on the insect remains unknown to me.

ACKNOWLEDGEMENTS

This book is a product of studies begun in the collecting season of 1944. Many people have aided and encourged me in this project during the 48 intervening years, and limitations on space, time, and memory allow me to name only a few. My early collecing efforts were greatly aided by Mr. D.W. Rice, Dr. W.E. Ricker, and Dr. F.N. Young. My knowledge of psocidology was much enhanced in its early stages by correspondence with colleagues, especially Dr. A.B. Gurney (dec.), Mr. J.V. Pearman (dec.), and Dr. K.M. Sommerman. Dr. L.E. Berner, my major professor on my master's work, was an invaluable source of ideas. During nearly two years while I was stationed in the U.S. Army at Fort Detrick, Maryland, Dr. Gurney opened his laboratory at the National Museum to me, and many weekends were spent there in pursuit of psocidology. Dr. H.H. Ross (dec.), my major professor on my doctoral work, provided valuable stimulation, sage advice, and a good foundation in the art and science of classification. At Illinois State University, Dr. R.O. Rilett, head of the Department of Biological Sciences at the time of my arrival, provided incentive and a suitable environment for research. Recent correspondence and conferences with colleagues, especially Dr. A. Badonnel (dec.), Dr. C. Lienhard, and Dr. C.N. Smithers, and with my former and present students, especially Dr. G.E. Eertmoed, Dr. A.N. García Aldrete, Dr. B.W. Betz, Dr. C.M. Schmidt, Mr. D.J. Schmidt, and Mr. D.M. Sullivan, have continued to broaden my outlook and deepen my insight in psocidology. Finally, the faculty and chairman of the Department of Biological Sciences at Illinois State University have been exceedingly generous in allowing me to use laboratory spece and facilities long after my retirement.

I thank the custodians of collections at the following institutions for the privilege of examining material: American Museum of Natural History (New York City), California Academy of Science (San Francisco), Canadian National Collection (Ottawa), Cornell University (Ithaca, New York), Field Museum (Chicago), Florida State Collection of Arthropods (Gainesville, Florida), Illinois Natural History Survey (Champaign, Illinois), Michigan State University (East Lansing, Michigan), Museum of Comparative Zoology (Cambridge, Massachusetts), Philadelphia Academy of Natural Science, United States National Museum (Washington, D.C.), University of California at Davis, and University of California at Riverside.

Glossary

Aedeagal arch. Sclerotic arch closing distally the Phallosome (principal male copulatory structure) and formed by fusion of external parameres.

Agamic (populations). Populations which reproduce asexually.

Anal vein(s). The most posterior vein(s) of the wing. In Psocoptera, two anal veins in the forewing are regarded as primitive state and one as a derived state.

Anteclypeus. In Psocoptera, a slender, transverse sclerite lying between the labrum and the large, bulging postclypeus.

Aptery. Adult condition of complete absence of wings, as in all adult *Liposcelis*.

Areola postica. Cell of hind margin of the forewing delimited by veins Cu1a and Cu1b.

Areoles. Regular, repeating geometric patterns in cuticle delimited by depressed lines or breaks between groups of nodules.

Baguettes. Term used by Badonnel (1931) to characterize certain short, cylindrical setae of the anal lobes in some Psyllipsocidae.

Beating. Mode of collecting by striking branches and foliage over a sheet or umbrella.

Berlese sample. A sample of specimens taken by extraction with a Berlese funnel, usually from soil or leaf litter.

Biotope. A specific stand of a habitat type.

Bi-seasonal. With population peaks in two seasons of the year.

Brachelytroptery. Possessing short, leathery winglets similar to short elytra.

Brachyptery. Possession of short wings, probably not suitable for flight, but not extremely short.

Bristle. A stout seta, usually blunt-tipped.

Chaetotaxy. Description of the arrangement of setae.

Cibarium. Region of the base of the buccal cavity usually functioning as a pump. In Psocoptera the floor is a heavy sclerite (sitophore sclerite) in the base of the hypopharynx bearing a median indentation which receives a rod-shaped structure from the roof operated by muscles originating on the large postclypeus. The apparatus extracts water from the lingual sclerites via the hypopharyngeal filaments (Rudolph, 1982b).

Ciliation. Synonym of Chaetotaxy.

Claspers. Paired projection arising one on each side of the hypandrium in the psocid genera Lachesilla and Trichadenotecnum.

Clunium. Fused and well sclerotized abdominal terga 8-10 or 9-10 in Psocoptera.

Clunial comb. Ornamentation, often consisting of a row of spines or teeth along the hind margin of the clunium in male psocids.

Clypeo-frontal suture. Suture separating the postclypeus from the frons in Psocids.

Coeloconic sensillum. A minute sense organ consisting of a pit containing a largely concealed, thin-walled hair.

Copeognatha. Earlier name for Order Psocoptera.

Coriaceous. Of a leathery consistency, pertaining to forewings.

Corrodentia. Earlier name for Order Psocoptera.

Cosmopolitan. As used here, of very wide distribution, usually involving several continents.

Coxal organ. An organ located on the inner face of each hind coxa in many adult psocids, consisting of a small tympanum ("mirror") and adjacent rasp. The rasps of opposite sides are thought to produce a sound by rubbing together, augmented by the mirrors.

Cretaceous. Last period of the Mesozoic Era, ending about 80 MYBP, from which several essentially modern psocid fossils are known.

Cubital loop. A loop formed by vein Cu1a in the forewing of many psocids.

Detritivore. As used here a feeder on small particles of organic debris.

Diapause. A programmed interruption of development.

Dimorphism. As used here, the existence of two different body forms of adults. The difference may be between sexes or within one sex. It may involve body size, or wing development, or some other feature.

Diploid. Possessing a single complete set of chromosomes of each parental type.

Discoidal cell. A cell in the forewing of many psocids closed by the junction of the cubital loop with vein M.

Domestic. Used here to mean species found in houses.

Dotted areas. Areas of the vertex which bear rounded, brown spots, probably marking points of muscle attachment.

Ecdysial lines. Lines of weakness in the cuticle along which the cuticle ruptures during ecdysis.

Ecdysis. Phase of molting in which the old cuticle is cast off.

Eclosion. Hatching from the egg.

Elytriform. In the form of a beetle's forewing (elytron), as seen in forewings of some psocids.

Endophallus. A membranous sac often with spines or denticles on its inner surface, located distally in the phallosome; it is everted during copulation.

xiii

Epiproct. A lobe or flap above the anus.

Epistomal suture. Synonym of clypeo-frontal suture, q.v.

Exopterygote. An insect in which wings develop externally.

Facultative parthenogenesis (or thelytoky). Reproduction without fusion of gametes in an organism which has the capacity for sexual reproduction.

Filiform antenna. Antenna in which the flagellomeres are cylindrical and slender.

Flagellomere. A non-musculated unit of the flagellum, or third antennal metamere.

Foliate claw. A broad, flattened pretarsal claw.

Frons. Region of the front of the head lying immediately above the postclypeus.

Fronto-clypeal region. Region formed by fusion of frons and postclypeus.

Fuscous. As used here, a dark brown color.

Galea. Distal outer lobe of the maxilla.

Gonapophyses. Copulatory and egg-laying appendages of the female eighth and ninth abdominal segments. See also ovipositor valvulae.

Harvestmen. Arachnids of the Order Opiliones.

Homonyms. Names in biological nomenclature spelled the same but referring to different taxa.

Hypandrium. As used here, in male psocids, a plate formed by the ninth abdominal sternum, or that plus one or more preceding sterna, underlying the phallosome (q.v.).

Hypopharyngeal filaments. A pair of sclerotized tubes, one from each lingual sclerite to the sitophore sclerite, or the two joining together before reaching the sitiphore sclerite.

Incisor region. A group of tooth-like or knife-blade-like projections located on the distal, inner margin of the mandible.

Instar. An individual between molts.

Interception. As used here, a capture at a port of entry, usually on material being brought into the country.

IO/d. Index of least distance between compound eyes in front view divided by horizontal diameter of an eye in same view.

Jurassic. Middle period of the Mesozoic Era, ending about 135MYBP, from which the earliest definite psocid fossil is known.

Jute. Coarse cloth made of fiber of a tiliaceous herb, *Corchorus* sp.

Labral sensilla (distal inner). A row of minute sense organs, usually consisting of placoids and trichoids, on the inner surface of the labrum near its distal margin.

Labral stylets. A pair of short projections located disto-laterally on the labrum in some psocids.

Lacinia. Median distal process of the maxilla. In Psocoptera, it has become a long, slender rod.

Lingual sclerites. A pair of ovoid sclerites located distally on the hypopharynx in Psocoptera. Their ventral surfaces can capture water molecules from the atmosphere at high relative humidity.

Macroptery. Possession of long wings; the normal adult condition in Psocoptera.

Mesepisternal suture. A suture running antero-posteriorly across the mesepisternum in some psocids.

Meso-metathorax. The fused mesothorax and metathorax in some psocids.

Meso-precoxal bridge. A sclerotic connection between mesothoracic pleuron and sternum anterior to the coxa.

Meso-trochantin. A process from the meso-pleuron running between the precoxal bridge and the coxa, providing the ventral articular point for the coxa.

Microptery. Possession by adults of extremely short wings.

Microspades. A group of minute organs, presumably sensory, which somewhat resemble short-handled spades, located on the antennal pedicel in some Psyllipsocidae.

Mirror. A minute tympanum on each hind coxa in many adult psocids. The two mirrors presumably reflect and amplify the sound produced by the rasps located adjacent to them.

Molar region. Basal medially-bulging region of mandible with ridged or denticulate surface.

Mymaridae. Family of minute wasps, some of which parasitize psocid eggs.

Neoteny. Possession of immature characters in an adult.

Nodulus. A point on the hind margin of the forewing in Psocomorpha and Psyllipsocidae at which veins Cu2 and 1A join. The in-flight wing coupler is located there.

Obligate parthenogenesis (or thelytoky). Parthenogenesis in a species (or population) in which males do not exist.

Ocellar field. In most psocids, a small, circumscribed area including the ocelli.

Ommatidium. The visual unit of a compound eye.

Ovipary. Reproduction by egg laying.

Ovipositor valvulae. Three pairs of reproductive appendages on female abdominal segments 8 and 9. The first valvula (v1) arises on segment 8, and the other two (v2 and v3) on segment 9. Some authors refer to these as the ventral (~v1), dorsal (~v2) and lateral (~v3) valvulae.

Oviruptor. A tooth-like, knife-like, or spinous structure on the head of the pronymph which is pushed against the egg chorion, facilitating its rupture during hatching.

Paralectotype. A specimen accompanying, and of the same species as a lectotype.

Paramere. One of a pair of rami forming the sclerotized distal end of the phallosome or principal male copulatory organ.

Paraproct. One of a pair of lobes lateral to the anus.

Parasitoid. A predatory organism which slowly consumes its living host or prey during its growth period.

Parthenogenesis. Reproduction without agency of sex, i.e., without fusion of gametes.

Pedicel. The second metamere, or musculated unit, of an antenna.

Phallosome. The principal male copulatory organ in psocids.

Phoresy. A relationship between two kinds of organisms in which a smaller one is transported on the body of a larger one.

Placoid sensillum. A minute, rounded sense organ with flattened or slightly depressed surface.

Polymorphism. Term used here to mean the existence of several forms of adults of the same species. In psocids, this usually involves several levels of wing development (such as macroptery, brachyptery, and microptery), with associated differences in development of thoracic notal lobes and ocelli.

Postclypeus. The bulging upper region of the clypeus housing the cibarial dilator muscles.

Preclunial region. Portion of the abdomen lying before the clunium, i.e., segments 1-7 or 1-8, usually with weakly-sclerotized cuticle.

Pretarsal claws. Pair of claws on the pretarsus, or terminal metamere of the leg, at the distal end of the tarsus.

Protein-electrophoresis. Process by which proteins are separated by passing them through an electrical field on a gel.

Psocina. Old name for the group comprising Psocoptera.

Pterostigma. A cell on the anterior margin of the forewing delimited by veins Sc and R1. In most Psocomorpha it is somewhat sclerotized.

Pulvillus. Term used in the Psocoptera for a non-setiform ventral appendage of the pretarsal claw.

Radula. Denticulate or spinose region of the endophallus, the eversible inner pouch of the phallosome.

Rasp. A denticulate field on the hind coxa of many adult psocids. The rasps of opposing coxae are thought to be scraped together to produce a sound.

Scape. The basal metamere, or musculated unit, of an antenna.

Sculpturing. Any sort of design visible in the cuticle.

Semidomestic. Referring to habitats associated with house, but not indoors, such as birds' nest under eves.

Sense cushion. A region on the paraproct consisting of closely set minute sense organs, each composed of a basal floret or rosette surrounding a trichobothrium, q.v.

Sitophore sclerite. The sclerite of the floor of the cibarium, q.v.

Spermapore. Opening of the spermathecal duct on the ninth abdominal sternum, the roof of the genital chamber in psocids.

Spermatophore. A capsule enclosing sperms, formed in the male reproductive system and used to convey sperms to the female's genital chamber. The spermatophore never becomes lodged in the spermatheca, as some accounts suggest.

Stridulation. A sound produced by scraping two rough surfaces together or scraping a sclerotized projection over a rough surface.

Struts. A pair of sclerotized rods, completely separate or fused basally, forming the skeletal base of the phallosome in Psocoptera.

Subgenital plate. A plate formed by sclerotization of the eighth abdominal sternum in female psocids, underlying the opening of the genital chamber.

Superspecies. A taxon consisting of a set of closely related, very similar species.

Syntype. A specimen forming part of a type series in which no holotype has been designated.

Tarsomere. A non-segmental division of the penultimate leg segment or tarsus.

Tenerality. Condition of cuticular softness and lack of pigmentation following a molt.

Tertiary. Earlier of the two periods of the Coenozoic Era, ending about 5 MYBP.

Thelytoky. Form of parthenogenesis (q.v.) in which only females are produced.

Trichobothrium. A slender, elongate seta, probably innervated. In psocids trichobothria usually arise from a specialized base; those of the paraproctal sense cushion each arises in the center of a "basal rosette", of which the 'petals' are depressed areas in the cuticle.

Trichoid sensillum. A minute sense organ consisting of a short, thin-walled sensory hair.

T-shaped sclerite. A sclerite of unknown function, in form of a T, located on the subgenital plate in several genera of Troctomorpha.

Ventral abdominal vesicles. Inflatable sacs apparently formed as expansions of intersegmental membranes on the ventral surface of the abdomen in several psocid taxa. They are apparently capable of secreting a sticky substance, enabling the insect to adhere to a surface.

Vivipary. Bearing live young in a relatively advanced state of development. In psocids the phenomenon is restricted to very few taxa, and the young emerge as typical first instars.

Chapter 1
Order Psocoptera

a) General Information

The insects of the order Psocoptera are commonly called psocids. Out-door forms on tree trunks and branches have been called bark lice, and in-door forms, sometimes found in old books, have been called book lice. A detailed description of these insects is given in the diagnosis. Suffice it to say here that they are small, usually soft bodied. The body and forewings are commonly shades of brown or gray. Some domestic and cave-dwelling species have no pigment in the body wall, the muscle and fat of the subcuticular layer showing through the cuticle as creamy white. The unique mouthparts (see diagnosis) set this order apart and indicate a close relationship to the Mallophaga, hence, probably in turn to the Anoplura. These three 'louse' orders are probably hemipteroids, as they share with the other hemipteroid orders a low number of tarsomeres — not more than three —, absence of cerci, and a well developed cibarial pump apparatus, which in psocids and Mallophaga appears to be involved in water uptake from the atmosphere (Rudolph, 1982a,b).

b) Biology

The biology of the Psocoptera was recently reviewed by New (1987). The following account adds information from personal observations, where possible, and stresses aspects peculiar to the North American fauna.

Fossil insects identified as psocids have been reported from the Permian (Carpenter, 1926, 1932, 1933, 1939, Tillyard, 1926, 1928, 1935, 1937). These are principally wing impressions in fine-grained substrates. Although they are quite psocid-like, the fact that small size tends to correlate with simple venation makes one shun a rigorous interpretation of these as psocids. Indeed, Vishniakova (1981) showed that the Lophioneuridae are thrips, and Carpenter's (1933) reconstructions of *Dichentomum tictum* Tillyard suggests that the insect is not even a hemipteroid, although it may be near the ancestry of the group. Even the Lower Jurassic Archipsyllidae continued to possess four tarsomeres (Vishniakova, 1976), so that the case for their being psocids remains doubtful. None of the middle-Jurassic forms assigned to Psocoptera by Hong (1983) (a paper in which homonyms were created on the genera *Mesopsocus* Kolbe, *Parapsocus* Scudder, *Pseudopsocus* Kolbe, and *Trichopsocus* Kolbe) appear to be psocids beyond reasonable doubt, although his '*Trichopsocus*' may be. Badonnel (*in litt.*) mentioned a Jurassic fossil wing with Caeciliid-like venation which he felt was definitely a psocid. Other than this, the earliest unquestionable fossil psocids known to me are those of Cretaceous amber, including those described by Vishniakova (1975) from Taimyr Peninsula, USSR, and those awaiting description from Cedar Lake, Manitoba. These Cretaceous forms are essentially modern in appearance and do not show any clear connections to the older fossils of doubtful ordinal assignment.

The life cycle of most psocids involves an adult period of sexual inactivity including, but slightly longer than, the teneral period; courtship and copulation (often only once in females, several times in males); oviposition; eclosion and four to six nymphal instars.

Courtship and copulation have been described by Pearman (1928d), Sommerman (1943b, 1943c, 1944, 1956a), Badonnel (1951), Broadhead (1952), Thornton and Broadhead (1954), Klier (1956), Mockford (1957a, 1977b), Broadhead and Wapshere (1966), Eertmoed (1966), and Betz (1983a). The works by Badonnel (1951) and Klier (1956) compare examples from several higher taxa. Broadhead (1952) and Mockford (1957a) compare closely related species (of Liposcelis and Archipsocidae respectively). Betz (1983a) presented a detailed study of *Trichadenotecnum alexanderae* Sommerman. He included a summary of evidence for a sex-attractant pheromone and noted a sideways gait of the male during courtship, speculating that this gait involves stridulation with the coxal organs. Passage of a solid spermatophore was noted by several of the above authors, while Klier (1956) observed that sperms are transferred in a solid spermatophore in some taxa and in a liquid medium in others. Mockford (1985a) noted a correlation in family Lachesillidae between rapid copulation and a solid spermatophore versus slow copulation and transfer of sperms in liquid. He speculated that the former was the more primitive state in this family.

Observations on psocid eggs are summarized by New (1987). Pearman (1928b) recognized four categories of egg deposition for British species which seem to apply universally:

1. Eggs laid bare: a) webbed, b) not webbed.
2. Eggs encrusted with material from the digestive tract:
    a) webbed, b) not webbed.

The material covering encrusted eggs appears to be a mixture of particles passed rapidly through the digestive tract and secretions of the digestive tract (pers. obs.). There is a systematic and an ecological relation to the category of egg deposition. Thus, the species of Psocidae and Peripsocidae, which are open bark inhabitants, lay unwebbed encrusted eggs. Leaf inhabiting members of the families Caeciliidae and Pseudocaeciliidae usually lay bare, overwebbed eggs (New, 1987 and pers. obs.).

Vivipary is rare in psocids. In North America it has been observed only in the species of *Archipsocopsis* (Mockford, 1957a).

Eclosion has been described by several authors (Pearman, 1928c, Badonnel, 1951, New, 1987). There is an oviruptor on the anterior end of the pronymphal cuticle in the form of a blade or tooth or group of spines. By rhythmically pushing this structure against the egg cuticle, the pronymph achieves a hatching orifice and pushes partially out. The first instar then emerges from the pronymphal cuticle, leaving that cuticle attached to the hatching orifice.

Nymphs generally resemble adults of their taxon in body form and markings and can, thus, often be identified to species. They lack functional ocelli, never have more than two tarsomeres, and early instars have fewer flagellomeres than adults of their species. There are usually four to six nymphal instars, and, as noted by New (1987) and other authors, the higher number is usually associated with macroptery and the lower number with extreme wing reduction. Broadhead (1947b) observed three nymphal instars in the small, apterous males of *Embidopsocus enderleini* (Ribaga). Nymphs of some species of *Psocomorpha* have gland hairs, short setae with flaring tips. These in some cases retain bits of debris, thus forming a camouflaging coat over the body surface. In other cases, the gland hairs may facilitate body contact among nymphs. Nymphs often aggregate in groups. In North America, nymphs of the species of *Cerastipsocus* form large, dense herds which remain together after adulthood is reached but disperse soon after adults lose tenerality (pers. obs.). Ecdysis was described by Pearman (1928c).

Parthenogenesis in Psocoptera was reviewed by Mockford (1971c). Subsequently, Badonnel and García Aldrete (1980) and Betz (1983b) demonstrated the existence of species complexes involving one sexual species and one or more parthenogenetic sister species. Parthenogenesis in psocids is usually obligate thelytoky. Facultative thelytoky is known only in a few species. Betz (1983c) found that in the facultatively parthenogenetic species *Trichadenotecnum alexanderae*, mating receptivity of virgin females dropped rapidly after the third day of adult life and reached zero on the seventh day. Jostes (1975) and C. Schmidt (unpublished doctoral dissertation) studied chromosome numbers in obligate parthenogenetic species and found them to be usually diploid, but C. Schmidt (ops. cit.) found near-triploidy in *Trichadenotecnum castum* Betz. C. Schmidt (ops. cit.) also studied protein-electrophoresis patterns in the *T. alexanderae* species complex. She found fixation at all loci in each of the obligate parthenogenetic species (but with interspecific differences in pattern), but variation at several loci in the sexual species.

Chromosome numbers in sexual species were treated by Boring (1913), Badonnel (1951), Wong and Thornton (1966), and Meinander et al. (1974). XO sex determination was found in every case. Diploid male chromosome numbers ranged from 14+1 to 18+1 (the lower number incorrectly stated as 4+1 by New, 1987), with a single species exhibiting 28+1. The most frequent chromosome number was 16+1.

Psocid habitats were treated as six series by Mockford (unpublished ms.) This system was adopted by García Aldrete (1990a). The series are 1) living foliage, 2) dead foliage, 3) ground litter, 4) bark of trees, subdivided into a) open-bark surfaces, and b) subcorticolous situations, 5) rock surface, and 6) human habitations (indoors). In a somewhat imprecise manner, the habitat series are related to taxonomic groups. Thus, in the study area, the inhabitants of series 1 are primarily members of families Caeciliidae and Amphipsocidae, with representatives of families Pseudocaeciliidae and Archipsocidae in southeastern Texas, the north Gulf Coast, and much of Florida. The inhabitants of series 2 are primarily lachesillids and ectopsocids, with some psoquillids and hemipsocids in southeastern United States. The inhabitants of series 3 are somewhat complex, and will be treated below. Those of sub-series 4a are primarily members of families Psocidae, Myopsocidae, Peripsocidae, Philotarsidae, and Mesopsocidae while subseries 4b is primarily inhabited by liposcelids. Inhabitants of series 5 are also primarily Psocidae, Myopsocidae, and Peripsocidae (but no Mesopsocidae and rarily Philotarsidae), and in parts of the study area representatives of Amphientomidae and Epipsocidae are included. Inhabitants of series 6 include primarily families Liposcelidae, Troglidae, and Psoquillidae.

Series 3, ground litter, was treated in general by New (1969) and for the study area by García Aldrete (1990a). New (1969) treated leaf-litter dwellers in three ecological groups: 1) primary: species which spend all of their life in litter and are not found in other habitats; 2) secondary: species with generations in two habitats: at least one in litter and one or more others in another habitat; 3) casual: species which do not normally breed in litter. The first category needs no comment. The second consists primarily of species which live on leaves of deciduous trees. Their eggs fall to the ground in autumn and the first generation of the following year is in the litter. The third category consists primarily of individuals which fall to the ground during violent weather. García Aldrete (1990a) recorded 60 species in leaf-litter in continental United States. Of these, 24 were primary, seven were secondary, and the rest casual.

Psocids have been reported from mammal nests by Badonnel (1969), Mockford (1971a), and García Aldrete (1984b,c). For ground-nesting mammals, the psocids are primarily leaf-litter forms, while for a tree-nesting species the psocids are those occurring usually on bark and in dead leaves. Psocids have also been reported from the fur of mammals by Gurney (1950), Pearman (1960), and Badonnel (1969).

Psocids in bird nests have been reported by Hicks (1937), Rapp (1961), Wlodarczyk (1963), Badonnel (1967), Thornton and Wong (1968), Wlodarczyk and Martini (1969), Thornton et al. (1972), New (1972), and García Aldrete (1988b). Most of the species seem to be those likely to be encountered in dead leaves or under bark. Mockford (1967) reported several species of psocids (families Troglidae, Liposcelidae, and Ectopsocidae) from plumage of live birds. He speculated that some species may have become widely distributed by phoresy on migrating birds.

Psocid phenology has been studied by Schneider (1955), Broadhead (1958b), Broadhead and Wapshere (1966), New (1969, 1970), Turner (1974), Turner and Broadhead (1974), Bigot et al. (1983), Wolda and Broadhead (1985), D. Schmidt (unpubl.), and Baz (unpubl.). Within the study area, most species overwinter as eggs in the north, but in southeastern Texas, around the northern Gulf Coast, and throughout most of Florida, many species occur as nymphs and adults throughout the year (pers. obs.). D. Schmidt (unpubl.) found a distinct summer fauna and a distinct winter fauna in the psocids of the Archbold Biological Station in southern Florida, with only a few species occurring bi-seasonally or non-seasonally. The winter fauna was the larger, including all of the species of families Caeciliidae and Psocidae. In southern California, several species have a single generation per year, active in the winter, apparently undergoing a long egg diapause throughout the hot, dry summer (Mockford, 1984b and pers. obs.).

Glinyanaya (1975) and Eertmoed (1978) demonstrated egg diapause in some north-temperate species, regulated primarily by effects of photoperiod on the parental female.

As noted by New (1987) the feeding habits of psocids are quite varied, but in general these insects are herbivores or detritivores feeding on microflora and organic debris on surfaces where they live. A few psocids are partially predators, including insect eggs and possibly scale insects in their diets. In general, the bark-inhabiting Psocomorpha appear to be specialists either on pleurococcine algae or on lichens. Both groups take in some bark material when feeding (pers. obs.). The bark-inhabiting Troglomorpha and Troctomorpha as well as the domestic members of these groups feed on molds and detritus and are often found to have cuticular parts of other psocids in their digestive tracts, presumably ingested dead (pers. obs.). Leaf-inhabiting psocids feed primarily on small leaf fungi and often ingest small amounts of leaf tissue with the fungal hyphae (pers. obs.).

Humidity relationships of psocids have been studied by Knulle and Spadafora (1969), and Rudolph (1982a,b). These authors have shown that psocids are able to absorb water vapor from the atmosphere at high relative humidity, and that this process takes place on the specialized lower surfaces of the lingual sclerites.

Information on parasitoids and predators of psocids was summarized by New (1987), who emphasized the importance of small sphecid wasps as psocid predators. In southeastern United States, *Rhopalum atlanticum* Bohart has been noted to provision its nests with numerous psocids representing eight species in six genera (Kislow and Matthews, 1977). Broadhead (1958a) lists species of harvestmen, spiders, mites, hemerobiid larvae, and carabid larvae found with psocid prey in England. Salticid spiders are sometimes seen to stalk psocids on the beating sheet and lose interest in their prey or reject it after capture (pers. obs.). Recently, R. Pape has sent me a photograph of a pseudoscorpion, *Albiorix* sp. feeding on a live psocid, probably *Psyllipsocus* sp., in an Arizona cave. Assassin bugs of the genus *Empicoris* are frequently found on the surfaces of webs of *Archipsocus* spp. in Florida (Mockford, 1957a). As noted by New (1987), the eggs of several species of psocids in North America are parasitized by mymarid wasps of the genus *Alaptus*. Also, nymphal psocomorph psocids of several genera are parasitized in North America by braconid wasps of the genus *Euphoriella*. The psocids involved are usually of medium to larger size, but small species of *Archipsocus* in Florida are sometimes parasitized by an unknown wasp. These psocids often reach adulthood, but their external genitalia are usually modified, so that superficial observation would suggest a distinct species (pers. obs.).

Records of psocids as prey of vertebrates are few. Several kinds of small birds are known to feed on psocids (Palmgren, 1932; Betts, 1956; Turner, 1979). Chapman (1930) recorded a specimen of *Caecilius 'aurantiacus'* (=*flavidus*) from the stomach of a tree frog.

Wing polymorphism in psocids was reviewed by Mockford (1965d) and New (1987). Its manifestations are discussed briefly below in the diagnosis. Unfortunately, this interesting aspect of psocid biology has not yet received the detailed study which it deserves.

In general, psocids are of little economic importance to man, but they may reach pest proportions in households, stored products, and collections of dried insect specimens. Turner (1987) and Turner and Maude-Roxby (1987, 1988) have studied aspects of *Liposcelis bostrychophilus* as a household and stored products pest. Mockford (1991) reported 50 species of psocids occurring in households and stored products. The dead bodies of some domestic psocids in household dust are thought to contribute to asthmatic attacks (Spieksma and Smits, 1975). Allen (1973) found several species of psocids to be probable intermediate hosts of the fringed tapeworm of sheep.

c) Classification

There have been several classifications of the Psocoptera which might be called modern, i.e., based on various aspects of morphology. These are in works by Pearman (1936a), Roesler (1944), Badonnel (1951), and Smithers (1972). The classification of Badonnel is followed here with minor modifications. This system employs three suborders, which had been proposed by Roesler (1944), and a series of 'groups' which had been proposed by Pearman (1936a). As the term 'group' is commonly used informally for taxa at any level, I have replaced it here with 'family group'. This category is regarded as being above the level of superfamily; thus, superfamilies may be established within a family group, as has been done in Caecilietae (Mockford and García Aldrete, 1976). There are about 3200 described species in the world distributed in about 230 genera and 33 families. In the present work, 287 species are treated in 78 genera and 28 families.

d) Synonymy and Diagnosis
Psocina. Burmeister, 1839:772.
Copeognatha. Enderlein, 1903a:424.
Psocoptera. Shipley, 1904:259.
Corrodentia (*s. str.*). Comstock, 1924:331

In the diagnosis which follows, general morphological terms are introduced and defined or figured. More specific morphological terms are introduced where needed in subsequent chapters.

Diagnosis. Neopterous, exopterygote insects ranging in body length from 1.0 to 6.0 mm, feeding on terrestrial algae, lichens, molds, and organic debris. Head with eyes variously developed, from large and globose, multi-faceted structures down to a single ommatidium in each eye. Antennae filiform, relatively long, 11 to 50 segmented, often with scattered coeloconic sensilla on first flagellomere (f1) one or more trichoid sensilla on distal ends of flagellomeres (Fig. 1). Anteclypeus short, transverse; postclypeus large and bulging, housing cibarial dilator muscles. Mandibles (Fig. 3) robust, with well developed molar region. Maxillary palpus four-segmented (mx1-mx4); lacinia (Fig. 4) elongate, slender ('maxillary pick' or 'maxillary fork'). Hypopharynx (Fig. 5) with large basal sclerite (cibarial or sitophore sclerite) forming floor of cibarium; more distally, two ovoid sclerites (lingual sclerites) connected to cibarial sclerite by pair of sclerotized tubes (hypopharyngeal filaments) often partially joined on midline. Labial palpi (Fig. 6) short, one- or two-segmented.

Prothorax reduced, neck-like; mesonotum greatly raised over pronotum in 'normal' body form, (Figs. 7, 834) only slightly raised in 'depressed' body form (Fig. 8). Tarsi two- or three-segmented (t1, t2); hind coxae usually each with rasp and adjacent tympanum ('coxal organ' or 'Pearman's organ' Fig. 9). Pretarsal claws usually with basal appendage, the pulvillus, subtending each claw (Fig. 36). Usually four membranous wings present, extended beyond tip of abdomen when closed (macroptery), but wings showing various states of reduction (brachyptery, microptery), sometimes absent (aptery); some forms dimorphic with macropterous individuals of one sex, apterous or micropterous individuals of opposite sex. Wing venation (Figs. 10, 11) usually simple with few crossveins; Rs of forewing usually two-branched; M of forewing usually three-branched; Cu of forewing two-branched; one or two anal veins; hindwing venation more reduced.

Abdomen generally membranous except last two or three segments usually well sclerotized. Sclerotized terga 8-10 or 9+10 forming incomplete ring called clunium; segments basal to these constituting preclunial segments. Occasionally all abdominal terga well sclerotized. Abdomen composed of 11 segments with first often fused with second segment; 11th segment reduced to three flaps: dorsal epiproct and lateral paraprocts (Fig. 12). Cerci absent but possibly represented by sense cushion bearing trichobothria usually present on base of paraproct (Fig. 12). Male external genitalia consisting of hypandrium (ninth sternum), phallosome, and auxiliary structures on epiproct and paraprocts. Phallosome (Fig. 13) usually consisting of pair of struts, often fused basally, once-branched distally, their median branches (internal parameres) fused distally, forming aedeagal arch; lateral branches often somewhat broadened, forming external parameres, often bearing pores; a sac, the endophallus, either membranous or with denticulate sclerites forming distal end of penis canal above and attached to basal struts and parameres, capable of eversion and extrusion over aedeagal arch during copulation. (Note: a very denticulate or granulose endophallus has been called a radula in certain groups). Hypandrium (Fig. 14) simple or variously sclerotized, sometimes highly asymmetrical. Female external genitalia consisting of a subgenital plate (eighth sternum, Fig. 15) and three pairs of ovipositor valvulae (= gonapophyses, Fig. 16): first valvulae (v1) usually slender, attached to eighth tergum; second valvulae (v2), variously developed, attached to ninth tergum; third valvulae (v3), usually a broad flap, attached to ninth tergum; ninth sternum, forming roof of genital chamber, including opening of spermathecal duct (spermapore, sometimes surrounded by a spermapore sclerite) and sometimes auxiliary sclerotizations. Subgenital plate often terminating posteriorly in pair of lobes or processes, or single median process, called egg guide in some groups.

Note. Scale bars are provided for some figures, primarily wings, facial views of heads, and habitus drawings. All scale bars are in mm. For those not marked with a number, a solid bar = 1.0 mm and a bar with a hatch mark in the middle = 0.5 mm. Generally, a scale bar for a particular figure applies to other figures of the same sort on the same plate unless otherwise noted.

Key to Suborders

1. Adults with more than 18 flagellomeres; hypopharyngeal fila-
    ments separate, never fused on midline (Fig. 5); labial palpus
    with minute basal segment and rounded distal segment (Fig.
    6); tarsi three-segmented .... Suborder Trogiomorpha
-- Adults with fewer than 18 flagellomeres; hypopharyngeal fila-
    ments fused on midline at least part of their length (Fig. 17);
    labial palpus as above or with no basal segment 2

2. Adults with 13 flagellomeres (rarely fewer); at least some
    flagellomeres annulated with cuticular sculpture (Fig. 18).
    Labial palpus usually with minute basal segment, rounded
    distal segment (Fig. as in Fig. 6). Tarsi usually three- seg-
    mented. Forewing, when present, lacking sclerotized
    pterostigma .... Suborder Troctomorpha
-- Adults with 11 flagellomeres (rarely fewer); no flagellomeres
    annulated with cuticular sculpture, but sometimes with re-
    ticulate sculpture pattern (Fig. 19). Labial palpus lacking basal
    segment, consisting of single rounded or triangular segment.
    Tarsi two- or three- segmented. Forewings usually present,
    with sclerotized pterostigma .... Suborder Psocomorpha

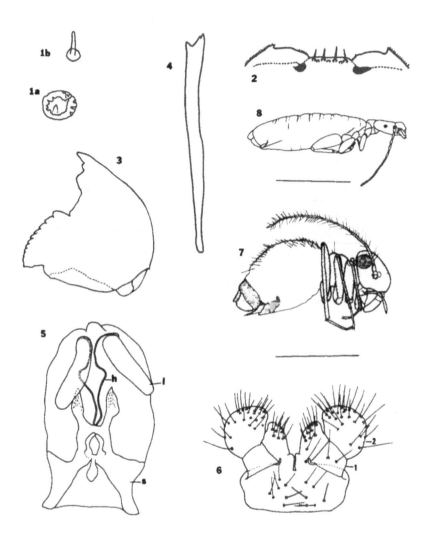

Figs. 1-8. 1a,b - typical coeloconic (1a) and trichoid (1b) sensilla of flagellomeres; 2 - distal inner labral sensilla (*Notipsocus* sp.); 3 - mandible (*Notiopsocus* sp.); 4 - lacinia (*Echmepteryx madagascariensis*); 5 - hypopharynx (*E. madagascariensis*, s - sitophore sclerite, l - lingual sclerite, h - hypopharyngeal filament); 6 - labium (*E. madagascariensis*; 1, 2 - first and second segments of labial palpus); 7 - normal body form (*Bertkauia crosbyana*); 8 - depressed body form (*Embidopsocus needhami*).

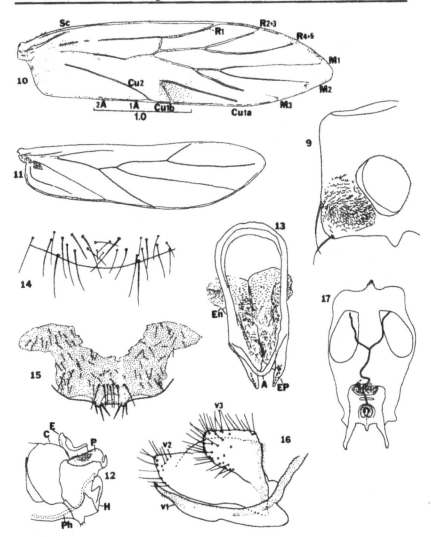

Figs. 9-17. 9 - coxal organ showing tympanum and rasp (*Indiopsocus bisignatus*); 10 - forewing of *Amphientomum hystrix* with veins named; 11 - hindwing of *A. hystrix*; 12 - male terminal abdominal segments (*Camelopsocus* sp.): C - clunium, E - epiproct, H - hypandrium, P - paraproct, Ph - phallosome; 13 - phallosome (*Elipsocus* sp.): A - aedeagal arch, EP - external paramere; En - endophallus; 14 - simple hypandrium (*Xanthocaecilius* sp.); 15 - subgenital plate (*Elipsocus guentheri*); 16 - ovipositor valvulae (*Peripsocus madidus*), v1-v3 - first to third valvulae; 17 - hypopharyx with partially fused hypopharyngeal filaments.

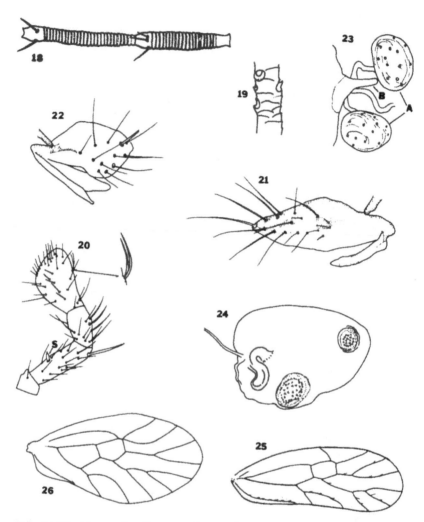

Figs. 18-26. 18 - two flagellomeres of *Liposcelis* sp. showing annulate sculpture; 19 - flagellomere with reticulate sculpture (*Hemipsocus* sp.); 20 - maxillary palpus (*Echmepteryx madagascariensis*), S - spur sensillum; forked sensillum of mx4 shown enlarged; 21 - ovipositor valvulae of Soa flaviterminata, typical of Family Group Atropetae; 22 - ovipositor valvulae of *Psocatropos microps*, typical of Family Psyllipsocidae; 23 - region of spermathecal sac at exit of duct (*Rhyopsocus bentonae*, typical of Family Psoquillidae) A - accessory bodies or glands, B - beak; 24 - spermathecal sac of *Lepinotus inquilinus* showing sessile maculae (denticulate structures), typical of Family Trogiidae; 25 - forewing of *Psyllipsocus oculatus*, macropterous male; 26 - forewing of *Speleketor flocki*.

Chapter 2
Suborder Trogiomorpha

Diagnosis. Adults generally with more than 18 flagellomeres; hypopharyngeal filaments separate, never fused on midline (Fig. 5); labial palpus with minute basal segment in addition to rounded distal segment (Fig. 6); tarsi generally three-segmented. Wings almost always present, though reduced to minute buttons in some forms; forewing (Fig. 407) lacking a sclerotized pterostigma, veins Cu_2 and IA usually ending separately on wing margin (exception: Family Psyllipsocidae). Ovipositor usually reduced to two valvulae (v2 and v3) or a single valvula (v3) on each side. Paraproct usually with a stout spine on free margin near middle (Fig. 406).

North American taxa: Family Group Atropetae (Families Lepidopsocidae, Psoquillidae, Trogiidae), Family Group Psocatropetae (Families Psyllipsocidae, Prionoglaridae).

Key to the Family Groups
and Families of Suborder Trogiomorpha

1. Spur sensillum always present on mx2 (Fig. 20). Forewing, when fully developed, with veins Cu2 and IA ending separately on wing margin. Ovipositor valvulae (Fig. 21): v3's elongate, partially joined together on midline by membrane; v2's small or absent; v1's absent Family Group Atropetae 2
-- Spur sensillum of mx2 present or absent. Forewing, when fully developed, with veins Cu2 and IA ending together on wing margin (nodulus, Fig. 26). Ovipositor valvulae: v3 never elongated but sometimes very broad; v2 usually present, slender; v1 frequently present, slender (Fig. 22) Family Group Psocatropetae 4

2. Wings usually pointed apically when fully developed but rounded when much reduced. Body and forewings covered with scales or (rarely) dense setae .... Family Lepidopsocidae
-- Wings rounded at apex when fully developed. Body and forewings never covered with scales or dense setae 3

3.   Wings, even when much reduced, with visible veins. Ovipositor
         consisting entirely of v3's. Two conspicuous accessory bodies
         at opening of spermatheca (Fig. 23) ....
         .... Family Psoquillidae
--   Wings greatly reduced (occasionally absent) lacking visible veins.
         Ovipositor generally including v2's and v1's. Spermathecal
         accessory bodies consisting of two denticulate plaques ('mac-
         ulae') attached to spermathecal wall (Fig. 24) ....
         .... Family Trogiidae

4.   Wings frequently reduced; when fully developed, basal segment
         of vein Sc in forewing short, not joining R (Fig. 25). Ovipositor
         valvulae: v3 ~ 3X as broad as v1 and v2 together, but not in
         form of a large rounded plate (Fig. 22) ....
         .... Family Psyllipsocidae
--   Wings fully developed; basal segment of vein Sc in forewing a long
         arc joining R (Fig. 26). Ovipositor valvulae: v3 developed as
         large, rounded plate with or without a small accessory valve on
         median margin (Fig. 473) .... Family Prionoglarldae

Family Group Atropetae

Diagnosis. Spur sensillum always present on mx2 (Fig. 20).
Forewings, when fully developed, never with Cu2 and 1A ending
together on wing margin. Ovipositor with at most two valvulae on the
side (v2 and v3, Fig. 409); v3 elongate, much longer than v2; v2 often
absent. Spermatheca with one or two partially sclerotized gland areas
('maculae') on surface of sac near duct attachment or as separate
bodies with ducts opening near spermathecal duct. Phallosome with
basal struts never fused anteriorly.

Family Lepidopsocidae

Diagnosis. Adults generally with wings pointed apically when fully
developed. Body and forewings generally covered with scales, but
occasionally with dense setae. Forked sensillum present on mx4 (Fig.
20). Vein R1 in forewing (Fig. 407) usually re-uniting with Rs distally by
a crossvein or a short fusion; R faint near wing base; Cu1 branching a
short distance beyond its separation from M. Hindwing (Fig. 27) with
two M veins, arising separately or together from Rs - M stem.
     North American subfamilies: Thylacellinae, Perientominae,
Lepidopsocinae, Echinopsocinae (new status).

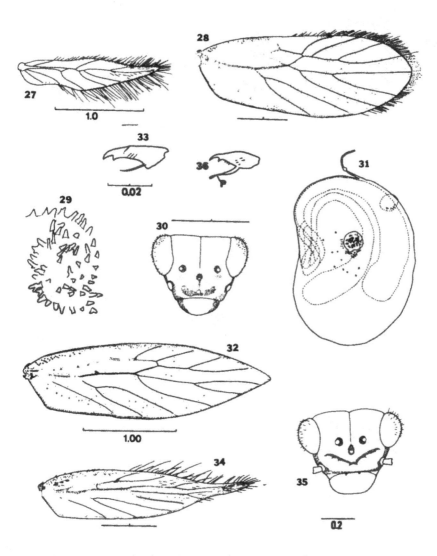

Figs. 27-36. 27 - *Thylacella cubana*, hindwing; 28 - *Soa flaviterminata*, forewing; 29 - field of dactyloid papillae of spermathecal wall of *Nepticulomima* sp.; 30 - facial markings of *Proentomum personatum*; 31 - spermathecal sac of *Neolepolepis occidentalis* (note teeth around maculae); 32 - forewing of *Echmepteryx hageni*; 33 - pretarsal claw of *E. hageni*; 34 - forewing of *E. madagascariensis*; 35 - head markings of *E. falco*; 36 - pretarsal claw of *E. madagascariensis* (P - pulvillus).

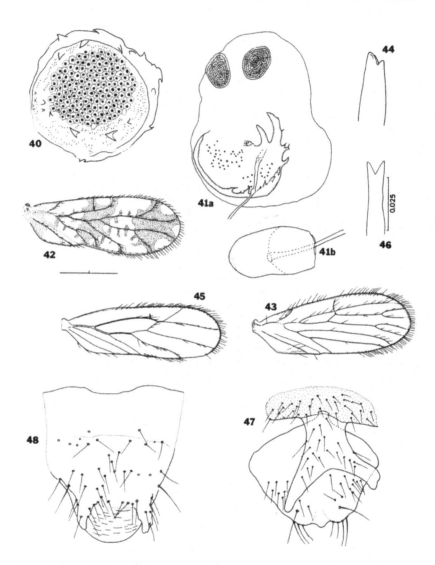

Figs. 40-48. 40 - single spermathecal macula with dentate semi-ring of *Neolepolepis occidentalis*; 41a - spermathecal sac of *Pteroxanium kellogi*; 41b - sheath of spermathecal duct of *P. kellogi*; 42 - *Psoquilla marginepunctata*, macropterous female, forewing; 43 - *Rhyopsocus texanus*, macropterous female, forewing; 44 - lacinial tip of *Balliella ealensis*; 45 - *B. ealensis*, forewing; 46 - lacinial tip of *Rhyopsocus* sp.; 47 - hypandrium and underlying process of *Rhyopsocus bentonae*; 48 - clunium and epiproct of *R. bentonae*.

5.   Adults macropterous. No dentate sclerotic rings on spermathe-
        cal sac ..... Subfamily Lepidopsocinae, Genus *Echmepteryx* 6
--   Adults usually brachypterous. One or two incomplete dentate
        sclerotic rings on spermathecal sac (Fig. 31) ....
        .... Subfamily Echinopsocinae 10

6.   Forewing (Fig. 32) relatively broad (length/greatest width ~ 3.77
        - 4.40); Rs stem in forewing ≤ length of vein R4+5. Pulvillus
        acuminate-tipped (Fig. 33) .... Subgenus *Echmepteryx* 8
--   Forewing (Fig. 34) relatively slender (length/greatest width ~
        4.15-4.72); Rs stem of forewing 1.5-2.0X length of R4+5;
        Pulvillus broad-tipped (Fig. 36) ...... Subgenus *Thylacopsis* 7

7.   Forewing (Fig. 34) with longitudinal streak of brown through
        middle; face unpatterned .... *E. (T.) madagascariensis* (Kolbe)
--   Forewing unmarked; face patterned (Fig. 35), with a 'falcon-
        wings' mark on frons below ocelli ....*E. (T.) falco* Badonnel

8.   Hind margin of vertex between eyes approximately straight.
        Parietal arched marks not enclosing a pale spot, continuous
        with a narrow pigment line parallel to margin of eye (Fig. 37)
        .... *E. (E.) hageni* (Packard)
--   Hind margin of vertex at least slightly downcurved on sides to eye
        margins. No narrow pigment band paralleling eye margin . 9

9.   Hind margin of vertex only slightly downcurved on sides to eye
        margins. Parietal arched marks over ocelli each usually en-
        closing a pale spot (Fig. 38) .... *E. (E.) intermedia* Mockford
--   Hind margin of vertex decidedly downcurved to eye margins.
        Parietal arched marks over ocelli not enclosing a pale spot (Fig.
        39) .... *E. (E.) youngi* Mockford

10.   Brachypterous form with fore winglets reaching only to end of
        basal one-third of abdomen. Wall of spermatheca, where
        known, with two flat macular areas each surrounded by
        sclerotic ring bearing curved, thorn-like projections (Fig. 40)
        .... Genus *Neolepolepis* 11
--   Brachypterous forms (only form known) with fore winglets reach-
        ing end of basal half of abdomen. Spermathecal sac with single
        incomplete sclerotic ring bearing pointed processes, not asso-
        ciated with the two flat macular areas (Fig. 41) ....
        .... Genus *Pteroxanium, P. kelloggi* (Ribaga)

11.   Facial portion of head unmarked ....*N. caribensis* (Turner)

--   Facial portion of head (Fig. 424) marked with elongate brown bar
       in each parietal region, triangular brown spot to each side of
       median ocellus and smaller brown spot before each triangular
       spot .... *N. occidentalis* (Mockford)

Subfamily Thylacellinae

Diagnosis. Antennae with at least 26 segments. Wings and body
without scales but with abundant setae, many long. Wings (Figs. 407,
27) elongate, slender; forewing length ~ 4.1X greatest forewing width.
Hindwing with a closed cell; R1 in hindwing arising far distad of Cu1
from common R - M - Cu1 stem. Both wings with marginal fringe of long
setae.

North American taxon: Genus *Thylacella* Enderlein.

Genus *Thylacella* Enderlein

Thylacella Enderlein, 1911:439.
Udamolepis Enderlein, 1912: 301.

    Diagnosis. As for the subfamily.
    Generotype. *T. eversiana* Enderlein.
    North American species: *T. cubana* (Banks).

Thylacella cubana (Banks)

Echmepteryx cubana Banks, 1941: 393.
Thylacella cubana (Banks). Mockford, 1974a: 106.

    Recognition features. Body dull yellow. Ocelli dark brown. A
reddish brown band on side of head from antennal base to hind margin,
continuing on thoracic pleura above leg bases to hind margin of thorax.
Forewing (Fig. 407) clear with three dusky transverse bands: one in
middle, one at distal two-thirds, one, sometimes lacking, near tip.
Lacinial tip (Fig. 408) with three short, blunt denticles. V3 relatively
short and broad (Fig. 409).

    Relationships. This is the only known species of its genus in the
Western Hemisphere. It appears to be most closely related to *T. fasciata*
Badonnel (1955), from Angola, differing in (1) complete middle band
through forewing versus elongate spot through and a little beyond
pterostigma in *T. fasciata*, (2) shorter, broader v2, (3) blunter denticles
of Lacinial tip.

Distribution and habitat. Within the study area the species occurs around Brownsville, Texas and in the southern two-thirds of peninsular Florida. Outside the study area it is known from Cuba, most of Mexico, Guatemala, and Belize. It occurs primarily in dead persistent leaves of various plants, but also on trunks and branches of trees.

Subfamily Perientominae

Diagnosis. Wings and body covered with scales. Antennae with relatively few (not over 24) relatively long segments. Hindwing venational characters as in Thylacellinae but closed cell sometimes very narrow (Fig. 410).

North American genera: *Nepticulomima* Enderlein, *Proentomum* Badonnel, *Soa* Enderlein.

Genus *Nepticulomima* Enderlein

Nepticulomima Enderlein, 1906f: 95.

Diagnosis. Lateral tine of lacinial tip with subapical tooth (Fig. 411). Forewing (Fig. 412) relatively slender (length/greatest width ~ 3.42), apex pointed. Sc usually not joined to R stem by crossvein. In hindwing (as in Fig. 410) R1 originating distad of M1. Scales of forewing medium brown. Third valvula short, relatively slender (Fig. 413). Field of dactyloid papillae on spermathecal sac (Fig. 29).

Generotype. *N. sakantala* Enderlein.

The genus is represented in the study area by a single undescribed species, which is found only in the southern two-thirds of peninsular Florida.

Genus *Proentomum* Badonnel

Proentomum Badonnel, 1949a: 109.

Diagnosis. Lateral tine of lacinial tip lacking subapical tooth. Forewing relatively slender (length/greatest width ~ 3.39). Sc rejoining R stem; distal end of wing acuminate (Fig. 414). In hindwing R1 originating distad of M1 (Fig. 410). Scales of forewing pale to medium brown. No field of dactyloid papillae on spermathecal sac. V3 short, relatively broad; v2 nearly half length of v3 (Fig. 415).

Generotype. *P. personatum* Badonnel.

North American species: *P. personatum* Badonnel.

Proentomum personatum Badonnel

Proentomum personatum Badonnel, 1949a: 23.

Recognition features. As for genus plus following. Head marked (Fig. 30) with dark v-shaped mark along median ecdysial line in posterior half of vertex, faint dark smudge below each lateral ocellus, dark anchor-shaped mark descending from median ocellus.

Relationships. This is the only described species of its genus.

Distribution and habitat. Within the study area, the species has been found only in south-peninsular Florida (Broward County). Outside the study area, the species has been found in Ivory Coast (type locality) Angola, Brazil, Cuba, French Guiana Guatemala, southern Mexico, Panama, Trinidad (W.I.), and Sri Lanka (Western Hemisphere and Sri Lankan records author's unpublished). It has been taken on dead persistent leaves, primarily of palms, musaceous, and bromeliadaceous plants, and in ground litter.

Note. Of 42 specimens examined, all but one had the M branches separate in the hindwing; one individual had them separate on one side and arising from a common stem on the other. It would appear, then, that the two type specimens are unique or represent a unique population regarding this character. All adult specimens seen to date are females.

Genus *Soa* Enderlein

Soa Enderlein, 1904: 109.

Diagnosis. Forewing (Fig. 28) broad (length/greatest width ~ 1.69). Sc joined to R stem by a crossvein. Dark scales covering most of forewing. V3 short and broad (Fig. 21).

Generotype. *S. dahliana* Enderlein.

North American species: *S. flaviterminata* Enderlein.

Soa flaviterminata Enderlein

Soa flaviterminata Enderlein, 1906f: 79.

Recognition features. Body and wings dark brown with slight reddish cast. Distal end of forewing beset with row of long yellow setae.

Relationships. Three other species have been described: *S. angolana* Badonnel (Angola), *S. dahliana* Enderlein (Bismark Archipelago and Guam), and *S. violacea* New (Mato Grosso). The other species lack the distal fringe of yellow setae on the forewing. Badonnel (1955) noted several other differences for *S. angolana*.

Distribution and habitat. Within the study area, the species has been found only in extreme southern Florida, where it has been taken in ground litter. Outside of the study area, it is widely distributed in the Tropics of both hemispheres and is occasionally introduced on plant material into England and Germany.

Subfamily Lepidopsocinae

Diagnosis. Antennal flagellum with numerous (30 - 50) short segments. Hindwing without a closed cell. Wings and body covered with scales, forewings also with numerous long setae.

North American genus: *Echmepteryx* Aaron; subgenera: *Echmepteryx* Aaron, *Thylacopsis* Enderlein.

Genus *Echmepteryx* Aaron

Echmepteryx Aaron, 1886: 17.

Diagnosis. Antennae with fewer than 40 segments. In forewing, R1 generally joined to Rs stem by a crossvein; anterior wing catch relatively short and thick.

Generotype: *E. hageni* (Packard).

Subgenus *Echmepteryx* Aaron

Subgeneric rank, Roesler, 1944:133.

Diagnosis. Forewings (Fig. 32) relatively broad, length/greatest width ~ 3.77 - 4.40. Rs stem in forewing equal to or shorter than R4+5. Pulvilli acuminate-tipped (Fig. 33). Spermathecal sac with single bulging, or slightly stalked macula (Fig. 416).

North American species: *E. (E.) hageni* (Packard), *E. (E.) intermedia* Mockford, *E. (E.) youngi* Mockford.

Echmepteryx (E.) hageni (Packard)

Amphientomum hageni Packard, 1870: 405.
Echmepteryx agilis Aaron, 1886: 17.
Echmepteryx hageni (Packard). Enderlein 1906f: 104.

Recognition features. Hind margin of vertex between eyes virtually straight. Parietal arched marks over ocelli continuous on each side with a narrow band close to and paralleling margin of compound eye (Fig. 37).

Relationships. On the basis of facial markings, this species appears to be closely related to the other two species of its subgenus included here, plus the Jamaican species *E. vara* Turner, *E. submontana* Turner, and *E. barba* Turner, and the Brazilian species *E. lealae* New.

Distribution and habitat. The species occurs throughout eastern United States west to central Iowa, Missouri, Arkansas, and eastern Texas; south to north-peninsular Florida, north to the Boundary Waters region of northern Minnesota. In eastern Canada it has been recorded in New Brunswick (Kouchibouguac National Park), southern Ontario and southern Quebec. It occurs on bark of tree trunks and branches, and on stone outcrops.

Echmepteryx (E.) intermedia Mockford

Echmepteryx intermedia Mockford, 1974b: 261.

Recognition features. Hind margin of vertex between eyes nearly straight, very slightly downcurved to eye margin on side. Parietal arched marks over ocelli each usually enclosing a pale spot; no narrow band paralleling eye margin (Fig. 38).

Relationships. As noted for *E. hageni.* It is possible the species will prove to be the same as *E. lealae* New, described from Pernambuco, Brazil.

Distribution and habitat. Within the study area, the species is known from the southern half of peninsular Florida, but not the Florida Keys south of northern Key Largo. Outside the study area it has been found in Trinidad (W.I.). It occurs on trunks and branches of trees. The male is unknown.

Echmepteryx (E.) youngi Mockford

Echmepteryx youngi Mockford 1974b: 259.

Recognition features. Hind margin of vertex decidedly downcurved to eye margins. Parietal arched marks over ocelli generally not including a pale spot; no narrow band paralleling eye margin (Fig. 39).

Relationships. As noted for *E. hageni.*

Distribution and habitat. The species occurs throughout peninsular Florida, near the Gulf Coast in Franklin and Wakulla Counties, West Florida, and at Crooked River State Park, Camden County, Georgia. It occurs on the trunks and branches of various trees.

Subgenus *Thylacopsis* Enderlein

Genus *Thylacopsis* Enderlein, 1911:348
Subgeneric rank, Roesler, 1944:133.

Diagnosis. Forewing relatively narrow, length/greatest width ~ 4.24 - 4.72. Rs stem of forewing much longer than (1.5 - 2.0X) R4+5. Pulvilli broad-tipped (Fig. 35). Spermathecal sac variable as to number (1 or 2) and disposition (attached or stalked) of maculae.

Type species: *E. (T.) mihira* Enderlein.

North American species: *E. (T.) falco* Badonnel, *E. (T.) madagascariensis* (Kolbe).

Echmepteryx (Thylacopsis) falco Badonnel

Thylacopsis falco Badonnel, 1949a:24.

Recognition features. Wings relatively long and slender, length/greatest width of forewing ~ 4.15. Forewing unmarked. Lacinial tip (Fig. 419) with three denticles on lateral tine, head with distinctive reddish brown markings on creamy yellow background (Fig. 35): a band on frons between antennal bases bordering postclypeus, continuing from antennal base to eye on each side; a mark in shape of spread falcon wings across frons below ocelli; a faint spot above each lateral ocellus. Spermathecal sac (Fig. 420) with single, stalked macula, complex sclerotizations in region of duct opening.

Relationships. An undescribed species with similar head markings is found on the Florida Keys.

Distribution and habitat. Within the study area, this species has been intercepted twice at Miami, Florida on plant material from Peru. Outside the study area, it is known from the southern Mexican states of Chiapas, Tabasco, and Veracruz, from Brazil and the Guianas in South America, the West Indian islands of Trinidad, Puerto Rico, and Cuba, and from the Ivory Coast and Madagascar in the Old World. In the field it is collected primarily on palm leaves, but has also been found on ferns and foliage of broad-leaf trees.

Echmepteryx (Thylacopsis) madagascariensis (Kolbe)

Thylax madagascariensis Kolbe, 1885: 184.
Thylacopsis madagascariensis (Kolbe), Enderlein, 1911: 348.
Echmepteryx costalis Banks, 1931: 349.
Lepidopsocus costalis (Banks), Zimmerman, 1948: 224.
Thylacopsis albidus Badonnel, 1949a: 25.
Echmepteryx albidus (Badonnel), Badonnel, 1962: 186.
Echmepteryx (Thylacopsis) madagascariensis (Kolbe), Smithers, 1967: 8.

Recognition features. Wings exceedingly long and slender, length/ greatest width of forewing ~ 4.72. Forewing marked with broad brown band running its entire length through middle (Fig. 34). Lacinial tip with only two denticles on lateral tine (Fig. 417). Head with facial region unmarked. Maculae of spermathecal sac forming two compact fields closely applied to wall of sac (Fig. 418).

Relationships. This species shares with *E. (T.) falco* Badonnel the three characters noted in the subgeneric diagnosis. It seems likely, however, that narrow-wingedness has arisen more than once in the genus *Echmepteryx*, and questions of relationships can only be resolved by analysis of more characters in numerous species.

Distribution and habitat. Within the study area, this species has been found only in south-peninsular Florida north to Brevard County on the Atlantic Coast. Outside the study area, it occurs very widely in the Tropics and has been found occasionally in greenhouses in northern Europe. It inhabits primarily dead persistent leaves, especially of palms and bananas.

Subfamily Echinopsocinae, revised status

Diagnosis. Antennae with 30 - 50 short flagellomeres. Lacinial tip with four teeth, the outermost pair sometimes very close together (Fig. 421). Front coxa with lateral flange forming decided angle on margin (Fig. 422). Wings usually reduced, the hindwing minute. If wings occasionally elongate, no closed cell present in either fore- or hindwing or at most a small one in forewing (Fig. 423). Forewings and body covered with scales; forewings also with stout, upright setae (Fig. 424). Spermathecal sac with one or two incomplete sclerotic rings on surface bearing irregular pointed processes (Fig. 31).

Note. Roesler (1944) placed the genera *Scolopama* Enderlein, *Echinopsocus* Enderlein, and *Pteroxanium* Enderlein in this taxon, placing *Lepolepis* Enderlein in a separate subfamily. It seems likely that he was correct in doing this. The single North American species

which has long been assigned to *Lepolepis*, *L. occidentalis* Mockford,
agrees with the above diagnosis and differs from the Old World species
of *Lepolepis* in lacinial tip, front coxal, and spermathecal sac characters. On that account, a new genus is set up to contain it.

North American genera: *Neolepolepis* n. gen., *Pteroxanium*
Enderlein.

Genus *Neolepolepis* new genus

Antennae with ~ 32 moderately short flagellomeres. Wings generally much reduced, with venation vague or absent. Fully winged
individuals (rare) with no closed cells in forewing or hindwing or a small
closed cell in forewing (Fig. 423), pterostigma absent. Front coxa with
lateral flange forming decided angle on margin (Fig. 422). Pretarsal claw
with no minute denticles before preapical tooth, a single setiform
appendage at base. Spermathecal sac with two flat macular areas, each
surrounded by a partial ring of curved thorn-like projections (Fig. 31).
Ovipositor valvulae relatively broad and stout (Fig. 424b). Phallosome:
endophallus (Fig. 425) with basal bodies elongate, traversed by rib-like
thickenings; phallosome otherwise typical of subfamily. Epiproct and
paraprocts normal for family.

Type species. '*Lepolepis*' *occidentalis* Mockford.

Other included species. '*Echmepteryx*' *xerica* Garcia Aldrete
(Mexico), '*Lepolepsis*' [sic] *caribensis* Turner.

Neolepolepis caribensis (Turner), new combination

Lepolepsis [sic] *caribensis* Turner, 1975:547.

Recognition features. Known from brachypterous form only;
forewings about equal in length to hind femur, therefore probably as
described for following species. Head brown, unmarked except for dark
brown stripe between eye and antennal base.

Note. I have not seen this species. I place it in the new genus on the
basis of characters of the phallosome, which agrees with the characters
noted above for the genus.

Relationships. Beyond the generic placement, these remain unknown.

Distribution and habitat. Within the study area, the species was
recorded from Key Largo, Monroe County, Florida, by Garcia Aldrete
(1984b), where it was taken in the nest of a pack rat, *Neotoma floridana
smalli* Sherman. Outside the study area, it is known only from the type
locality, Hellshire Hills, St. Catherine's Parish, Jamaica.

Neolepolepis occidentalis (Mockford), new combination

Lepolepis occidentalis Mockford, 1955b: 436.

Recognition features. Brachypterous form with fore winglets reaching only to end of basal one-third of abdomen. Macropterous form (rare) with wings exceeding tip of abdomen. Head marked (Fig. 424a) with elongate brown bar in each parietal region, frons with triangular brown spot to each side of median ocellus, smaller brown spot before each triangular one. Body in general straw brown. Numerous dark, slender, upright scales on winglets in living and carefully preserved specimens.

Relationships. In addition to the two species known from the study area, the only other known species of the genus is *N. xerica* (García Aldrete) (new combination), from southeastern Mexico, in which wings are somewhat reduced but the forewings reach nearly to the tip of the abdomen. Relationships among these three species cannot be determined at present.

Distribution and habitat. The species occurs throughout eastern United States and north to Kouchibouguac National Park, New Brunswick, west to Nebraska and spottily in the southwest to south-central New Mexico. It inhabits woodland ground litter, especially over well-drained sandy soil.

Genus *Pteroxanium* Enderlein

Lepidilla Ribaga, 1905: 100 (preoccupied).
Pteroxanium Enderlein, 1922: 103.
Tasmanopsocus Hickman, 1934: 78.

Diagnosis. Only brachypterous forms known; forewings extending about to end of basal half of abdomen. Spermathecal sac with single incomplete sclerotic ring bearing pointed processes (Fig. 41a) and two flattened macular regions (without sclerotic rings). Sheath of spermathecal duct short and broad (Fig. 41b). Pretarsal claw with two basal setiform appendages, no minute denticles before preapical tooth.

Generotype. *P. kelloggi* (Ribaga).
North American species: *P. kelloggi* (Ribaga).

Pteroxanium kelloggi (Ribaga)

Lepidilla kelloggi Ribaga, 1905: 100.
Hyperetes brittanicus Harrison, 1916: 108.
Pteroxanium squamosum Enderlein, 1922: 103.

Tasmanopsocus litoralis Hickman, 1934: 78.
Pteroxanium kelloggi (Ribaga), Roesler, 1943: 13.

Recognition features. Head marked (Fig. 426) with longitudinal purplish brown bar through each parietal region, the bars pale in middle; lateral to each bar a fine line of same color paralleling eye margin; frons with pair of triangular marks outlined in purplish brown, pale in middles, converging on median ocellar spot; pair of spots of same color anterior to triangles. Clypeus and genae largely purplish brown. Rest of body straw brown to creamy yellow except all tibiae with two broad purplish brown rings.

Relationships. Five other species have been described in the genus: P. funebris Badonnel (Chile), P. ralstonae Smithers and Thornton (Norfolk Island), P. evansi Smithers and Thornton (Norfolk Island), P. insularum Smithers and Thornton (Norfolk Island), and P. forcepetus Garcia Aldrete (Mexico). This species and the Chilean and Norfolk Island species appear to form a group of close allies on the basis of similar state of wing reduction, short and wide sheath of the spermathecal duct, and similarity of the phallosome. The Mexican species, based on a single, damaged specimen, is not well enough known to permit determination of its relationships, but the phallosome suggests other affinities.

Distribution and habitat. Within the study area, this species occurs along the Pacific Coast from central California (San Francisco Bay) north to the Olympic Peninsula of Washington. Outside the study area, it has been found in France, United Kingdom, Tasmania, New Zealand, Chile, and Argentina. It has been found on trunks and branches of coniferous and broad-leaf trees, in dead fern leaves, and in dead plant material under stones.

Family Psoquillidae

Diagnosis. Adults with wings rounded apically. Body and forewings without scales, but often with abundant setae. Forked sensillum present or absent on mx4. Vein Cu1 in forewing separating from R + M at about one-fourth to one-sixth distance from wing base to tip and not rejoining R + M (Fig. 42). Vein M3 leaving from common Rs + M stem separately from M1+2 (probably in both fore- and hindwing; if so interpreted, Rs simple in hindwing). Coxal organ absent. Pretarsal claws lacking preapical tooth. Phallosome with two basal struts never fused anteriorly. Ovipositor formed entirely of third valvulae. Two conspicuous spermathecal glands ('accessory bodies') at opening of spermathecal sac (Fig. 427).

North American genera: *Balliella* Badonnel, *Psoquilla* Hagen, *Rhyopsocus* Hagen.

Key to the North American Genera and Species of Family Psoquillidae

1. Body largely dark brown. Wings heavily patterned clear and dark brown. No closed cell in forewing (Fig. 42) .... Genus *Psoquilla, P. marginepunctata* Hagen
-- Body largely pale brown; wings clear. At least a small closed cell (not readily visible where wing greatly reduced) in forewing (Fig. 43) 2

2. Lacinial tip with lateral tine much larger than median (Fig. 44). Vein Sc absent in forewing thus pterostigma open basally (Fig. 45) .... Genus *Balliella, B. ealensis* Badonnel
-- Lacinial tip with two approximately equal tines (Fig. 46). Vein Sc present in forewing; pterostigma closed basally .... Genus *Rhyopsocus* 3

3. Male clunium with three lobes on hind margin in middle. At least vestigial ocelli present in brachypterous forms 4
-- Male clunium lacking three lobes on hind margin in middle. Ocelli totally absent in brachypterous forms 5

4. Male with a flap in form of asymmetrical fish tail arising before and underlying hypandrium (Fig. 47) lateral lobes of male clunium slender, finger-like (Fig. 48). Female clunium expanded as broad, rounded area before epiproct (Fig. 49) .... *R. bentonae* Sommerman
-- Male without a flap underlying hypandrium; lateral lobes of male clunium about equal in width to median lobe (Fig. 50). Female clunium not expanded before epiproct .... *R. eclipticus* Hagen

5. Only brachypterous forms known. Sexual dimorphism in wing length slight (fore winglets of male reaching end of basal half of abdomen; those of female slightly beyond basal one-third). Compound eyes reduced with not over 20 facets .... *R. micropterus* Mockford
-- Either both macropterous and brachypterous forms known, or sexual dimorphism in wing length marked. Compound eyes not reduced, with numerous facets 6

6.   Sexual dimorphism in wing length marked: male fore winglets
        reaching base of clunium; those of female reaching only second
        abdominal tergum. Spermapore plate not obvious ....
        .... *R. disparilis* (Pearman)
--   Both macropterous and brachypterous forms known. Sexual
        dimorphism in wing length not perceptible: brachypterous
        forms of both sexes with fore winglets reaching about basal
        one-third of abdomen. Spermapore plate heavily sclerotized, in
        form of elongate frame (Fig. 51) .... *R. texanus* (Banks)

Genus *Balliella* Badonnel

Balliella, Badonnel, 1949b:9.

Diagnosis. Lacinial tip with two tines, lateral much larger than
median. Body largely pale brown, wings clear. Sc absent in forewing
(Fig. 45), hence pterostigma open basally. R1 leaving R + M + Cu stem
in basal one-fifth of forewing, rejoining Rs in distal three-fifths, thus
forming large closed cell in wing. Rs beyond separation with R1 simple
or with short distal fork. M1+2 unbranched in forewing. In hindwing
(Fig. 428) R1 absent, a small closed cell present along R + M + Cu stem,
with Cu1 arising from hind margin of cell. Pretarsal claws lacking basal
seta but with pulvillus broadened at tip. Spermathecal glands mush-
room-shaped (Fig. 429). Female ninth sternum with small spermapore
sclerite. Phallosome relatively short, about one-third length of abdomen.
    Generotype. *B. ealensis* Badonnel.
    North American species: *B. ealensis* Badonnel.

Balliella ealensis Badonnel

Balliella ealensis Badonnel, 1949b:10.

    Recognition features. As described in generic diagnosis.
    Relationships. This is the only described species of its genus.
Originally described from Zaire, it has not been recorded elsewhere
previously. The two specimens, taken as introductions in North America,
differ from the Zairian specimens in details of the lacinial tip, viz. the
lateral tine is not denticulated (Fig. 44), but otherwise appear to be the
same.

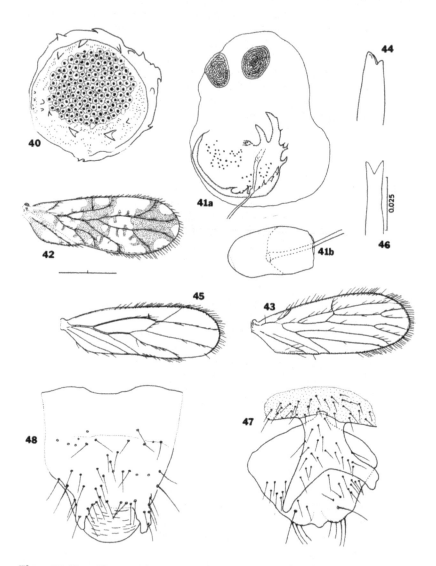

Figs. 40-48. 40 - single spermathecal macula with dentate semi-ring of *Neolepolepis occidentalis*; 41a - spermathecal sac of *Pteroxanium kellogi*; 41b - sheath of spermathecal duct of *P. kellogi*; 42 - *Psoquilla marginepunctata*, macropterous female, forewing; 43 - *Rhyopsocus texanus*, macropterous female, forewing; 44 - lacinial tip of *Balliella ealensis*; 45 - *B. ealensis*, forewing; 46 - lacinial tip of *Rhyopsocus* sp.; 47 - hypandrium and underlying process of *Rhyopsocus bentonae*; 48 - clunium and epiproct of *R. bentonae*.

Distribution and habitat. Within the study area, one specimen was taken at Anchorage, Alaska, the other at Indianapolis, Indiana, both on plant materials originating in the Oriental region. Outside the study area, the species has been recorded only at Eala, Zaire.

Genus *Psoquilla* Hagen

Psoquilla Hagen, 1865: 123.
Heteropsocus Verrill, 1902: 817.

Diagnosis. Lacinia with two nearly equal distal tines. Mx4 with one or two large forked sensilla on side. Body largely dark-pigmented and forewing (Fig. 42) extensively marked with dark brown areas; several narrow brown spots touching veins containing small spur veins. Forewing lacking closed cell. Pretarsal claws with basal seta and pulvillus, the latter somewhat broadened at tip. Female ninth sternum with conspicuous spermapore sclerite (Fig. 430). Spermathecal glands rounded.

Generotype. *P. marginepunctata* Hagen.
North American species: *Psoquilla marginepunctata* Hagen.

Psoquilla marginepunctata Hagen

Psoquilla marginepunctata Hagen, 1865:123.
Heteropsocus dispar Verrill, 1902:817.

Recognition features. Only species of its family in study area with body and wings (even when latter much reduced) very contrastingly marked dark and light.
Relationships. Only one other species of *Psoquilla* is known: *P. infuscata* Badonnel, from Ivory Coast, West Africa.
Distribution and habitat. Within the study area, the species seems to be established only in peninsular Florida, where it occurs primarily in buildings. Outside the study area it is known from most of the wet Tropics, Bermuda, and Hawaii, and has been taken in greenhouses in Germany and warehouses in England. In the Tropics it occurs primarily on and under bark of trunks and branches of trees.

Genus *Rhyopsocus* Hagen

Rhyopsocus Hagen, 1876:55.
Deipnopsocus Enderlein, 1903f:358.
Rhyopsocopsis Pearman, 1929:107.

Diagnosis. Lacinial tip with two tines, approximately equal in size. A single forked sensillum on side of mx4. Body largely pale brown. R1 in forewing (Fig. 43) separating from R-M stem at about basal one-third of wing, rejoined to Rs by crossvein at about one-half wing length, thus forming small closed cell. Wing venation otherwise normal for family. Pretarsal claws lacking basal seta, pulvillus broadened at tip. Spermathecal glands rounded. A relatively large spermapore sclerite present.

Generotype. *R. eclipticus* Hagen.

North American species: *R. bentonae* Sommerman, *R. disparilis* (Pearman), *R. eclipticus* Hagen, *R. micropterus* Mockford, *R. texanus* (Banks).

Rhyopsocus bentonae Sommerman

Rhyopsocus bentonae Sommerman, 1956a:146.

Recognition features. Both sexes dimorphic; macropterous form with wings extending beyond tip of abdomen; brachypterous form with wings not quite reaching tip of abdomen. Macropterous form with head and thorax dark brown; brachypterous form with these regions much paler brown. Ocelli present, at least in vestigial state. Mx4 relatively wide (Fig. 431). Male abdomen ventrally with flat, asymmetrical fish-tail-like flap arising before and underlying hypandrial plate (Fig. 47); dorsally, posterior margin of clunium with broad, rounded lobe over (and hiding) epiproct (Fig. 48), flanked by two finger-like lobes. Female clunium expanded as broad, rounded area before epiproct (Fig. 49); anterior edges of clunium on sides heavily sclerotized. Spermathecal accessory bodies rounded; beak of spermathecal sac of medium length, tapering to distal blunt point (Fig. 23).

Relationships. *R. bentonae* belongs to a small complex of closely related species of which one undescribed species occurs on the Florida Keys and another undescribed species occurs in southern Mexico.

Distribution and habitat. Within the study area, the species occurs throughout peninsular Florida, north along the Atlantic Coast to Brunswick, Georgia, and around the Gulf Coast to southeastern Texas. Outside the study area, García Aldrete (1987) recorded it from several localities in southern and southeastern Mexico. It has been taken primarily in dead leaves of palms, yuccas, and Typha.

Rhyopsocus disparilis (Pearman)

Deipnopsocus spheciophilus var.*disparilis* Pearman, 1931b:96.
Rhyopsocus disparilis Pearman. Badonnel, 1949a:29.

Recognition features. Sexually dimorphic. Males with wings extending approximately to base of clunium; females with wings much shorter, forewings extending only to second abdominal tergum. Both sexes lacking ocelli completely. Mx4 relatively wide (Fig. 432). Male abdomen with no pre-hypandrial ornamentation and no dorsal clunial lobes. Female clunium broad before epiproct, the widened area slightly indented in middle on its anterior margin (Fig. 433). Sides of clunium narrowly reinforced. Spermathecal accessory bodies rounded, their ducts elongate; beak of spermathecal sac slender, relatively long (Fig. 434).

Relationships. It is not possible to assess the relationships of this species at present.

Distribution and habitat. Within the study area, the species has been taken only as an introduced form at ports of entry, principally New York City. Materials on which it has been taken originated in Guyana, Philippines, West Indies, Nigeria, and Japan. Outside the study area, the species was found introduced in England on materials originating in Ghana.

Rhyopsocus micropterus Mockford

Rhyopsocus micropterus Mockford, 1971a:129.

Recognition features. Both sexes brachypterous; forewings of male extending nearly to half length of abdomen; those of female extending slightly beyond one-third length of abdomen. Ocelli totally absent. Compound eyes reduced (Fig. 435). Mx4 slender (Fig. 436). Male abdomen with no ventral flap or dorsal clunial lobes. Female clunium only slightly and narrowly reinforced. Spermathecal accessory glands somewhat elongate, one side slightly indented. Beak of spermathecal sac relatively short, tapering (Fig. 437).

Relationships. The species is probably close to several other brachypterous litter-inhabiting forms of southwestern United States and Mexico, including *R. maculosus* Garcia Aldrete and the dimorphic species *R. texanus* (Banks).

Distribution and habitat. The species has been found only in southern California (San Diego County) in the nest of a pack rat, *Neotoma fuscipes* Howell.

Rhyopsocus eclipticus Hagen

Rhyopsocus eclipticus Hagen, 1876b:52.
Rhyopsocus phillipsae Sommerman, 1956a:146, new synonym.

Note. The type (MCZ type no. 10123) is a brachypterous male slide-mounted in balsam. It clearly shows the genitalic features of the species.

Recognition features. Both sexes dimorphic; wings in brachypterous female reaching slightly beyond middle of abdomen; those of brachypterous male slightly longer. Macropterous individuals with wings extending beyond tip of abdomen. Macropters and brachypters differing in color as noted for *R. bentonae*. Ocelli present, at least vestigially, in both sexes. Mx4 broad (Fig. 438). Male abdomen with small process extending forward from hypandrium on midline (Fig. 439); clunium dorsally bearing three broad lobes (Fig. 50); median lobe covering base of epiproct. Female clunium slightly expanded before epiproct (expansion more obvious and slightly bilobed anteriorly in macropterous form, Fig. 440); anterior edges of clunium on sides slightly sclerotized. Spermathecal accessory bodies rounded, beak of spermathecal sac elongate, rounded at tip (Fig. 427).

Relationships. This species shares with the *R. bentonae* complex the three posterior lobes in the male clunium and the same pattern of wing dimorphism. Tentatively, then, its closest relationships appear to be with the *R. bentonae* complex.

Distribution and habitat. The species occurs throughout Florida, north on the Atlantic Coastal Plain to Maryland, around the Gulf Coast to southeastern Texas and up the Mississippi Embayment to southern Illinois and southern Missouri. It inhabits primarily persistent dead leaves, also Spanish Moss (*Dendropogon usnioides*), conifer foliage, and ground litter. It was originally described from Kerguelen Island, indoors, in a room where instrument boxes from Washington, D.C. containing much packing straw had been unpacked. It was undoubtedly introduced there.

Rhyopsocus texanus (Banks)

Deipnopsocus texanus Banks, 1930b:223.
Rhyopsocus (Deipnopsocus) texanus (Banks), Sommerman, 1956a:145.
Rhyopsocus squamosus Mockford and Gurney, 1956:357.
Rhyopsocus pescadori Garcia Aldrete, 1984a:25.

Note. Mockford and Garcia Aldrete (1991) established the synonymy of the names *R. squamosus*, *R. pescadori*, and *R. texanus*, thus the distribution as noted below.

Recognition features. Both sexes dimorphic. Wings in brachypters of both sexes extending to about basal one-third of abdomen. Wings of macropters reaching or exceeding tip of abdomen. Compound eyes somewhat larger in macropters than brachypters. Ocelli present in macropters, absent in brachypters. Preclunial abdominal segments each with a partial ring of subcuticular brown pigment, the rings incomplete ventrally. Male abdomen with no ventral flap or dorsal clunial lobes. Female clunium slightly expanded before epiproct; sides of clunium only slightly sclerotized (Fig. 441). Spermathecal accessory glands somewhat elongate. Beak of spermathecal sac broad and flattened at tip (Fig. 442). Spermapore plate heavily sclerotized, in form of elongate frame (Fig. 51).

Relationships. As noted for *R. micropterus.*

Distribution and habitat. Within the study area, the species has been found only in the lower Rio Grande Valley of Texas. Outside the study area, it has been found in the northern Mexican states of Nuevo León and Tamaulipas, also in the central and southern Mexican states of Jalisco, Morelos, Nayarit, Puebla, and San Luis Potosí. It occurs in ground litter and low vegetation in semi-arid to mesophytic woodlands.

Family Trogiidae

Diagnosis. Adults with wings reduced to small flattened pads lacking venation, (brachelytroptery of Günther, 1974a) or minute buttons, or (rarely) absent. Body and winglets without scales but often with abundant setae. Forked sensillum present or absent on mx4. Coxal organ present or absent. Pretarsal claws lacking preapical tooth. Ovipositor generally including second and third valvulae. Spermathecal accessory bodies in form of two denticulate plaques attached to spermathecal wall.

North American genera: *Cerobasis* Kolbe, *Lepinotus* Heyden, *Myrmicodipnella* Enderlein, *Trogium* Illiger.

Key to the North American Genera and Species of Family Trogiidae

1. Winglets entirely lacking ....
    .... Genus *Myrmicodipnella, M. aptera* Enderlein
-- Winglets present as small flattened pads, scales, or small buttons 2

2. Mx4 with forked sensillum (Fig. 52). No spurs on hind t1 ....
    .... Genus *Lepinotus* 5
-- Mx4 lacking forked sensillum. Three to nine spurs on hind t1
    3

3.   Coxal organ present as bulge or bulge plus rasp. Six to nine spurs
         on hind tl Genus *Cerobasis* 4
--   Coxal organ absent. Three to four spurs on hind tl ....
         .... Genus *Troglum, T. pulsatorium* (Linnaeus)

4.   Fore winglets developed as scale-like structures reaching second
         abdominal tergum. Rasp of coxal organ present ....
         .... *C. annulata* (Hagen)
--   Fore winglets developed as minute buttons, just reaching meta-
         thorax. Rasp of coxal organ absent ....
         .... *C. guestfalica* (Kolbe)

5.   Winglets with reticulate pattern (Fig. 53) Spermathecal sac
         absent ....*L. reticulatus* Enderlein
--   Winglets without reticulate pattern. Spermathecal sac present
         6

6.   Lacinial tip with large lateral and smaller median tine with no
         tooth between (Fig. 54). Both spermathecal maculae with
         tubercles oriented in sunburst pattern and surrounded by
         wreath of minute papillae (Fig. 24) .... *L. inquilinus* Heyden
--   Lacinial tip with well developed tooth between median and lateral
         tines (Fig. 55). Both spermathecal maculae uniformly covered
         with minute papillae (Fig. 56) .... *L. patruelis* Pearman

Genus *Cerobasis* Kolbe

Hyperetes Kolbe, 1880:132 (preoccupied).
Cerobasis Kolbe, 1882b:212.
Tichobia Kolbe, 1882b:212.
Albardia Jacobson and Bianchi, 1904:496.
Myopsocnema Enderlein, 1905:17.
Zlinia Obr. 1948:93.

Diagnosis. Lacinial teeth normal. Mx4 lacking forked sensillum.
Fore winglets developed as flattened scales, or small buttons, or
absent. Two to four large apical spurs on hind tibia plus two on inner
side before apex in some species. Six to nine large spurs on hind tl.
     Generotype. *C. guestfalica* (Kolbe).
     Genus *Cerobasis* appears to be divisible into two species groups.
These are diagnosed and their constituent (world) species listed below.

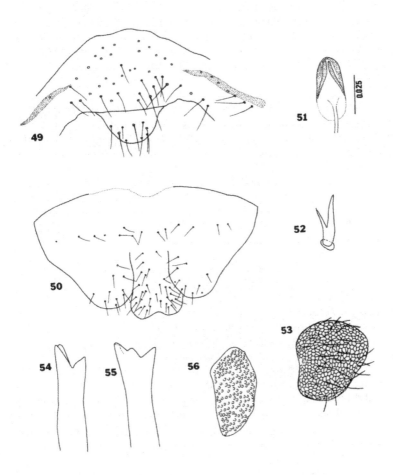

Figs. 49-56. 49 - female clunium and epiproct of *Rhyopsocus bentonae*; 50 - male clunial lobes of *Rhyopsocus eclipticus*; 51 - spermapore sclerite of *Rhyopsocus texanus*; 52 - forked sensillum of mx4 of *Lepinotus reticulatus*; 53 - fore winglet of *L. reticulatus*; 54 - lacinial tip of *Lepinotus inquilinus*; 55 - lacinial tip of *Lepinotus patruelis*; 56 - spermathecal macula of *L. patruelis*.

Group I

Diagnosis. Fore winglets present, at least as minute buttons. Pulvilli well developed, curved, slightly expanded at tip (Fig. 443). Coxal organ developed as rasp and bulge. Hind tibiae with two to four (usually two) preapical spurs on inner face.

Included species (world): *C. alfredi* Lienhard, *C. annulata* (Hagen), *C. bundyi* Turner, *C. canariensis* (Enderlein), *C. guestfalica* (Kolbe), *C. lapidaria* Badonnel, *C. maderensis* Lienhard, *C. multispinosa* Obr. Tentatively: *C. maculiceps* Badonnel, *C. recta* Thornton and Woo.

Group II

Diagnosis. Fore winglets totally absent or greatly reduced. Pulvilli reduced to short setae. Coxal organ represented only by a bulge. Hind tibiae with two apical spurs, zero to three (usually zero) preapical spurs on inner face.

Included species (world): *C. caboverdensis* Lienhard, *C. harteni* Lienhard, *C. intermedia* Lienhard, *C. papillata* Garcia Aldrete, *C. treptica* Thornton and Woo. Tentatively: *C. lambda* Thornton and Woo.

North American species (all group I): *C. annulata* (Hagen), *C. guestfalica* (Kolbe).

Cerobasis annulata (Hagen)

Clothilla annulata Hagen, 1865:122.
Atropos annulata (Hagen). Kolbe, 1880:135.
Myopsocnema annulata (Hagen). Enderlein, 1905:17.
Cerobasis annulata (Hagen). Smithers, 1967:11.

Recognition features. Fore winglets present, brown spotted, reaching second abdominal tergum. Abdominal terga rust-brown with only slight indication of spotting pattern (Fig. 441). Pulvilli about as long as pretarsal claws. Hind tibia with three apical spurs, two preapical spurs on inner face. Spermathecal maculae each with well developed papillar wreath (Fig. 445).

Relationships. This species appears to be most closely related to *C. bundyi* Turner (Morocco and Israel). Lienhard (1984) noted that *C. multispinosa* (Obr) may be a synonym of this species.

Distribution and habitat. Within the study area, the species appears to be scarce. I have only seen specimens from San Francisco and an intercepted specimen from Greece at New York City. It has been found only in domestic situations and is sometimes taken in human commerce on stored food materials. Hence, it is widely distributed, occurring throughout Europe, in Morocco, the island of St. Helena, and possibly the Canary Islands.

Cerobasis guestfalica (Kolbe)

Hyperetes guestfalicus Kolbe, 1880:132.
Hyperetes pinicola Kolbe, 1881:227.
Tichobia alternans Kolbe, 1882b:212.
Cerobasis muraria Kolbe, 1882b:212.
Hyperetes tessulatus Hagen, 1883:316.
Albardia alternans (Kolbe). Jacobson and Bianchi, 1904:496.
Cerobasis guestfalica (Kolbe). Roesler, 1943:13.

Recognition features. Fore winglets minute, just reaching metathorax. Head markings (Fig. 446): anchor mark on frons, longitudinal stripe through each parietal region. Abdominal terga gray with checkered pattern of brown spots at least on some posterior terga. Pulvillar, coxal organ, and hind tibial characters as in *C. annulata*. Spermathecal maculae (Fig. 447) without wreath of papillae, but each with sclerotic ring.

Relationships. This species appears to be most closely related to *C. alfredi* Lienhard (Tunisia).

Distribution and habitat. Within the study area, this species occurs on the Atlantic Coast from South Carolina to central Florida, west on the Gulf Coast to Texas, up the Mississippi Embayment to southern Illinois, and on the Pacific Coast in British Columbia, Washington, and Oregon. Outside the study area, it occurs throughout western Europe, in Morocco, St. Paul Island, Bermuda, South Africa, Argentina, Australia, and at two Mexican localities. The species occurs on a variety of trees and shrubs, on stone outcrops, and stone walls. Most populations are parthenogenetic. Badonnel (1943) noted that it sometimes enters houses in France.

Genus *Lepinotus* Heyden

Lepinotus Heyden, 1850:84.
Paradoxides Motschulsky, 1851:510.
Paradoxerus Molschulsky, 1852:19.
Cuixa Navás, 1927:151.
Heterolepinotus Obr, 1948:105.

Diagnosis. Lacinial teeth normal or with median tooth fused into lateral. Mx4 with forked sensillum. Fore winglets developed as flat scales. Coxal organ absent. Two large apical spurs on hind tibia; generally none on inner face before apex. No spurs on hind t1.

Generotype. *L. inquilinus* Heyden.

North American species: *L. inquilinus* Heyden, *L. patruelis* Pearman, *L. reticulatus* Enderlein.

Lepinotus inquilinus Heyden

Termes pulsatorium Scopoli, 1763:380.
(not *Termes pulsatorium* Linnaeus, 1761:474).
Lepinotus inquilinus Heyden, 1850:84.
Paradoxides psocoides Motschulsky, 1851:510.
Paradoxenus psocoides (Motschulsky). Motschulsky, 1852:19.
Clothilla inquilina (Heyden). Brauer, 1857:32.
Clothilla picea Hagen, 1861:8.
Atropos inquilina (Heyden). Kolbe, 1880:136.
Atropos sericea Kolbe, 1883:86.
Lepinotus piceus (Hagen). Hagen, 1883:314.
Atropos distincta Kolbe, 1888b:190.
Atropos picea (Hagen). Kolbe, 1888b:190.
Clothilla distincta (Kolbe). Tetens, 1891:372.
Lepinotus sericeus (Kolbe). Tetens, 1891:373.
Cuixa canaria Navás, 1927:151.

Recognition features. Color ranging from medium brown with pale frons to uniformly pale straw brown. Lacinial tip (Fig. 54) with large lateral and somewhat smaller median tine with no tooth between tines. Both maculae of spermathecal wall with tubercles oriented in sunburst pattern and surrounded by wreath of minute papillae (Fig. 24).

Relationships. This species appears to be closest to *L. stellatus* Badonnel of Angola, in which alary and spermathecal characters are the same, and there is asymmetrical reduction of the middle lacinial tooth.

Distribution and habitat. Within the study area, this domestic species has appeared primarily as an introduction on dried food material and dried plant parts at southwestern ports of entry. It is not known to be established in the study area. Outside the study area, it is common throughout Europe, several African localities and Madagascar. It has apparently not been recorded from South America, Central America, or Mexico. Usually, it occurs in human dwellings and warehouses, but it also occurs in bee hives, dwellings of animals, and (rarely) on shrubbery.

Lepinotus patruelis Pearman

Lepinotus patruelis Pearman, 1931a:47.

Recognition features. Color similar to that of *L. inquilinus*; always a dark brown band from eye to antennal base, not visible in darkest specimens. Lacinial tip (Fig. 55) with well developed tooth between lateral and median tines. Both maculae of spermathecal wall uniformly covered with minute papillae (pores?); pyriform sperm packet in spermatheca encased in an outer brown body.

Relationships. Badonnel (1969) noted that this species stands apart from other species lacking a reticulate pattern in the winglet in possession of the brown body in which the sperm packet comes to lie, and in having no large median tubercles of the spermathecal maculae.

Distribution and habitat. This is another largely domestic species which occurs within the study area primarily as an introduction on dried plant material. Essig (1940) reported it as having been found established in a household in California. Outside the study area, this species is known from several European localities and from Argentina. It appears to be restricted to domestic biotopes.

Lepinotus reticulatus Enderlein

Clothilla inquilina (Heyden). Hagen, 1882:526.
Atropos inquilina (Heyden). Kolbe, 1888b:190.
Lepinotus reticulatus Enderlein, 1905:31.

Recognition features. Color pale straw brown. Lacinial tip with large lateral tine, much smaller median tine, no tooth between tines (Fig. 448). Winglets with distinct reticulate pattern (Fig. 53). Spermathecal sac absent.

Relationships. The absence of a spermathecal sac in European, African (Badonnel, 1969, 1971d), North American, and Southeast Asian (personal observation) forms of this species indicates that it is an obligate parthenogen, despite Roesler's (1935) find of a single male in Germany. The parthenogenetic species may have been derived from a sexual form recorded by Smithers (1969) under the same name, from New Zealand, which may prove to be the form described by Hickman (1934) from Tasmania, as *L. tasmaniensis*. The latter form is clearly distinct, having a spermatheca with maculae developed as remarkable stellate bodies.

Distribution and habitat. The species is nearly world wide in distribution. It occurs in granaries, bird and mammal nests, ground litter on well-drained soil, and occasionally in the plumage of birds.

Within the study area it has been recorded from Nova Scotia, Quebec, Manitoba, Saskatchewan, Alberta, British Columbia, Utah, California, Arizona, New Mexico, Illinois, Indiana, and Michigan.

Genus *Myrmicodipnella* Enderlein

Myrmicodipnella Enderlein, 1909a:335.

Diagnosis. Wings lacking entirely. Lacinial teeth normal (tridentate). Mx4 only slightly enlarged apically. Two apical spurs on hind tibia and three on inner surface before apex, two of these side by side. Ten spurs on hind t1. Second valvula about half length of v3.

Generotype. *M. aptera* Enderlein.
North American species: *M. aptera* Enderlein.

Myrmicodipnella aptera Enderlein

Myrmicodipnella aptera Enderlein, 1909a:337.

Recognition features. The only species of its genus.
Relationships. The genus is probably close to *Cerobasis*.
Distribution and habitat. The species was found at San Francisco, California in a nest of an ant, *Leptothorax nitens* Emery. It is only known from the single type.

Genus *Trogium* Illiger

Trogium Illiger, 1798:500.
Atropos Leach, 1815:139.
Clothilla Westwood, 1841:480.

Diagnosis. Lacinial tip tridentate, teeth short. Mx4 lacking forked sensillum. Fore winglets present, developed as flat scales reaching base of abdomen. Coxal organ absent. Two apical spurs on hind tibia, none before apex. Hind t1 with three to four spurs.

Generotype. *T. pulsatorium* (Linnaeus).
North American species: *T. pulsatorium* (Linnaeus).

Trogium pulsatorium (Linnaeus)

Termes pulsatorium Linnaeus, 1758:610.
Hemerobius pulsatorius Fabricius, 1775:331.
Termes lignarium de Geer, 1778:314.

Trogium pulsatorium (Linnaeus). Illiger. 1798:500.
Atropos lignaria (de Geer). Leach. 1815:139.
Psylla pulsatoria Billburg. 1820:94.
Psocus pulsatorius (Linnaeus). Nitzch. 1821:276.
Troctes pulsatorius (Linnaeus). Burmeister. 1839:773.
Clothilla studiosa Westwood. 1841:480.
Clothilla pulsatoria (Linnaeus). Hagen. 1866:122.
Atropos pulsatoria (Linnaeus). Kolbe. 1880:91.
Clothilla ocelloria Weber. 1906:886.

Recognition features. Body in general creamy yellow with reddish brown band through middle of head and another from compound eye to antennal base. Abdomen with reddish brown spots mostly bordering anterior margins of segments.

Relationships. This almost exclusively domestic species has several congeners which occur in outdoor situations on dead grass and palm foliage.

Distribution and habitat. Within the study area, this species is not common, occurring in households and mills primarily in northeastern United States. It probably occurs in southeastern Canada, but there are no records at present. Outside the study area, it is widespread in Europe, and also occurs in Japan, Taiwan, Australia, and New Zealand. Although it occurs primarily within buildings, it is known in Europe from bee hives, wasp nests, and bird nests associated with buildings. It has been known to cause damage to herbaria and insect collections.

Family Group Psocatropetae

Diagnosis. 'Microspades' organ frequently present on antennal pedicel (Fig. 485). Spur sensillum of mx4 present or absent. Forked sensillum lacking on mx4. Forewing, when fully developed (Fig. 462), with apex rounded. Cu2 and 1A ending together on wing margin. Ovipositor with two or three (rarely one) valvulae on the side: v3 never very elongate, but in some cases very broad, always much broader than other valvulae. Spermatheca lacking gland area attached on sac but sometimes with vesicle opening on sac near duct attachment. Phallosome with basal struts usually separate basally, occasionally fused basally, or struts not discernible.

North American families: Psyllipsocidae, Prionoglaridae.

Family Psyllipsocidae

Diagnosis. Body and wings without scales but often with numerous setae. Mx2 without sensory spur. Distal inner sensillar field of labrum with five placoid sensilla each heavily ringed (Fig. 449). Vein Cu1 in forewing (Fig. 462), together with M, separating from R near wing base, then Cu1 separating from M about one-third distance from wing base to wing tip. Coxal organ present as rasp or absent. Pretarsal claw with preapical tooth but no basal setae or pulvillus. Phallosome with two basal struts never fused anteriorly. Ovipositor formed of v1, v2, and v3, or v1 absent. Spermatheca frequently with sac-like gland opening on spermathecal sac near exit of duct. Sperms not enclosed in vesicles within spermatheca.

North American genera: *Dorypteryx* Aaron, *Pseudorypteryx* Garcia Aldrete, *Psocatropos* Ribaga, *Psyllipsocus* Selys-Longchamps.

Key to the North American Genera and Species of Family Psyllipsocidae

1. Forewings reduced to slender straps with simple venation; R and M veins unbranched (Fig. 57). 'Microspades' organ of pedicel absent. Coxal organ absent .... Genus *Dorypteryx* 2
-- Forewing wider, or if a slender strap, with veins Rs and M both branched distally. Forewings sometimes greatly reduced. 'Microspades' organ present on pedicel (Fig. 58). Coxal rasp present 3

2. Forewing with a closed cell (Fig. 57a) in basal half; veins R1 (usually) and Cu1 (always) present. Preclunial abdominal segments dorsally well pigmented ....
.... *D. domestica* (Smithers)
-- Forewing lacking a closed cell in basal half; veins R1 and Cu1, never present (Fig. 57b). Preclunial abdominal segments with little or no pigmentation .... *D. pallida* Aaron

3. Forewing slender, at least length/greatest width ~ 3.85. 'Microspades' organ consisting of seven to eight units ....
.... Genus *Pseudorypteryx*, *P. mexicana* Garcia Aldrete
-- Forewing broader, at most length/greatest width ~ 2.76, or forewing greatly reduced, without venation 4

4. Vertex relatively narrow: distance between eyes/greatest width of postclypeus ~ 1.25. Meso- and metanotum considerably raised above level of head (Fig. 60) ....
.... Genus *Psocatropos*, *P. microps* (Enderlein)

--   Vertex relatively broad: distance between eyes/ greatest width of postclypeus ~ 1.41. Meso- and metanotum scarcely, if at all, raised above level of head .... Genus *Psyllipsocus* 5

5.   Mx4 widest near distal end. Spermapore sclerite a short sheath (Fig. 61) .... *P. oculatus* Gurney

--   Mx4 widest in middle, tapering towards distal end. Spermapore sclerite a rounded frame (Fig. 62) ....
.... *P. ramburii* Selys-Longchamps

Genus *Dorypteryx* Aaron

Dorypteryx Aaron, 1883:37.
Doloptexyx Smithers, 1958:113.

Diagnosis. Forewings (Fig. 59) reduced to slender straps with simple venation; R and M veins unbranched. Mx4 with one large sensillum associated with a curved seta (Fig. 450). 'Microspades' organ absent on pedicel. Coxal organ absent. Struts of Phallosome relatively far apart (Fig. 451). Three heavy setae distally on v3 (Fig. 452). Elaborate sclerotizations around opening of spermatheca and its accessory vesicle (Fig. 453). Paraproctal spine present and unmodified in both sexes.

Generotype. *D. pallida* Aaron.

North American species: *D. domestica* (Smithers), *D. pallida* Aaron.

Dorypteryx domestica (Smithers)

Doloptexyx domestica Smithers 1958:113.
Dorypteryx domestica (Smithers). Lienhard 1977:434.

Recognition features. Forewing (Fig. 59) with a closed cell in basal half; veins R1 (usually) and Cu1 (always) present; Veins Rs and M separating in distal half of wing. Preclunial abdominal segments dorsally well pigmented. 'Wings' of spermathecal sclerotizations abruptly tapered to slender tips (Fig. 453). Phallosome and hypandrium (Fig. 454) essentially same as in *D. pallida*.

Relationships. The species is close to its congener, *D. pallida*.

Distribution and habitat. Within the study area this species is represented by two specimens, one, taken by the author at Felmley Hall, Illinois State University, Normal, Illinois, the other from a house in New York City. Outside the study area, the species is known from

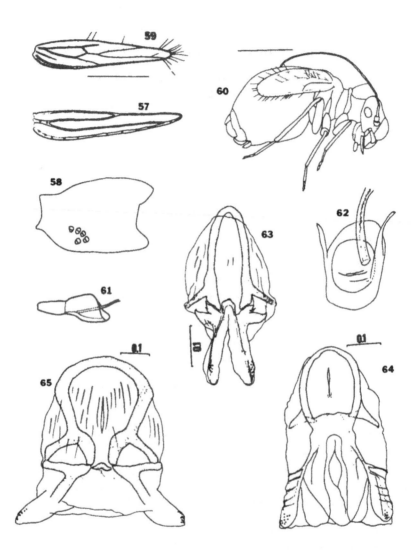

Figs. 57-65. 57 - forewing of *Dorypteryx pallida*; 58 - microspades organ of pedicel of *Psocatropos microps*; 59 - forewing of *Dorypteryx domestica*; 60 - habitus of *P. microps*; 61 - sheath of spermathecal duct of *Psyllipsocus oculatus*; 52 - spermapore plate of *Psyllipsocus ramburti*; 63-65 - phallosomes of *Speleketor* spp.: 63 - *S. irwini*; 64 - *S. pictus*; 65 - *S. flocki*.

several localities in central Europe, southern England, and from Zimbabwe. It is only known from domestic situations.

Note. The wing figured by Gurney (1950, Fig. 62c) is of this species, although the figure was meant to apply to *D. pallida*. Vein R1 is missing in both forewings of (presumably) this specimen.

Dorypteryx pallida Aaron

Dorypteryx pallida Aaron, 1883:38.
Dorypteryx albicans Ribaga (1907:1).

Recognition features. Veins R1 and Cu1 absent in forewing (Fig. 57). Preclunial abdominal segments with little or no pigmentation. 'Wings' of spermathecal sclerotization (Fig. 455) gradually tapered to slender tips. Phallosome and hypandrium (Fig. 451).

Relationships. As noted for *D. domestica*.

Distribution and habitat. Within the study area, the species has been taken at the type locality (Philadelphia Academy of Natural Sciences) and at the United States National Museum, Washington, D.C. Outside the study area, it is known from several localities in central Europe (France, Germany, Italy), and from a cave in Egypt. Aside from the latter find (specimens received by the author from A.A. El Meguid), all collections have been from domestic situations.

Genus *Pseudorypteryx* Garcia Aldrete

Pseudorypteryx Garcia Aldrete, 1984a:53.

Diagnosis. Forewings usually reduced to slender straps (Fig. 456) with simple venation with R1 absent, Rs and M both one-branched distally; or forewing somewhat broader with R1 present and M three-branched. Mx4 with single large sensillum plus smaller ones near distal end (Fig. 457). 'Microspades' organ present on pedicel, consisting of seven to eight units. Coxal rasp present. Struts of phallosome fused distally in middle (Fig. 459). Two heavy setae and one somewhat lighter distally on v3 (Fig. 460). Elaborate sclerotizations present around opening of spermatheca and its accessory vesicle (Fig. 461). Paraproctal spine present and unmodified in both sexes.

Generotype. *P. mexicana* Garcia Aldrete.

North American species: *P. mexicana* Garcia Aldrete.

Pseudorypteryx mexicana Garcia Aldrete

Pseudorypteryx mexicana Garcia Aldrete. 1984a:55.

Recognition features. Only species of its genus.

Relationships. This monotypic genus probably stands closest to *Dorypteryx*, and the two genera probable form a sister clade with *Psocatropos*.

Distribution and habitat. Within the study area, the species is known from two localities: Tucson, Arizona (indoor wall of a motel), and Great Sippewissett Marsh, Cape Cod, Massachusetts (presumably outdoor collection). Outside the study area, the species is known primarily from outdoor collections on dead persistent leaves at several localities in the Mexican states of Nuevo León and Jalisco, and in an herbarium at the University of Mexico, Mexico City.

Genus *Psocatropos* Ribaga

Psocathropos Ribaga, 1899:156. (amended spelling: Badonnel, 1931:254.)
Psocinella Banks, 1900a:431.
Axinopsocus Enderlein, 1903a:2.
Vulturops Townsend, 1912:267.
Gambrella Enderlein, 1931:221.

Diagnosis. Vertex relatively narrow: distance between eyes/greatest width of postclypeus 1.25. Most adults brachypterous with forewings (Fig. 60) extending little, if at all, beyond tip of abdomen, hindwings reduced to small flaps with little or no venation. Macropterous adult (rare), with forewings (Fig. 462) extending well beyond tip of abdomen, hindwings(Fig. 463) normal, with venation: closed cell in basal half, Rs branched, M simple. Forewing of macropterous form with two closed cells, second formed by r1 - rs crossvein (present often in brachypterous form), Cu1 branched (simple in brachypterous form). Both forms with meso- and metanotum considerably raised above level of head (Fig. 60). Mx4 with several small sensilla on outer margin. 'Microspades' organ present on pedicel, consisting of five to six units. Coxal rasp present. Struts of phallosome fused distally in middle. Ovipositor including three pairs of valvulae (Fig. 465). Only one heavy seta distally on v3. Only slight development of sclerotizations around opening of spermatheca and its accessory vesicle. Paraproctal spine absent in female, absent or modified in male.

Generotype. *P. lachlani* Ribaga.

North American species: *P. microps* (Enderlein).

Psocatropos microps (Enderlein)

Axinopsocus microps Enderlein, 1903a:3.
Psoquilla microps (Enderlein). Enderlein, 1908:776.
Psocatropos lesnei Badonnel, 1931:254.
Psocatropos microps (Enderlein). Badonnel, 1944:59.
Psocatropos lachlani Ribaga. Gurney, 1949:63 (? not Ribaga 1899:157).
Psocinella slossonae Banks, 1900a:432.
Vulturops floridanus Corbett and Hargreaves, 1915:142, new synonym.

Recognition features. Brachypterous forms with forewings not reaching tip of abdomen, hindwings very short, lacking venation. Macropterous forms (rare) with forewings far exceeding tip of abdomen, venation of fore- and hindwing as described in generic diagnosis. Both forms lacking ocelli. Hypandrium much broader than long (Fig. 466). Paraproctal spine present but modified in male (Fig. 467). Females with 'cylinder setae' (baguettes of Badonnel, 1931) on epiproct and paraproct(Fig. 468).

Relationships. The only other recognizable species at present is *P. pilipennis* (Enderlein). Other names in the literature (*P. lachlani* Ribaga, the generotype, and *P. territorum* Townsend) will probably prove to be synonyms of one of these two. *P. pilipennis* appears to have only one wing form, a rather long-winged brachypter, in which the forewings slightly exceed the tip of the abdomen and the hindwings have a closed cell from which arise a branched Rs, simple M, and simple Cu 1. The male of *P. pilipennis* has the hypandrium much longer than broad and lacks the paraproctal spines; the female lacks cylinder setae on the anal flaps.

Distribution and habitat. Within the study area the species has been found in several localities in Florida. It has been found almost entirely within buildings, but once on the north-facing outer wall of a building and once on natural vegetation in the Florida Keys. Outside the study area, the species has been recorded from Mexico, Cuba, Colombia, India, Madagascar, and several African localities.

Genus *Psyllipsocus* Selys-Longchamps

Psyllipsocus Selys-Longchamps, 1872:145.
Nymphopsocus Enderlein, 1903d:76.
Ocelloria Weber, 1906:858.
Ocellataria Weber, 1907:189.
Fita Navás, 1913:332.
Fabrella Lacroix, 1915:194.

Diagnosis. Vertex relatively broad: distance between eyes/greatest width of postclypeus - 1.41 or greater. Adults ranging from micropterous to macropterous. Macropterous adults generally with forewings extending beyond tip of abdomen; hindwing with closed cell, branched Rs and branched M. Forewing venational characters as noted for Psocatropos. Meso- and metanotum scarcely, if at all, raised above level of head. Mx4 with several small sensilla on outer margin. 'Microspades' organ of pedicel present but small, with two to three units. Coxal rasp present. Ovipositor including three pairs of valvulae or v1 absent. No heavy setae distally on v3. Only slight development of sclerotization around spermathecal opening; spermatheca lacking accessory vesicle. Paraproctal spine present or absent.

Generotype. *P. ramburi* Selys-Longchamps.

North American species: *P. oculatus* Gurney, *P. ramburi* Selys-Longchamps.

Psyllipsocus oculatus Gurney

Psyllipsocus oculatus Gurney, 1943:214.
Not *Psyllipsocus oculatus* Gurney. Mockford and Gurney, 1956:357.

Recognition features. Males frequent. Adults macropterous with forewings extending well beyond tip of abdomen or brachypterous, with forewings extending not quite to tip of abdomen. Wings unmarked (Fig. 25). Vein IA in hindwing branched (n - 17). Head, thorax, and terminal abdominal segments orange brown; preclunial abdominal segments pale yellowish white. Wing setae sparse, none on wing margin. Phallosome (Fig. 469) with basal struts fused in middle, forming angle of about 60°; median fused structure slightly extended anteriorly and producing small pair of jaw-like structures before and to sides of distal slender paired (fused?) forcep-tips. Ovipositor of three pairs of valvulae (Fig. 470). Spermapore sclerite short, rounded (Fig. 61). Paraproctal spine present. Cylinder setae absent on anal flaps.

Relationships. This species is a member of a complex of three species, including an undescribed one in the Monterrey region of Mexico and another undescribed one in southern Arizona.

Distribution and habitat. Within the study area, the species has been found in South Texas (Cameron County) and West Texas (Jeff Davis County). It has also been found at Laredo as an introduction from Mexico. Outside the study area it has been taken in the Mexican states of Michoacan, Puebla, Jalisco, Chiapas, and Oaxaca, also in Guatemala. Specimens cited by García Aldrete (1984) from Nuevo León and northern San Luis Potosí should be re-examined, as a similar species may be involved. It is usually found in dead persistent leaves of yucca, but has also been collected in bromeliads and Sabal palms.

Psyllipsocus ramburii Selys-Longchamps

Psocus pedicularius Rambur, 1842:323.
Psyllipsocus ramburii Selys-Longchamps, 1872:145.
Ocelloria gravonympha Weber, 1906:885.
Ocellataria gravinympha Weber, 1907:189.
Nymphopsocus troglodyta Enderlein, 1909b:536.
Nymphopsocus troglodyta var. *algericus* Enderlein, 1909b:538.
Fita vestigator Navás, 1913:333.
Fabrella convexa Lacroix, 1915:194.
Psyllipsocus (*Nymphopsocus*) *troglodytes* Enderlein. Badonnel, 1935b:201.
Psyllipsocus ramburii brachypterus Badonnel, 1943:131.
Psyllipsocus ramburii destructor (Enderlein). Badonnel, 1943:132.
Psyllipsocus ramburii troglodytes (Enderlein). Badonnel, 1943:132.
Dorypteryx hageni Banks, 1897:382, new synonym.

Recognition features. Males unknown in the study area. Adults micropterous (forewings not quite or barely reaching base of abdomen), brachypterous (forewings extending to about mid-length of abdomen), or macropterous (forewings extending beyond tip of abdomen). All forms with at least traces of ocelli. Pigmentation of short-winged individuals largely restricted to reddish-brown pigment of compound eyes and ocelli and slightly grayish-brown pigment of terminal abdominal segments. Macropterous individuals with head, thorax, and preclunial abdominal segments beyond about second pale reddish-brown. Ovipositor of three pairs of valvulae. Spermapore sclerite a circular structure (Fig. 62). Spermathecal sac with small sclerotic cap around exit of duct. Paraproctal spine present, very short in male.

Relationships. Lienhard (1977) has suggested that *P. ramburii* and most of the other described species of the genus stand rather close together, with *P. oculatus* somewhat apart, but data are sparse.

Distribution and habitat. Within the study area, the species occurs throughout the United States in caves, cellars, and occasionally stored collections of insects. It also occurs occasionally on shaded rocks outcrops. This is a widespread domestic species found throughout Europe and in many parts of the world where human commerce has taken it.

Note. The population from Mammoth Cave, Kentucky, named 'Dorypteryx' *hageni* Banks, appears to have appendages slightly more elongate than in other populations but otherwise does not differ.

Family Prionoglaridae

Diagnosis. Adults macropterous. Body and wings with only sparse setae. Mx2 with or without sensory spur. Sensilla of distal margin of labrum five tricholds or placoids with heavy rings. Antennal flagellum with sculpture annulations. Relationships of veins Cu1, M, and R as described for Psyllipsocidae. Sc in forewing (Fig. 472) forming large arc and fusing with R1 for short distance. Coxal organ present or absent. Pretarsal claw usually with preapical tooth; basal seta present or absent. Phallosome largely enclosed in membranous sac; basal struts, when discernible, fused anteriorly. Ovipositor consisting of pair of very broad, setose third valvulae (Fig. 473) with two to four heavy setae on distal end; a small accessory valvula (v1 or v2) present or not. Paraproctal spine absent.

North American genus: *Speleketor* Gurney.

Key to the North American Species of Family Prionoglaridae, Genus *Speleketor*

1.   Abdomen marked dorsally with transverse bands of reddish brown pigment on terga 2-8. Posterior (pore-bearing) lobes of phallosome touching on midline (Fig. 63) ....
.... *S. irwini* Mockford
--   Abdomen marked dorsally with spots, not transverse bands. Posterior lobes of phallosome separated on midline 2

2.   Vertex and postclypeus marked with dark brown blotches. Dorsal spots of abdomen extremely irregular in size and shape; those of midline mostly bisected by a clear line. Posterior lobes of phallosome directed posteriorly (Fig. 64) ....
.... *S. pictus* Mockford
--   Postclypeus unmarked; vertex at most with pale cloudy brown spots near hind margin. Dorsal spots of abdomen mostly quadrate, mostly of uniform size; those of midline undivided. Posterior lobes of phallosome directed laterally (Fig. 65) ......
.... *S. flocki* Gurney

Genus *Speleketor* Gurney

Speleketor Gurney, 1943:197.

Diagnosis. Mx2 with sensory spur. Sensilla of distal inner field of labrum five tricholds (Fig. 474). Adult lacking lacinial teeth (Fig. 475), with slender incisor region of mandibles (Fig. 476). M branches in hindwing arising separately from Rs-M stem (Fig. 477). Front femur with row of articulated spines (Fig. 478). Coxal organ absent. Pretarsal claw without basal seta. Phallosome with basal struts present and fused anteriorly, with well developed disto-lateral lobes but usually no disto-median lobes. Ovipositor with v3 bearing numerous setae peripherally, very few elsewhere; a minute auxiliary valvula (v1 or v2) present on median side in some species.

Generotype. S. *flocki* Gurney.

North American species: S. *flocki* Gurney, S. *irwini* Mockford, S. *pictus* Mockford.

Speleketor flocki Gurney

Speleketor flocki Gurney, 1943:197.

Recognition features. Head unmarked except for few cloudy brown spots on posterior margin of vertex. Abdomen marked with discrete reddish-brown spots on terga 3-8 forming five longitudinal series (Fig. 479). Phallosome with posterior lobes directed laterally (Fig. 65).

Relationships. The head shape and abdominal markings appear to be closer to those of S. *pictus*, but details of the phallosome appear closer to those of S. *irwini*.

Distribution and habitat. The species is known from two caves, one in the Tucson Mountains of southern Arizona and one (Gypsum Cave) in southeastern Nevada.

Speleketor irwini Mockford

Speleketor irwini Mockford, 1984a:173.

Recognition features. Head unmarked. Abdomen marked (Fig. 480) with complete or partial band of reddish-brown pigment across each of terga 2-8. Phallosome with posterior lobes directed posteriorly (Fig. 63).

Relationships. See under S. *flocki*.

Distribution and habitat. The species has been found only in southern California (Riverside and San Diego Counties), where it occurs on the skirts of dead leaves of the palm *Washingtonia filifera* (Linden) Wendl in native stands.

Speleketor pictus Mockford

Speleketor pictus Mockford, 1984a:177.

Recognition features. Head extensively marked with brown blotches (Fig. 481). Abdomen dorsally (Fig. 482) marked with seven longitudinal series of irregular dark reddish-brown blotches forming rough transverse rows on segments 2 through 8. Phallosome with posterior lobes directed posteriorly; a series of membranous lobes between outer more sclerotized lobes (Fig. 64).

Relationships. See under *S. flocki.*

Distribution and habitat. The species is known only from the type, collected at the Boyd Deep Canyon Desert Research Center, Riverside County, California, at a light. Hence, the habitat remains unknown.

Chapter 3
Suborder Troctomorpha

Diagnosis. Adults generally with 13 flagellomeres; at least some flagellomeres annulated with cuticular sculpture; hypopharyngeal filaments fused on midline much of their length, separate only in distal half (as in Fig. 17); labial palpus usually with minute basal segment, rounded distal segment, sometimes with single rounded segment. Tarsi of adult usually three-segmented. Wings present or not; when present, sometimes reduced. Forewing generally lacking sclerotized pterostigma; Cu2 and 1A usually ending separately on wing margin; IIA sometimes present. Subgenital plate large, often with t-shaped or triangular sclerite. Ovipositor generally with three valvulae on each side. Paraproct usually without a stout spine on free margin.

North American taxa: Family Group Nanopsocetae (Families Pachytroctidae, Sphaeropsocidae, Liposcelididae), Family Group Amphientometae (Family Amphientomidae).

Key to the Family Groups and Families
of Suborder Troctomorpha

1.   Small forms, rarely over two mm in length. Wings, when present and fully developed (females only), without scales, with not over two M branches in forewing, vein IIA never present. Coxal organ absent or represented by slight bulge in cuticle ....
....Family Group Nanopsocetae 2
Larger forms, generally three to five mm in length. Wings present in both sexes; forewings often covered with scales. In forewing, M three-brached, vein IIA usually present. Coxal organ repre-sented by mirror and usually small rasp ....
.... Family Group Amphientometae, Family Amphientomidae

2.   Body flattened. Coxae of opposite sides widely separated by sternal plates. Forewings, when present (some females), with two parallel veins occupying main body of wing ....
.... Family Liposcelididae

Body not flattened. Coxae of opposite sides only narrowly separated. Forewings, when present (some females), with several branching veins occupying main body of wing 3

3. Females with elytriform forewings (Fig. 496), no hindwings. Males micropterous; phallosome with basal struts joined on midline forming anterior apodeme .. Family Sphaeropsocidae
Females macropterous or apterous, rarely micropterous. Males apterous. Macropterous females with forewings membranous, folding flat over back at rest; hindwings present, developed as forewings. Phallosome with basal struts fused at midline only at anterior extreme, not forming anterior apodeme (Fig. 66) .... Family Pachytroctidae

Family Group Nanopsocetae

Diagnosis. Small forms, rarely over 2 mm in length. Wings, when present (females only), without scales, with not over two M branches in forewing, never with IIA. Coxal organ absent or at most represented by slight bulge in cuticle. Ovipositor of three valvulae but v2 and v3 fused together in some forms.

Family Pachytroctidae

Diagnosis. Body not flattened; coxae of opposite sides narrowly separated. Adults macropterous, brachypterous, or apterous. Eyes of numerous ommatidia. Sensilla of distal inner labral margin three placoids with heavy rims alternating with two trichoids with short setae (Fig. 483). Mx4 with six to eight claviform sensilla (Fig. 484): two to four pointed apically, four rounded apically. Pretarsal claws without basal seta or pulvillus. Wings, when present (females only), and fully developed, membranous, folded flat over back at rest. Phallosome with basal struts fused on midline but only at anterior extreme. Ovipositor with v3 simple, not bilobed (Fig. 485).

North American genera: *Nanopsocus* Pearman, *Pachytroctes* Enderlein, *Tapinella* Enderlein.

Key to North American Genera and Species of Family Pachytroctidae

1. Subgenital plate with T-shaped sclerite. Male clunium with two slender processes protruding over epiproct (Fig. 100) 3

--   Subgenital plate lacking T-shaped sclerite. Male clunium lacking two slender processes protruding over epiproct Genus *Pachytroctes* 2

2.   Head and abdomen dusky brown; thorax and legs pale yellowish brown (both sexes). Phallosome with pair of incurved anterior processes (Fig. 103) .... *P. aegyptius* Enderlein

--   Male with head, meso- and metathorax, and abdominal terga 4-10 reddish brown; pronotum and abdominal terga 1-3 pale yellowish brown. Female more uniformly reddish brown. Phallosome (Fig. 104) without anterior processes ....
                           P. neoleonensis García Aldrete

3.   Posterior pretarsal claw of each foot foliate, anterior claw normal (Fig. 106) .... Genus *Nanopsocus*, *N. oceanicus* Pearman

--   The two pretarsal claws of each foot equal, both normal ....
           Genus *Tapinella*, *T. maculata* Mockford and Gurney

Genus *Nanopsocus* Pearman

Nanopsocus Pearman, 1928a:134.
Onychotroctes Badonnel, 1969:73.

Diagnosis. Compound eyes relatively large, reaching relatively straight posterior head margin. Hind pretarsal claw of each foot foliate (Fig. 103). Subgenital plate with t-shaped sclerite (Fig. 486). Ovipositor v2 and v3 fused together. Phallosome (Fig. 66) rounded or truncate anteriorly. Male clunium with two slender processes protruding over epiproct (as in Fig. 100).

    Generotype. *N. oceanicus* Pearman.
    North American species: *N. oceanicus* Pearman.

Nanopsocus oceanicus Pearman

Nanopsocus oceanicus Pearman, 1928a:134.
Tapinella africana Badonnel, 1948:276.
Tapinella pallida Badonnel, 1949a:36.
Onychotroctes africanus (Badonnel), Badonnel, 1971d:23.
Tapinella oceanica (Pearman), Smithers, 1967:28.
Nanopsocus africanus (Badonnel), Broadhead and Richards, 1982:213.

Recognition features. Females macropterous or apterous. Macropterous females with head, thorax, and terminal abdominal segments medium brown; preclunial abdominal segments pale brown; wings with slight brown wash. Apterous females creamy white except pale brown along posterior margin of vertex. Males (apterous) colored as apterous females but with a reddish brown strip along anterior margin of each of abdominal terga 2-7.

Relationships. Three other species of *Nanopsocus* have been described: *N. trifasciatus* Badonnel, from Angola (seemingly the same species occurs in Uganda), *N. falsus* Badonnel, from Angola, and *N. longicornis* Badonnel, from Madagascar. An undescribed species occurs in Malaysia. Only brachypterous females of *N. trifasciatus* and the Malaysian species are known (ten individuals), both with the forewing banded. Females of *N. oceanicus* have not been found as brachypters, and the wing of the macropter is unmarked, suggesting that *N. oceanicus* stands somewhat apart from the African and Malaysian species. Females of the Madagascan species and *N. falsus* remain unknown.

Distribution and habitat. Within the study area, the species occurs in southeastern United States, north to Gallatin County, Illinois. At the northern end of its range, it has been found only on a long-established sawdust pile, which it shared with *Zorotypus hubbardi* Caudell. Farther south it occurs on palms, bamboos, grasses, and in houses. Outside the study area, it occurs in Mexico, Central America, West Indies, West and Central Africa, Japan (domestic), and some of the South Pacific islands, generally in non-domestic biotopes.

Genus *Pachytroctes* Enderlein

Pachytroctes Enderlein, 1905:46.

Note. This genus is treated here in the strict sense, with the taxa treated as subgenera by Roesler (1944) regarded as distinct genera. *Psyllotroctes* Roesler cannot be a synonym of *Pachytroctes*, s. str., (synonymy of Smithers, 1967) as the type species, *P. plaumanni* Roesler, is described as having a t-shaped sclerite of the subgenital plate.

Diagnosis. Compound eyes relatively small, generally not reaching hind head margin (Fig. 487). Posterior pretarsal claw of each foot same as anterior, a slender claw with preapical denticle. Subgenital plate lacking t-shaped sclerite. Ovipositor v2 and v3 not fused together. Phallosome variable anteriorly. Male clunium lacking two slender processes protruding over epiproct.

Generotype. *P. aegyptius* Enderlein.

North American species: *P. aegyptius* Enderlein, *P. neoleonensis* García Aldrete.

Pachytroctes aegyptius Enderlein

Pachytroctes aegyptius Enderlein, 1905:46.

Recognition features. Male body length ~ 1.00 mm. Head and abdomen dusky brown; thorax and legs pale yellowish brown; abdomen with quadrate spots somewhat clearer than surrounding cuticle along anterior margins of terga 3-7. Phallosome (Fig. 101) with elongate, incurved anterior processes.

Relationships. The body color and form of the phallosome suggest relationship to *P. dichromoscelis* Badonnel, a species known from West Africa and Brazil, and *P. ealensis* Badonnel, also a West African species.

Distribution and habitat. The specimens studied, two males, were taken at Gainesville, Florida, on a shipment of yams from West Africa, probably Ivory Coast. The species was originally found at Cairo, Egypt.

Pachytroctes neoleonensis García Aldrete

Pachytroctes neoleonensis García Aldrete, 1986b:9.

Recognition features. Known from males and apterous females. Male ~ 1.11, females ~ 1.60 mm in body length. Males medium reddish brown on head, meso- and metathorax, legs and abdominal terga 4-10; pale yellowish brown on prothorax, abdominal terga 1-3, and anal flaps. Antenna reddish brown from base through f2, fading to colorless on f3, f4, colorless on f5 to tip. Females with head and thorax reddish brown, preclunial abdominal segments somewhat paler, clunium and anal flaps same shade or somewhat darker than head and thorax. Phallosome lacking anterior processes, with long u-shaped anterior sclerite of endophallus (Fig. 102).

Note. Arizona specimens are in general duskier than the Texas and Mexican specimens but do not appear to differ otherwise.

Relationships. Lack of anterior processes on the phallosome and presence of a long anterior endophallic sclerite suggest close relationship to *P. pacificus* García Aldrete of western Mexico. The median marginal bulge of the subgenital plate suggests close relationship to *P. maculosus* García Aldrete, known from northern, central, and southwestern Mexico.

Distribution and habitat. Within the study area, the species has been found at Davis Mountains State Park, Jeff Davis County, Texas, and at one locality each in Gila and Graham Counties, Arizona. At these localities, it occurs in ground litter, primarily of oak leaves. Outside the study area, the species has been found at several localities in Nuevo León State, Mexico, where it was taken primarily in oak leaf litter.

Genus *Tapinella* Enderlein

Tapinella Enderlein, 1908:772.

Diagnosis. Compound eyes relatively large, reaching relatively straight posterior head margin (Fig. 488). Posterior pretarsal claw of each foot slender, same as anterior claw. Subgenital plate with t-shaped sclerite (Fig. 489). Ovipositor v2 and v3 not fused together (Fig. 485). Phallosome rounded or truncate anteriorly. Male clunium with two slender processes protruding over epiproct.

Generotype. *T. formosana* Enderlein.

North American species: *T. maculata* Mockford and Gurney.

Tapinella maculata Mockford and Gurney

Tapinella maculata Mockford and Gurney, 1956:360.

Recognition features. Known from males, macropterous and apterous females. Both sexes with ventro-lateral longitudinal series of dark brown spots on abdominal segments 1-6, occasionally forming continuous band. Male with lateral series of reddish brown spots each shaped as backward-directed u on abdominal segments 3-8; terga 5-7 each with a transverse dark sclerotized strip along anterior margin in middle (Fig. 488). Female with same lateral series of spots as in male plus two longitudinal series of quadrate spots on terga 2-7, or these spots continuous as band across each tergum (Figs. 490, 491).

Relationships. This species appears to be closest to *T. olmeca* Mockford of Mexico, Central America, and the West Indies. In the latter species, the ventro-lateral series of brown abdominal spots is absent, and females occur as macropters and brachypters, but not apters.

Distribution and habitat. Within the study area, the species occurs in the lower Rio Grande Valley of Texas, where it is found on palm leaves. Outside the study area, the species occurs throughout eastern Mexico and in the western Mexican states of Sonora, Sinaloa, Nayarit, and Jalisco. It is widespread in Guatemala, and also occurs in Belize. It is known from several kinds of palms, also from foliage of banana, yucca, *Heliconia*, and *Typha*.

Family Sphaeropsocidae

Diagnosis. Body not flattened; coxae of opposite sides narrowly separated. Adult males micropterous, females with elytriform forewings, no hindwings. Eyes of three to ten ommatidia. Sensilla of distal inner labral margin consisting of all tricholds, three more basal with heavy, rounded sockets and short setae alternating with 1 - 2 - 2 - 1 or 3 - 1 - 1 - 3 more distal with quadrate sockets and longer setae (Fig. 492). Mx4 with three to six trichoid and two to three claviform sensilla (Fig. 493). Pretarsal claws without basal seta or pulvillus. Phallosome with basal struts fused anteriorly forming slender apodeme. T-shaped sclerite always present on subgenital plate (Fig. 494). Ovipositor with v3 rounded to deeply bilobed; all valvulae from a common stem (Fig. 495).

North American genera: *Sphaeropsocopsis* Badonnel, *Sphaeropsocus* Hagen (?).

Genus *Sphaeropsocopsis* Badonnel

Sphaeropsocopsis Badonnel, 1963:322.

Diagnosis. Sensilla of distal labral margin consisting of eight longer tricholds alternating with three shorter tricholds in numbers 3 - 1 - 1 - 1 - 1 - 3. Forewing with not more than five veins, lacking Cu2; wings tucked under abdomen on sides. Ovipositor with v3 deeply bilobed (Fig. 495).

Generotype. *S. chilensis* Badonnel.

North American species: *S. argentinus* Badonnel.

Sphaeropsocopsis argentinus (Badonnel)

Sphaeropsocus argentinus Badonnel, 1962:218.
Sphaeropsocopsis argentinus (Badonnel). Badonnel, 1963:323.

Recognition features. Forewing with four veins, all ending freely. Each vein with double row of short club-shaped sensilla along its length (Fig. 496). Short club-shaped sensilla scattered on other body parts, especially abundant along anterior margin of mesonotum, hind margin of vertex. Body and forewings medium brown in color. Ovipositor valvulae (Fig. 495).

Relationships. Most of the species of *Sphaeropsocopsis*, including this one, are South American (one Tasmanian, one West African). This species seems to stand closest to *S. microps* Badonnel, of Chile, in wing shape and venation (all veins free from margin), but differs markedly from all others in having abundant club-shaped sensilla on wing and body surfaces.

Distribution and habitat. Within the study area, a single specimen was taken at J. F. Kennedy International Airport in New York on *Nothofagus* stem galls from Argentina. Outside the study area, the species is known from three individuals, one a nymph, one a wingless adult, and one an isolated wing from three Argentine localities in the region of Esquel.

Genus *Sphaeropsocus* Hagen (?)

Sphaeropsocus Hagen, 1882:225.
Paleotroctes Enderlein, 1911:350.

The type of this genus, *S. kunowii* Hagen, is a fossil from the Baltic amber. Little is known of its morphology beyond wing shape and venation. Three female specimens from California, possibly representing two species, appear to be closer to the type of this genus than to any described species of *Sphaeropsocopsis* in wing shape and venation. All have six veins, including Cu2, but Cu1 is simple and joins M2 distally, unlike *S. kunowii*. The forewing appears to be wider than in the known species of *Sphaeropsocopsis*, but not quite as wide as suggested by the figure of *S. kunowii* (Enderlein, 1911, Fig. 97). These will be studied in greater detail in a later work.

Family Liposcelididae

Diagnosis. Body flattened. Coxae of opposite sides widely separated by sternal plates. Adult males apterous; females apterous or macropterous; macropterous females with wings folded flat over back at rest; forewings with not over four veins. Apterous forms with two to eight ommatidia in eye, winged forms with larger number. Sensilla of distal inner labral margin as described for Family Sphaeropsocidae. Hind femora swollen. Pretarsal claws without pulvillus, with or without basal setae. Phallosome with basal struts fused on midline forming slender apodeme anteriorly. Ovipositor valvulae with common stem; v3 rounded apically. T-shaped sclerite of subgenital plate present or absent.

North American taxa: Subfamily Embidopsocinae, Subfamily Liposcelidinae.

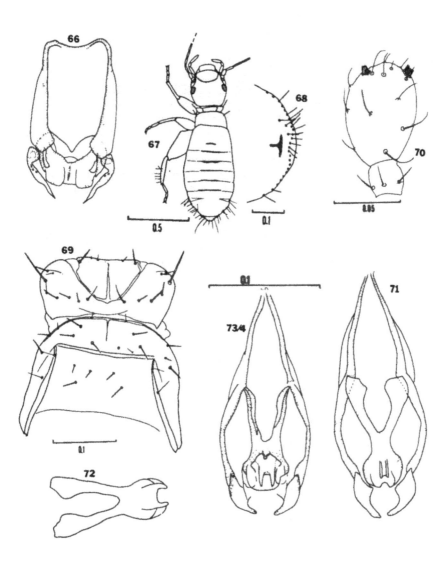

Figs. 66-74. 66 - phallosome of *Nanopsocus oceanicus*; 67 - habitus of *Liposcelis brunnea*; 68 - subgenital plate of *Liposcelis* sp.; 69 - thoracic nota of *Belaphotroctes alleni*; 70 - mx4 of *Belaphotroctes badonneli*; 71 - phallosome of *B. badonneli*; 72 - endophallus of *Belaphotroctes alleni*; 73/74 - phallosome of *Belaphotroctes simberloffi*.

Key to North American Genera and
Species of Family Liposcelididae

1. Hind femur lacking lateral protuberance. Females apterous or macropterous, without t-shaped sclerite on subgenital plate .... Subfamily Embidopsocinae 2
-- Hind femur with lateral protuberance (Fig. 67). Females apterous, with t-shaped sclerite on subgenital plate (Fig. 68) .. .. Subfamily Liposcelidinae, Genus *Liposcelis* 16

2. Mx4 at least slightly, sometimes considerably, wider than mx3. Meso-metanotum of apterous forms with slender transverse sclerite before middle (Fig. 69). Preclunial abdominal terga lacking slender transverse bands of heavy sclerotization .... Genus *Belaphotroctes* 3
-- Mx4 not or scarcely wider than mx3. Meso-metanotum of apterous forms lacking slender transverse sclerite before middle. Preclunial abdominal terga frequently each with a slender transverse band of heavy sclerotization .... Genus *Embidopsocus* 10

3. Mx4 ~ 2.0x width of mx3 in both sexes; no groups of closely-set rod-like sensilla on mx4 .... *B. ghesquierei* Badonnel
-- Mx4 ~ 1.5X width of mx3, or, if wider, with two groups of closely-set rod-like sensilla on ventral surface 4

4. Mx4 ~ 2.0x width of mx3 and with two groups of closely- set rod-like sensilla on ventral surface (Fig. 70) ....*B. badonneli* Mockford, female
-- Mx4 ~ 1.5x width of mx3, with or without groups of closely-set rod-like sensilla on ventral surface 5

5. Males (phallosome obvious). Mx4 lacking any conspicuous sensilla on ventral surface 6
-- Females (no phallosome). Mx4 with conspicuous sensilla on ventral surface 8

6. Phallosome with middle piece simple distally bearing two parallel thickened lines (Fig. 71) .... *B. badonneli* Mockford, male
-- Phallosome with middle piece more complex, bearing two posteriorly directed processes on hind margin 7

7. Processes of middle piece of phallosome separated by more than length of a process (Fig. 72) ..... *B. hermosus* Mockford, male

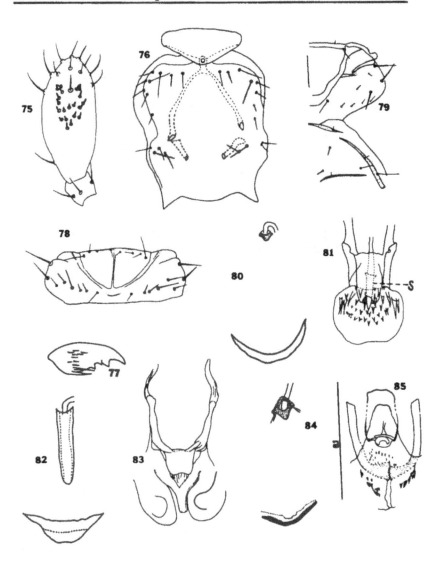

Figs. 75-85. *Belaphotroctes* and *Embidopsocus* spp. 75 - mx2 and mx3 of *B. alleni*; 76 - thoracic sterna of *E. bousemani*; 77 - pretarsal claw of *E. bousemani*; 78 - pronotum of *E. thorntoni*; 79 - pro- and mesonota of *E. needhami*; 80 - spermapore sclerite and u-shaped sclerite of *E. citrensis*; 81 - distal end of endophallus of *E. citrensis*; 82 - spermapore sclerite and u-shaped sclerite of *E. thorntoni*; 83 - endophallus of *E. thorntoni*; 84 - spermapore sclerite and u-shaped sclerite of *E. bousemani*; 85 - distal end of endophallus of *Embidopsocus laticeps*.

--   Processes of middle piece of phallosome separated by less than
         length of process (Fig. 73) .... *B. simberloffi* Mockford, male

8.   Ventral surface of mx4 with numerous scattered, short claviform
         sensilla but no groups of rod-like sensilla (Fig. 75) ....
         .... *B. alleni* Mockford, female
--   Ventral surface of mx4 with at least one group of rod-like sensilla
         9

9.   Ventral surface of mx4 with two groups of rod-like sensilla (as in
         Fig. 70) .... *B. simberloffi* Mockford, female
--   Ventral surface of mx4 with one group of rod-like sensilla ......
         .... *B. hermosus* Mockford, female

10.   Sclerotized bands of mesosternum forming closed arc joined to
         pro-mesothoracic spina (Fig. 76). Pretarsal claws with small
         denticles basal to larger preapical denticle (Fig. 77) ....
         .... Group I 14
--   Sclerotized bands of mesosternum, when present, not joined to
         pro-mesothoracic spina. Pretarsal claws with or without small
         denticles basal to larger preapical denticle ....
         .... Groups II and III 11

11.   Sclerotized bands of mesosternum absent. Pretarsal claws lack-
         ing minute denticles basal to preapical denticle ....
         .... Group III, *E. femoralis* Badonnel
--   Sclerotized bands of mesosternum present, forming closed arc or
         open in middle. Pretarsal claws with minute denticles basal to
         preapical denticle or not. .... Group II 12

12.   Lateral sclerotized bands delineating median pronotal lobe con-
         tinuing in a curve to form an arc (Fig. 78). Pretarsal claws with
         minute denticles basal to larger preapical denticle ....
         .... Subgroup A 13
--   Lateral sclerotic bands delimiting median pronotal lobe meeting
         hind margin of pronotum at a distinct angle (Fig. 79). Pretarsal
         claws lacking minute denticles basal to preapical denticle ...
         .... Subgroup B, *E. needhami* (Enderlein)

13.   Sclerotized bands of mesosternum not meeting in middle.
         Spermapore plate minute, triangular (Fig. 80). Radula of
         numerous acuminate spines (Fig. 81) ....
         .... *E. citrensis* Mockford

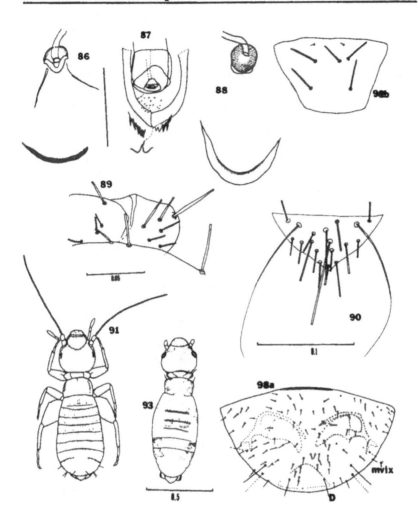

Figs. 86-91, 93, 98. *Embidopsocus* and *Liposcelis* spp. 86 - spermapore sclerite and u-shaped sclerite of *E. mexicanus*; 87 - distal end of endophallus of *E. mexicanus*; 88 - spermapore sclerite and u-shaped sclerite of *E. laticeps*; 89 - bristles of pronotum and anterior edge of mesonotum of *L. hirsutoides*; 90 - epiproct of *Liposcelis brunnea* showing pair of long, slender setae; 91 - habitus of *Liposcelis formicaria*; 93 - body color pattern of *Liposcelis ornata*; 98a - bristles of terminal abdominal segments of *Liposcelis corrodens*; 98b - bristles of prosternum of *L. bostrychophila*.

--   Sclerotized bands of mesosternum meeting in middle. Spermapore
        plate a long, tapering sheath (Fig. 82). Radula of a few minute
        denticles (Fig. 83) .... *E. thorntoni* Badonnel

14.   Spermapore sclerite in form of two triangular structures joined
        at bases, forming diamond-shaped structure around sperma-
        pore (Fig. 84). Radula of many minute denticles and a few
        larger ones (Fig. 507) .... *E. bousemani* Mockford
--   Spermapore sclerite more rounded. Radula with few minute
        denticles and 10 - 14 larger ones (Fig. 85) 15

15.   Spermapore sclerite of rounded anterior piece and broad- trian-
        gular posterior piece (Fig. 86). Orifice of penis canal slightly
        sunk into end of cylindrical terminal sclerotization (Fig. 87)
        .... *E. mexicanus* Mockford
--   Spermapore sclerite a single rounded piece (Fig. 88). Orifice of
        penis canal somewhat raised on end of cylindrical terminal
        sclerotization (Fig. 85) .... *E. laticeps* Mockford

16.   Abdominal terga 3-4 not presenting, at least in middle, a clearer
        posterior membranous band with sculpture different from the
        anterior part. All prosternal bristles anterior to middle ....
        .... Section I 17
--   Abdominal terga 3-4 presenting a clearer posterior membranous
        band with sculpture different from anterior part (Fig. 91). With
        or without a pair of prosternal bristles posterior to middle ..
        .... Section II 34

17.   Prothoracic humeral bristle relatively strong and long, a trans-
        verse row of two to five bristles (rarely only one) on lateral
        pronotal lobe (Fig. 89). Female usually with eight ommatidia in
        each eye .... Group A 18
--   Prothoracic humeral bristle relatively strong and long, but no
        other long bristles on lateral pronotal lobe. Female compound
        eye frequently with fewer than eight ommatidia . Group B 27

18.   A single supplementary long bristle on each lateral lobe of
        pronotum. Two long, curved, acuminate setae on epiproct (Fig.
        90) .... Subgroup Aa, *L. brunnea* Motschulsky
--   Each lateral lobe of pronotum with transverse row of two to five
        bristles (exceptions: only one supplementary bristle in species
        with uniformly pale body). Epiproct without pair of long,
        curved, acuminate setae .... Subgroup Ab 19

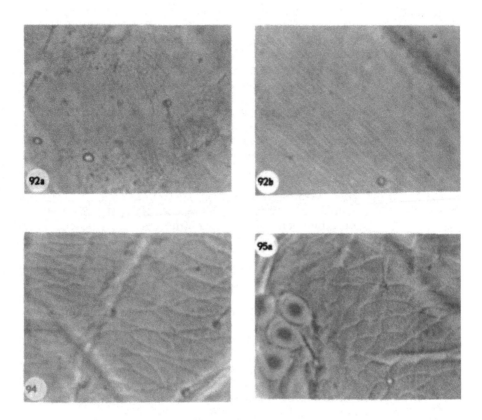

Figs. 92, 94, 95a. 92a, b - sculpture of abdominal terga of *Liposcelis pallida* (a) and *L. villosa* (b); 94 and 95a - sculpture of vertex of *L. rufa* (94) and *L. decolor* (95a).

19. A single long, strong supplementary bristle on lateral lobe of pronotum in addition to humeral bristle 20
-- A transverse row of long, strong bristles on each lateral lobe of pronotum in addition to humeral bristle 21

20. Antennae ~ 1.3X length of body. Sculpture of abdominal terga: irregular minute reticulations showing weak tendency to form areoles (Fig. 92-a) .... *L. pallida* Mockford
-- Antennae about same length as body. Sculpture of abdominal terga transverse areoles enclosing granules (Fig. 92-b) .... *L. villosa* Mockford

21. Abdomen marked with a conspicuous color pattern: brown or reddish-brown marks on a creamy white background 22
-- Abdomen uniform in color: pale, brown or yellow; at most, basal two segments contrastingly paler than others 24

22. Abdomen marked with repeating pattern of spots or transverse bands. Head unmarked 23
-- Abdominal pattern complex, not repeating (Fig. 93); head with y-shaped brown mark on vertex and frons .... *L. ornata* Mockford

23. Abdomen with conspicuous purplish brown spot on each side on terga 3-10 with faint transverse band of same color between spots .... *L. deltachi* Sommerman
-- Abdomen marked with transverse bands of purplish brown fading in middle along hind margins of terga 2 and 3 and fore margins of terga 6-9 .... *L. entomophila* (Enderlein)

24. Head entirely medium brown, contrasting with yellow remainder of body .... *L. fusciceps* Badonnel
-- Head in its entirety not contrasting in color with rest of body (but 'nasus' darker in one species) 25

25. Body color pale brown except white on abdominal terga 2 and 3. Short, truncate hairs abundant on all abdominal terga .... *L. hirsutoides* Mockford
-- Body color paler -- some shade of yellow. Hairs of abdomen slender, acuminate 26

26. Body color buffy yellow except anterior half of head gradually darkening to rusty brown on clypeus and labrum .... *L. nasa* Sommerman
-- Body uniformly pale ocraceous yellow with no darkening of head anteriorly .... *L. pallens* Badonnel

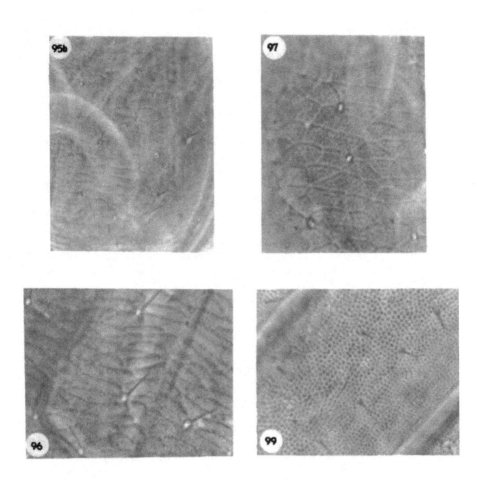

Figs. 95b, 96, 97, 99. 95b - sculpture of vertex of *L. decolor* (lesser magnification than in Fig. 95a); 96,97 - sculpture of vertex of *L. nigra* (96) and *L. silvarum* (97); 99 - sculpture of abdominal terga of *L. bostrychophila*.

27. Female eye with fewer than seven ommatidia 28
--   Female eye with seven ommatidia 29

28. Female eye with five to six ommatidia .... *L. pearmani* Lienhard
--   Female eye with three ommatidia .... *L. triocellata* Mockford

29. Body color neither dark brown nor with a striking pattern . 30
--   Body color either dark brown or with a striking pattern 32

30. Lacinial tip with more than three teeth ....
     ....*L. lacinia* Sommerman
--   Lacinial tip with three teeth 31

31. Color reddish brown. Head sculpture distinct (transverse areoles
     with granular surfaces) but without tubercles (Fig. 94) ....
     ....*L. rufa* Broadhead
--   Color ocraceous yellow. Head sculpture faint but with small
     tubercles forming wavy longitudinal lines (Fig. 95) ....
     .... *L. decolor* Pearman

32. Body color a striking pattern: head and abdomen brown, thorax
     yellow .... *L. bicolor* Banks
--   Body color dark brown 33

33. Areoles of vertex smooth or finely granulate (Fig. 96) ....
     .... *L. nigra* (Banks)
--   Areoles of vertex with nodules (Fig. 97) ....
     .... *L. silvarum* Kolbe

34. Prosternal setae all in front of middle of prosternum Group C 35
--   Two prosternal setae behind middle of prosternum (Fig. 98b)
     Group D 36

35. Color medium brown. Eight ommatidia in female eye ....
     ....*L. formicaria* (Hagen)
--   Color paler brown. Seven ommatidia in female eye ....
     ....*L. mendax* Pearman

36. Female eye with three ommatidia .... *L. paeta* Pearman
--   Female eye with seven ommatidia 37

37. Setae D and MvIX differentiated (Fig. 98a) ....
     .... *L. corrodens* Heymons
--   Setae D and MvIX not differentiated 38

38.   Body length 0.87 - 0.97 mm. Granules of abdominal terga
       beyond tergum 2 not forming areoles ....
       .... *L. prenolepidis* Enderlein
--   Body length 1.10 - 1.25 mm. Granules of abdominal terga
       beyond tergum 2 forming distinct areoles (Fig. 99) ....
       .... *L. bostrychophila* Badonnel

Subfamily Embidopsocinae

Diagnosis. Females apterous or macropterous. Hind femur lacking lateral protuberance. Pretarsal claw with or without basal seta. Subgenital plate without t-shaped sclerite.

North American genera: *Belaphotroctes* Roesler, *Embidopsocus* Hagen.

Genus *Belaphotroctes* Roesler

Eutroctes Ribaga, 1911:165 (preoccupied).
Belaphotroctes Roesler, 1943:13.

Diagnosis. Mx4 at least slightly, sometimes considerably wider than mx3. Meso-metanotum of apterous forms with slender transverse sclerite before middle (Fig. 69). Preclunial abdominal terga lacking slender transverse bands of heavy sclerotization. Phallosome with basal struts only touching at anterior end, or at most fused a very short distance.

Generotype. *B. traegardhi* (Ribaga).

North American species: *B. alleni* Mockford, *B. badonneli* Mockford, *B. ghesquierei* Badonnel, *B. hermosus* Mockford, *B. simberloffi* Mockford.

Belaphotroctes alleni Mockford

Belaphotroctes hermosus Mockford (in part), 1963:27.
Belaphotroctes alleni Mockford, 1978b:558.

Recognition features. Known only from apterous female. Mx4 ~ 1.5X as wide as mx3, with scattered, short claviform sensilla on ventral surface (Fig. 75). Pretarsal claw (Fig. 497) with short, slender basal seta, short preapical denticle. Spermapore sclerite slender, tapering toward posterior opening (Fig. 498).

Relationships. On the basis of shape of the spermapore sclerite and the nature and distribution of ventral sensilla of mx4, this species appears to be closest to *B. major* Badonnel from the Mato Grosso of Brazil.

Distribution and habitat. The species has been found at three localities: Davis Mountains State Park, Texas, Diamond A Ranch near Roswell, New Mexico, and Oak Creek Canyon, Yavapai County, Arizona. It occurs in leaf litter, principally of oak and juniper debris.

Belaphotroctes badonneli Mockford

Belaphotroctes okalensis Mockford (in part), 1963:31.
Belaphotroctes badonneli Mockford, 1972:155.

Recognition features. Known from male, macropterous and apterous female. Female mx4 at least 2X as wide as mx3 and bearing on its lower surface distally two groups of closely-set rod-like sensilla (Fig. 70). Pretarsal claw with short, slender basal seta, short preapical denticle. Spermapore sclerite very slender, tapering towards posterior opening (Fig. 499). Male mx4 ~ 1.5X width of mx3, lacking closely-set rod-like sensilla. Phallosome with middle piece simple distally, bearing two parallel thickened lines (Fig. 71).

Relationships. *B. vaginatus* Badonnel, from the Mato Grosso, Brazil, is very similar in shape of the spermapore plate and in having two distal groups of closely set rod-like sensilla on mx4 but differs in that mx4 is not quite as broad and it possesses a field of placoid sensilla laterally on the frons near the antennal base.

Distribution and habitat. The species has been found in Alachua County, Florida, and on the Texas coast in Matagorda County. It occurs on bark of trees and in ground litter.

Belaphotroctes ghesquierei Badonnel

Belaphotroctes ghesquierei Badonnel, 1949b:20.
Belaphotroctes okalensis Mockford (in part), 1963:31.
Belaphotroctes similis Mockford, 1969b:1268.

Recognition features. Known from male, macropterous female, and apterous female. Female mx4 2X as wide as mx3, with no groups of closely-set rod-like sensilla ventrally but with numerous short, scattered setae distally. Pretarsal claw with preapical denticle in about middle of claw. Spermapore sclerite (Fig. 500) rounded. Male mx4 slightly over 2X width of mx3. Phallosome with middle piece showing lateral lobes and central thickening distally, in middle with two heavy prongs slightly diverging (Fig. 501).

Relationships. On the basis of the nature of mx4 and its sensilla, spermapore sclerite, pretarsal claw, and phallosome, or some of these features, this species appears to be close to *B. striatus* Badonnel, *B. brunneus* Badonnel, and *B. ocularis* Badonnel, all from Sao Paulo State, Brazil.

Distribution and habitat. Within the study area the species occurs throughout most of Florida, where it is found under loose bark of dead tree trunks and branches. Outside the study area, it has been found in southern and western Brazil, West Africa, Madagascar, Mexico (present day) and the tertiary amber from Chiapas, Mexico.

Belaphotroctes hermosus Mockford

Belaphotroctes hermosus Mockford, 1963:27.

Recognition features. Known from male and apterous female. Mx4 in both sexes ~ 1.5X width of mx3: female with one group of close-set rod-like sensilla on ventral surface of mx4 (Fig. 502). Pretarsal claw with short, slender basal seta, short preapical denticle. Spermapore sclerite bell-shaped (Fig. 503). Phallosome with middle piece bearing two posteriorly-directed processes on hind margin at sides (Fig. 72).

Relationships. This species appears to be close to *B. simberloffi* Mockford in details of the middle piece of the phallosome and in proportions of mx4. It differs in having the distal processes of the middle piece of the phallosome farther apart, having only one group of closely-set rod-like sensilla on mx4 of the female, and in having a relatively much broader spermapore sclerite.

Distribution and habitat. Within the study area, the species occurs in southern Texas, where it is found in woodland ground litter. Outside the study area, it has been taken at Monterrey, Nuevo León, Mexico, in the same habitat.

Belaphotroctes simberloffi Mockford

Belaphotroctes simberloffi Mockford, 1972:159.

Recognition features. Known from male, apterous and macropterous female. Mx4 in both sexes ~ 1.5X width of mx3; female with two groups of closely-set rod-like sensilla on ventral surface of mx4 (as in Fig. 70). Pretarsal claw with short basal seta, distal preapical denticle. Spermapore sclerite relatively slender, tapering posteriorly (Fig. 504).

Phallosome with middle piece bearing posterior processes separated by slightly more than width of a process at its base (Fig. 73).

Relationships. As noted for *B. hermosus.*

Distribution and habitat. The species has been found only on small islands in the Florida Keys, where it occurs on red mangrove (*Rhizophora mangle* L.).

Genus *Embidopsocus* Hagen

Embidopsocus Hagen, 1866:170.
Tropusia Hagen, 1883:296.
Stenotroctes Enderlein, 1905:43.
Embidotroctes Enderlein, 1905:48.
Trigonosceliscus Enderlein, 1910:75.

Diagnosis. Mx4 not or scarcely wider than mx3. Meso-metanotum of apterous forms lacking slender transverse sclerite before middle. Preclunial abdominal terga often each with a slender transverse band of heavy sclerotization. Phallosome with basal struts variable: fused for a distance or only touching at anterior end.

Generotype. *E. luteus* Hagen.

Badonnel (1955, 1969, 1972) has elaborated a system of classification of this genus which is followed here. The categories, their characterization, and included species from the study area are as follows:

Group I. Sclerotized bands of mesosternum forming closed arc joined to pro-mesothoracic spina (Fig. 76). Pretarsal claws with small denticles basal to larger preapical denticle (Fig. 77).

North American species: *E. bousemani* Mockford, *E. laticeps* Mockford, *E. mexicanus* Mockford.

Group II. Sclerotized bands of mesosternum forming a closed or open arc not joined to pro-mesothoracic spina (Fig. 505). Pretarsal claws with or without minute denticles basal to larger preapical denticle.

Subgroup A. Pretarsal claws with small denticles basal to larger preapical denticle (as in Fig. 77). Lateral sclerotized bands delimiting median pronotal lobe continuing in a curve to form closed arc (Fig. 78).

North American species: *E. citrensis* Mockford, *E. thorntoni* Badonnel.

Subgroup B. Pretarsal claws lacking small denticles basal to larger preapical denticle. Lateral sclerotic bands delimiting median pronotal lobe meeting hind margin of pronotum at a distinct angle (Fig. 79).

North American species: *E. needhami* Enderlein.

Group III. Sclerotized bands of mesosternum absent. Pretarsal claws without small denticles basal to larger preapical denticle.
North American species: *E. femoralis* (Badonnel).

Group I

The species from the study area, all of which belong to subgroup B (abdominal terga with narrow transverse sclerotic bands; those of terga 4-6 flexuous) also share the following characters: (1) base and tip sclerotization around penis orifice firmly joined together, (2) sculpture of vertex, thoracic and abdominal terga minute-granulate; (3) male tenth abdominal tergum with field of trichoid sensilla on each edge behind seta M10 (Fig. 506).

Embidopsocus bousemani Mockford

Embidopsocus laticeps Mockford, 1974a:111 (in part).
Embidopsocus bousemani Mockford, 1987b:852.

Recognition features. Known from male and apterous female. Spermapore sclerite in form of two triangular pieces joined at bases around spermapore (Fig. 84). U-shaped sclerite slender, curved. Sclerotization of penis tip slightly bilobed posteriorly. Denticles of radula: large number of minute ones, a few scattered larger ones (Fig. 507).

Relationships. The group I-B species of the study area form a close complex together with *E. cubanus* Mockford (Cuba), *E. brazilianus* Badonnel (Brazil), and *E. virgatus* (Enderlein) of southern Brazil and Argentina.

Distribution and habitat. This species has been found only in ridge-top situations in the Ozark Uplift area of southern Illinois and in the western edge of the Appalachians in northern Alabama. It is found under loose bark of dead trees.

Embidopsocus laticeps Mockford

Embidopsocus laticeps Mockford, 1963:33.
Not *E. laticeps* Mockford, 1974b:111.

Recognition features. Known from males, macropterous and apterous females. Spermapore sclerite a single rounded piece; u-shaped sclerite slender, deeply curved (Fig. 88). Orifice of penis canal somewhat raised in middle of cylindrical terminal sclerotization (Fig.

85). Radula containing few scattered minute denticles in addition to two rows of larger denticles (Fig. 85).

Relationships. These are as noted for *E. bousemani.*

Distribution and habitat. The species occurs throughout Florida, north on the coastal plain to Bullock County, Georgia, and west on the coastal plain to New Orleans. It is found principally under bark of dead branches.

Embidopsocus mexicanus Mockford

Embidopsocus mexicanus Mockford, 1987b:858.

Recognition features. Known from male, macropterous and apterous female. Spermapore plate developed as front and hind semicircular pieces, the hind piece slightly triangular (Fig. 86). U-shaped sclerite slender, slightly curved. Orifice of penis canal slightly sunk into end of cyclindrical terminal sclerotization (Fig. 87). Radula containing relatively few minute denticles in addition to two rows of larger ones (Fig. 587).

Relationships. These are as noted for *E. bousemani.*

Distribution and habitat. Within the study area, this species is known only from two localities in Jim Wells County, southeastern Texas, where it was taken by beating shrubs. Outside the study area, it is known from the Mexican states of Chiapas, Oaxaca, and Veracruz, where it has been taken by beating vegetation in various kinds of woodlands, and under bark of dead branches.

Group II - A

Embidopsocus citrensis Mockford

Embidopsocus citrensis Mockford, 1963:35.

Recognition features. Known from male and apterous female. Sclerotized bands of mesosternum open in front. Spermapore plate minute, triangular, blunted posteriorly (Fig. 80). U-shaped sclerite curved, crescentic. Penis canal capped by a sheath with terminal point (Fig. 81); radula of numerous acuminate spines (Fig. 81).

Relationships. The spiny radula of this species is somewhat reminiscent of those of the *E. flexuosus* complex and *E. frater* Badonnel of South America.

Distribution and habitat. The species is known only from north-peninsular Florida, where it was collected under loose bark of a dead tree.

Embidopsocus thorntoni Badonnel

Embidopsocus thorntoni Badonnel, 1971b:325.

Recognition features. Known from male and apterous female. Sclerotized bands of mesosternum joined in middle. Spermapore plate a long, tapering sheath; u-shaped sclerite in form of shallow bowl (Fig. 82). Penis canal capped by broad sheath tapering to acuminate point (Fig. 83); radula of a few minute denticles.

Relationships. The species seems to stand decidedly apart from other known members of its section.

Distribution and habitat. Within the study area the species has been found at Miami, Florida as an introduction from Ecuador and established in a store room at Lake Buena Vista, Orange County, Florida. Outside the study area, the species is known only from the single type, collected on Santa Cruz (= Indefatigable) Island, Galapagos Archipelago.

Group II - B

Embidopsocus needhami (Enderlein)

Troctes needhami Enderlein, 1903f:360.
Stenotroctes needhami (Enderlein). Enderlein 1905:44.
Embidopsocus needhami (Enderlein). Mockford, 1952:198.

Recognition features. Known from male, apterous and macropterous female. Spermapore plate long-triangular with base (anterior margin) heavily sclerotized; u-shaped sclerite bowl-like (Fig. 508). Phallosome with internal parameres broad, rounded and longitudinally striated (or plicated?) at their apices (Fig. 509). Radula of scattered short denticles.

Relationships. Only two other known species are assigned to section II B: *E. enderleini* Ribaga of Europe, Argentina, and South Africa, and *E. trichurensis* Menon of southern India. *E. needhami* differs markedly from both of these in shape of the spermapore sclerite (bell-shaped in both). It differs from *E. enderleini* in nature of the internal parameres, which are pointed apically in that species. Thus, it seems not to be close to either of these species.

Distribution and habitat. This species occurs throughout the United States wherever woodland is well developed, north of the Gulf Coast and Texas, but south in the Southwest through much of New Mexico and Arizona, also in southeastern Canada and southern Saskatchewan. It is found primarily under loose bark, but is often taken in beating, suggesting that individuals frequently wander out on open bark surface.

Group III

Embidopsocus femoralis (Badonnel)

Stenotroctes femoralis Badonnel, 1931:252.
Embidopsocus femoralis (Badonnel). Badonnel, 1955:88.

Recognition features. Known from male, apterous, and macropterous female. Fore femur with large spine near middle of inner face. Fore tibia with six to 11 spines on inner face in distal half, most distal spine longer than others (Fig. 510). Hind femur exceptionally broad, with a few acuminate spines on inner face near distal end.

Relationships. This is the only known member of its species group and appears to stand rather far apart from any other member of the genus.

Distribution and habitat. Within the study area the species has been found only at two localities in St. Lucie County, Florida (lower Peninsula) where it may have become established following introduction. At both localities it was taken under bark of *Casuarina* sp. Outside the study area, the species is known from Guerrero and Veracruz States in Mexico and from Mozambique and Angola in Africa.

Subfamily Liposcelidinae

Diagnosis. Female apterous. Hind femur with a lateral protuberance (Fig. 67). Pretarsal claw without basal seta. Subgenital plate with t-shaped sclerite (Fig. 68).
North American genus: *Liposcelis* Motschulsky.

Genus *Liposcelis* Motschulsky

Liposcelis Motschulsky, 1852:19.

Note. The confusing synonymy of this genus was ably summarized by Smithers (1967:24), whose work should be referred to as a guide to the literature on this subject.
Diagnosis. As for the subfamily.
Generotype. *L. brunnea* Motschulsky.

The infrageneric classification of Badonnel (1962, 1963) is employed here for this large and difficult genus. This involves use of categories section, group, and subgroup. Lienhard (1990a) showed that the generic name, which had been treated as masculine, is of feminine gender. Specific name endings are changed here, where appropriate, to agree with this finding.

Section I

Diagnosis. Abdominal terga 3-4 not presenting, at least in middle, a posterior membranous band with sculpture different from the anterior part. All prosternal bristles anterior to middle.

Group A

Diagnosis. Prothoracic humeral bristle (S1) strong and long; a transverse row of two to five bristles (exceptionally one) on lateral pronotal lobe (Fig. 89). Female with normally eight ommatidia in each eye.

Subgroup Aa

Diagnosis. A single supplementary bristle on each lateral lobe of pronotum. Two long, curved, acuminate setae on epiproct (Fig. 90).
North American species: *L. brunnea* Motschulsky.

Liposcelis brunnea Motschulsky

Liposcelis brunneus Motschulsky, 1852:19.
Liposcelis liparus Broadhead, 1947:42.
Liposcelis brunnea Motschulsky. Lienhard, 1990a:145.

Recognition features. The only known member of taxon I-Aa in the study area. Color ranging from uniform pale yellowish brown through uniform medium brown, with intermediates medium brown on postclypeus and terminal abdominal segments, pale yellowish brown between. Abdominal terga always with darker, more heavily sclerotized strip along anterior edges of segments 4-8 (Fig. 67). Sculpture of vertex and abdominal terga transverse areoles with finely granulate surfaces separated by fine lines (Fig. 511). Female compound eye with seven ommatidia.

Relationships. Three other species are known in taxon IAa: *L. kyrosensis* Badonnel (Isle of Cypress), *L. liparoides* Badonnel (Argentina and Chile), and *L. setosa* Badonnel (Chile). *L. brunnea* does not seem to be very close to any of the others.

Distribution and habitat. Within the study area, this species occurs commonly throughout Arizona, New Mexico, southern Colorado, and western Texas, northeast to Ottawa, Ontario. It is collected primarily by beating branches and foliage of conifers and oaks, and by sifting ground litter under these trees. It has also been found in a domestic situation, among fiber slipsheets in a warehouse in St. Louis, Missouri. At Ottawa, it was taken in a bird nest in a window. Outside the study area it has been found in domestic sites and out-door situations in many parts of Europe (Lienhard, 1990a), and in a termitarium in South Africa.

Subgroup Ab

Diagnosis. Each lateral lobe of pronotum with transverse row of two to five bristles (only one bristle in addition to S1 in forms with body uniformly pale). Epiproct without pair of long, curved, acuminate setae.

North American species: *L. deltachi* Sommerman, *L. entomophila* Enderlein, *L. fusciceps* Badonnel, *L. hirsutoides* Mockford, *L. nasa* Sommerman, *L. ornata* Mockford, *L. pallens* Badonnel, *L. pallida* Mockford, *L. villosa* Mockford.

Liposcelis deltachi Sommerman

Liposcelis delta-chi Sommerman, 1957:127.

Recognition features. Known from the female only. Body and legs pale brown, slightly darker on postclypeus. Abdomen marked with purplish-brown spot on each side on segments 3-10 and on paraprocts. Very faint cross band of purplish-brown pigment between lateral spots on terga 3-7. Female compound eye with eight ommatidia.

Relationships. Other members of the subgroup somewhat similarly marked are *L. entomophila* (Enderlein) (cosmopolitan), and *L. marginepunctata* Badonnel (Angola).

Distribution and habitat. Within the study area, the species is known from southern and western Texas and from southern New Mexico. Outside the study area, it has been taken near Matehuala, San Luis Potosi, Mexico. It has been collected on foliage of junipers, yuccas, mesquite, and in ground litter.

Liposcelis entomophila (Enderlein)

Troctes entomophilus Enderlein, 1907a:34.
Liposcelis bakeri Pearman, 1928a:133.
Liposcelis virgulatus Pearman, 1929:106.
Liposcelis entomophilus (Enderlein). Broadhead, 1947a:109.

Recognition features. Known from both sexes. Body and appendages creamy yellow; abdominal terga marked with obscure bands of purplish-brown, fading to extinction, or nearly so, in middle, along hind margins of tg2 and tg3, fore margins of tg6, 7, 8, and 9. Female compound eye with eight ommatidia.

Relationships. As noted for *L. deltachi.*

Distribution and habitat. Within the study area, the species has been taken primarily in domestic situations -- stored grain and collections of biological material -- in midwestern and southeastern United States. There is one field collection from a small mangrove island in the Florida Keys. Outside the study area, the species is widely distributed, being known from most continents, primarily from domestic situations. It was found in forest ground litter in Guadeloupe (West Indies), a cave in Cuba, shrubs in a forest in Senegal, and from plumage of a fledgling bird in Mindanao, Philippine Islands. Field collections, then, appear to be entirely tropical and subtropical.

Liposcelis fusciceps Badonnel

Liposcelis fusciceps Badonnel, 1968:541.

Note. The specimens on hand agree with the original description in the distinctive coloration, chaetotaxy, and sculpture of the abdominal terga. They differ in that most of the areoles of the vertex (Fig. 512) bear moderate-sized nodules instead of relatively fine granulations, although some of the postero-central areoles bear fine granulations. The common stem of the gonapophyses appears somewhat shorter and thicker, but this is probably a question of relative flatness of the preparation. For the present, these specimens are regarded as representing *L. fusciceps.*

Recognition features. Known only from females. Two to three bristles in addition to humeral seta in transverse row of lateral pronotal lobe. Head medium brown with a slight reddish tinge; antennae slightly paler; palpi colorless. Rest of body, including legs creamy yellow. A sclerotized strip obvious along anterior margin of each of abdominal tg. 3-8. Sculpture of head as noted above. Meso-metathoracic and ab-

dominal terga uniformly covered with small, compact nodules not
forming areoles (Fig. 513). Female compound eye with eight ommatidia.

Relationships. The species does not appear to be very close to any
other species of its subgroup.

Distribution and habitat. Within the study area, the species was
found as an introduction on Mexican grass and hay at Nogales, Arizona.
Outside the study area, it is known from 30 females collected at
Barueri, Sao Paulo, Brazil in the nest of an ant, *Camponotus rufipes*.

Liposcelis hirsutoides Mockford

Liposcelis hirsutoides Mockford, 1978b:562.

Recognition features. Known from both sexes. Two bristles in
addition to humeral seta in transverse row of lateral pronotal lobe. Body
and appendages generally medium grayish brown; abdominal terga
slightly darker along sides than in middle; tg 1+2 paler than rest of
abdomen. Short, truncated setae abundant on all abdominal terga.
Female compound eye with eight ommatidia.

Relationships. It appears to be very close to *L. hirsuta* Badonnel,
from Zaire, differing in smaller size and in having the larger bristles of
the pro- and mesonotum scarcely or not bulging.

Distribution and habitat. The species is known only from southeastern Texas north to Matagorda County and from Highlands County
in central-peninsular Florida. It has been collected by beating woody
vegetation and by searching under bark of tree trunks.

Liposcelis nasa Sommerman

Liposcelis nasus Sommerman, 1957: 128.

Recognition features. Known from both sexes. One or two bristles
in addition to humeral seta in transverse row of lateral pronotal lobe.
Body and appendages pale buffy yellow; postclypeus and labrum pale
rusty brown. Abdominal terga with slender, short, sparse setae.
Female compound eye with eight ommatidia.

Relationships. Not enough is known to allow speculation at
present.

Distribution and habitat. The species is known only from southeastern Texas north to San Antonio. It has been taken on dead leaves
of palms and yuccas, branches of thorny trees, and in ground litter of
grass cuttings and oak leaves.

Liposcelis ornata Mockford

Liposcelis ornatus Mockford, 1978b:565.

Recognition features. Known from female only. Three bristles in addition to humeral seta in transverse row of lateral pronotal lobe. Marked with complex pattern of brown on white background: head with brown y-shaped mark with stem along midline of vertex, branches covering most of frons; postclypeus brown; meso-metanotum brown laterally, white in middle posteriorly, pale brown anteriorly; abdomen with irregular brown transverse band covering most of tg3 and tg4 and extending into tg5; tg7 and tg8 brown laterally, tg9, and tg10 brown along midline; epiproct brown (Fig. 93). Female compound eye with eight ommatidia.

Relationships. The body markings are somewhat similar to those of *L. marginepunctata* Badonnel of Angola.

Distribution and habitat. Within the study area, the species has been found throughout peninsular Florida, at New Orleans, Louisiana, and in southeastern Texas (Matagorda and Nacogdoches Counties). It has been taken by beating a great variety of trees and shrubs. Outside the study area, it has been found in the Mexican states of San Luis Potosí and Tabasco, and has been tentatively identified from Colombia.

Liposcelis pallens Badonnel

Liposcelis pallens Badonnel, 1968:537.

Recognition features. Known from both sexes. Two or three bristles in addition to humeral seta in transverse row of lateral pronotal lobe. Body and appendages uniformly pale ocraceous yellow except head a slightly darker shade. Sculpture of vertex (Fig. 514): areoles with smooth surfaces separated by arched ridges; abdominal terga (Fig. 515) with small nodules showing no tendency to form areoles on well sclerotized areas. Female compound eye with eight ommatidia.

Note. The species was originally described from material in a laboratory culture. Field collected specimens are somewhat darker in general or at least have the head darker with, in some examples, relatively abundant purplish-brown subcuticular pigment granules on the vertex, frons, thorax, and sides of the abdomen. Most field collected specimens have discrete granules of the size seen on the abdomen over the head areoles.

Relationships. *L. pallens* somewhat resembles the other pale species of its subgroup, *L. pallida* Mockford (Texas) and *L. villosa* Mockford (California) in sculpture of the vertex, but differs markedly from either species in smaller size, chaetotaxy of the lateral pronotal lobes, and sculpture of the abdominal terga.

Distribution and habitat. The original laboratory culture was started from field collected material at Wooster, Ohio. The species is common on bark of trees in central Illinois. A single specimen was taken in a nest of house sparrow (*Passer domesticus* L.) in Indiana County, Pennsylvania.

Liposcelis pallida Mockford

Liposcelis pallidus Mockford, 1978b:567.

Recognition features. Known from both sexes. Antennae ~ 1.3X length of body. One bristle in addition to humeral seta in transverse row of lateral pronotal lobe. Body and appendages dull ochre yellow with scattered reddish brown subcuticular pigment granules over vertex and sides of thorax and abdomen. Sculpture of abdominal terga (Fig. 92a) irregular minute reticulations showing very weak tendency to form areoles.

Relationships. The species is similar in color and chaetotaxy to *L. villosa* Mockford (southwestern U. S.), differing markedly in sculpture of abdominal terga.

Distribution and habitat. The species is known from the Davis Mountains of Texas and the Catalina Mountains of Arizona. It has been found in dead persistent leaves of yucca.

Liposcelis villosa Mockford

Liposcelis villosus Mockford, 1971a:134.

Recognition features. Known from both sexes. Antennae about same length as body. One bristle in addition to humeral seta in transverse row of lateral pronotal lobe. Body pale yellowish brown, somewhat darker on postclypeus and abdominal terga 8-11; reddish brown subcuticular pigment granules sparse on vertex, concentrated between eye and antennal base, relatively heavy on thoracic and abdominal terga, especially on sides. Sculpture of vertex (Fig. 516) narrow, transverse areoles separated by depresses lines and bearing exceedingly minute granulations. Sculpture of abdominal terga (Fig. 92b) transverse areoles enclosing relatively large granules. Female compound eye with eight ommatidia.

Relationships. These are as noted for *L. pallida.*

Distribution and habitat. The species is known from southern Colorado, New Mexico, and southern California. It has been found on foliage of ponderosa pine, Douglas fir, several species of junipers, and in a pack rat nest. Outside the study area, it was tentatively identified from Colombia (Badonnel, 1986d).

Group B

Diagnosis. Prothoracic humeral bristle (S1) relatively strong and long; no transverse row of bristles on lateral pronotal lobe.

Subgroup Bb

Diagnosis. Female with seven or fewer ommatidia in each eye.
North American species: *L. bicolor* (Banks), *L. decolor* Pearman, *L. lacinia* Sommerman, *L. nigra* (Banks), *L. pearmani* Lienhard, *L. rufa* Broadhead, *L. silvarum* Kolbe, *L. triocellata* Mockford.

Liposcelis bicolor (Banks)

Troctes bicolor Banks, 1900c:559.
Liposcelis bicolor (Banks), Roesler, 1939:138.

Note. There are two species in North America showing the color pattern described by Banks for the species. Examination of the type shows it to be a species of Group I-B and that the species designated by Pearman (1925) from England is the true *L. bicolor* Banks. The other North American species, an un-named member of group II D, has only short hairs plus two of intermediate length near the tip of the abdomen while *L. bicolor* has several long hairs in that region. Whether *L. bicolor* Enderlein, also described from North America, is the same as Banks' species, remains uncertain but seems likely in view of Broadhead's (1950) find that Enderlein's type specimens are the same as the European species.

Recognition features. Known from both sexes. Color pattern striking: head and abdomen brown, thorax yellow. Abdominal terga unsculptured except tg 5-7 in posterior, membranous portion of tergum with narrow transverse smooth areoles enclosed by granulose lines. Female compound eye with seven ommatidia.

Relationships. The species is thought to be close to *L. decolor* Pearman, although the color is very different.

Distribution and habitat. Within the study area, this species has been found at Falls Church, Virginia, and Normal and Virginia, Illinois (type locality of Enderlein's species). At Falls Church individuals were found running over dry boards. At Normal one specimen was taken on the outer wall of a house and another in the kitchen of the same house. At Virginia, Illinois, specimens were taken under bark of maple. Outside the study area, the species has been found in England and much of central and western Europe, primarily in outdoor sites.

Liposcelis decolor (Pearman)

Troctes bicolor var. *decolor* Pearman, 1925:126.
Liposcelis terricolis Badonnel, 1945:35.
Liposcelis divinatorius (Müller). Pearman, 1946:238.
Liposcelis luridus Broadhead, 1947a:45.
Liposcelis simulans race B Broadhead, 1950:356.
Liposcelis simulans Broadhead. Smithers 1967:27.
Liposcelis silvarum palpalis Badonnel, 1971a:1216.
Liposcelis terricolis monniotae Badonnel, 1971a:1220.
Liposcelis macedonicus Günther, 1980:4.
Liposcelis decolor Pearman. Badonnel, 1986a:72.

Recognition features. Known from both sexes. Body grayish white to medium brown, with scattered brown subcuticular pigment granules on vertex and along sides of thorax. Sclerotized strips along anterior margins of abdominal terga more marked on tg 6-8 than on more anterior terga. Sculpture of vertex (Fig. 95) transverse areoles (fairly wide antero-posteriorly) separated by ridges, bearing small nodules forming antero-posteriorly oriented wavy lines; abdominal terga sculptured with nodules of the same size to slightly larger weakly grouped into areoles separated by clear lines. Female compound eye with seven ommatidia.

Relationships. Lienhard (1990a) regards this as a superspecies, which may include several species-level taxa, currently regarded as synonyms. According to the same author, (loc. cit.), the superspecies is in a complex including *L. silvarum* Kolbe and *L. rufa* Broadhead.

Distribution and habitat. Within the study area, the species has been found in the states of New York, Indiana, Missouri, Kansas, Colorado, Texas, New Mexico, California, Idaho, and Washington. Collections have been made from stored grain, the basement of a house, fiber slipsheets in a warehouse, in beach debris (Lake Ontario shore), in a bee nest (*Osmia* sp., California), under tree bark, on cow dung (Pawnee Grasslands, Colorado), and in the nest of an ant (*Formica* sp.). Outside the study area, the species has been found in numerous localities in Europe, also in temperate South America and West Africa, in both domestic and a variety of outdoor situations.

Note 1. A slide was prepared of the type of '*Atropos*' *purpurea* Aaron. It is a female *Liposcelis* of section I, agreeing with *L. decolor* in number of ommatidia and sculpture of the head and abdominal terga. The deep reddish color of the thorax and abdomen reported by Aaron was probably due to a large quantity of reddish brown material in the digestive tract of the animal. I hesitate to propose synonymy of *L. decolor* with *L. purpurea* because of the poor condition of this type.

Note 2. It seems possible that two weakly differentiated species may exist in North America corresponding to *L. decolor* and *L. simulans*. A darker colored form has larger nodules of the abdominal tergal sculpture, with only a single row of nodules in areoles near the front margin of tg 8.

Liposcelis lacinia Sommerman

Liposcelis lacinia Sommerman, 1957:125.

Recognition features. Known from female only. Body uniformly light brown. Lacinial tip with inner tine slightly divided at tip, a minute denticle between inner and outer tine, middle tooth clearly on outer tine. Vertex and more heavily sclerotized portions of abdominal terga sculptured with transverse areoles enclosing nodules. Compound eye with seven ommatidia.

Relationships. These remain unknown.

Distribution and habitat. The species is known only from the holotype, taken at Kerrville, Texas, under bark of sycamore (*Platanus* sp.).

Liposcelis nigra (Banks)

Troctes niger Banks, 1900c:560.
Liposcelis niger (Banks). Mockford, 1950:194.
Liposcelis krullei Broadhead, 1971:264. New synonym.

Recognition features. Known from both sexes. Body shiny blackish brown (paler in long-preserved material); appendages paler. Vertex sculptured with fine-granulate-surfaced transverse areoles (Fig. 96); well sclerotized regions of abdominal terga with vague areoles enclosing groups of small nodules (Fig. 517). Female compound eye with six to seven ommatidia.

Relationships. *L. nigra* is similar in color and chaetotaxy to *L. keleri* Günther of Germany but differs decidedly in surface sculpture.

Distribution and habitat. The species is found in eastern, midwestern and southwestern Unites States, and southeastern Canada, principally under bark of trees.

Note. The synonymy of *L. knullei* with this species is based on comparison of the types of *L. nigra* with topotypic material of *L. knullei* received from Dr. Knulle and presumably collected along with the types of the species. Identity was found in color, chaetotaxy, surface sculpture, and nature of the stems of the ovipositor valvulae. No differences were found. Within the species, some variation is seen in the surface sculpture of the head. Some individuals have the areoles smooth while in others they are finely granulate.

Liposcelis pearmani Lienhard

Troctes kidderi (Hagen). Pearman, 1951:85.
Liposcelis kidderi (Hagen) auct. (e.g. Günther, 1974a).
Not *Atropos divinatoria* var. *kidderi* Hagen, 1883:293.
Liposcelis simulans Race A Broadhead. Pearman, 1951:85.
Liposcelis simulans Broadhead. New, 1974:41(in part).
Not *Liposcelis simulans* Broadhead. Pearman, 1952:150 (implied).
Liposcelis pearmani Lienhard, 1990a:157.

Recognition features. Known from both sexes. Body medium grayish brown to medium brown, somewhat paler on appendages, and paling with age in preserving fluid. Sculpture as described for L. decolor. Female compound eye with five to six ommatidia. Setae S1 and S2 about equal in length, ~ 2.5X length of next closest bristle.

Relationships. The species appears to be very close to *L. decolor*, differing in fewer ommatidia in the female compound eye and in relative length of seta S2.

Distribution and habitat. In the study area, the species has been found in Michigan, Indiana, and northern Florida, always in domestic situations, including the nest of a house sparrow, *Passer domesticus* L. Outside the study area, the species has been found in various parts of Europe (England, France, Switzerland, Germany, Finland, Luxembourg, Yugoslavia) and in Japan, mostly in domestic situations but also in birds' nests and under bark in the Mediterranean region (Lienhard, 1990).

Liposcelis rufa Broadhead

Liposcelis rufus Broadhead, 1950:366.

Recognition features. Known from both sexes. Head orange-yellow, slightly darker on postclypeus; thorax, abdomen, and appendages pale to medium reddish brown. Sclerotized strips of anterior margins of abdominal terga prominent on tg. 3-8. Sculpture of vertex (Fig. 94) clearly differentiated transverse areoles with granulate surfaces, separated by depressed lines; well sclerotized portions of abdominal terga (Fig. 518) with polygonal areoles bearing relatively large, diffuse nodules, delimited posteriorly by continuous rows of nodules. Compound eye of female with seven ommatidia.

Relationships. The species is presumably close to *L. decolor*.

Distribution and habitat. Within the study area, the species has been found in the states of Ohio, Indiana, Texas, and California, generally under bark of trees. Outside the study area, the species is known from England, France, Switzerland, Greece, Spain, Israel, West Africa (Angola), and Chile, and has been reported from both domestic and out-door situations.

Liposcelis silvarum (Kolbe)

Troctes silvarum Kolbe, 1888a:234.
Liposcelis silvarum (Kolbe). Enderlein, 1927:12.

Note. Specimens agreeing in color, sculpture of the vertex, and abdominal terga, and chaetotaxy with published descriptions of this species (Günther, 1968, 1974, Lienhard, 1977) have been collected on sagebrush (*Artemisia tridentata*) at two localities in central Oregon. The specimens agree with figures of head width (279 - 282 µm) given for European specimens, as well as for f2 length (58 - 62 µm) and length of longest epiproctal bristle (54 µm). Disagreement is seen in the following measurements: (1) one specimen exceeds the range given for humeral seta length (33 µm vs. .02 - .028 mm); (2) both measured specimens fall short of the range given for greatest width of the hind femur (152 and 156 µm vs. 168 - 176 µm). Although the habitat is not in agreement with European collections, Günther (1968) records finding this species on shrubs in arid steppe in Mongolia. Despite the morphometric differences noted above, the Oregon material is for the present regarded as representing this species.

Recognition features. Known from both sexes. Body and appendages in general dusky brown, somewhat paler on head and legs in some specimens. Sculpture of vertex (Fig. 97) areoles mostly transversely lengthened bearing somewhat diffuse nodules and separated by distinct depressed lines; abdominal terga (Fig. 519) with larger, distinct nodules forming vague polygonal areoles. Female compound eye with seven ommatidia.

Relationships. The species is probably close to *L.decolor* and *L. rufa*, differing principally in color and head width.

Distribution and habitat. The species has been found in the study area only in two localities in central Oregon (Lake and Grant Counties), at both on sagebrush. Outside the study area, it is found throughout Europe on and under bark of various trees, in bird nests, and in the leaf sheaths of dead reed (*Phragmites* sp.) (Günther, 1974b), also in Mongolia on arid-land shrubs.

Liposcelis triocellata Mockford

Liposcelis triocellatus Mockford, 1971a:137.

Recognition features. Known from female only. Body and appendages pale straw brown. Sculpture of vertex (Fig. 520) arched areoles separated by depressed lines, bearing minute nodules; thoracic and abdominal terga (Fig. 521) with large nodules not forming areoles except medially on meso- metanotum. Female with three ommatidia in compound eye.

Relationships. The species may be close to *L.pearmani*, which it resembles in chaetotaxy, sculpture, and reduction of number of ommatidia in the compound eye.

Distribution and habitat. This species has been found only in San Diego, Kern, and Los Angeles Counties in southern California, where it inhabits ground litter and pack rat nests.

Section II

Diagnosis. Abdominal terga 3-4 presenting a posterior membranous band (Fig. 91) with sculpture different from the anterior, more sclerotized part (i.e., abdomen of annulate type).

Group C

Diagnosis. Prosternal setae all in front of middle of sternum.
North American species: *L.formicaria* (Hagen), *L. mendax* Pearman.

Liposcelis formicaria (Hagen)

Atropos formicaria Hagen, 1865:121.
Troctes formicarius (Hagen). Rostock, 1878:93.
Liposcelis formicarius (Hagen). Broadhead, 1947a:36.

Recognition features. Known only from the female. Body medium brown with slight reddish hue. Head and thorax with scattered purplish brown subcuticular pigment granules. Sculpture of vertex (Fig. 522) arched areoles separated by depressed lines, their surfaces mostly fine-granulate, but some with a few nodules, especially in middle of each parietal region; well sclerotized regions of abdominal terga transverse-areolate, the surfaces fine-granulate but many areoles with diffuse nodules along their posterior edges (Fig. 523). Compound eyes with eight ommatidia.

Relationships. It is impossible to speculate on these at present.

Distribution and habitat. Within the study area, the species has been found in the states of Montana, Colorado, New Mexico, Arizona, and California, generally in litter of needle-and broad-leaf trees and in shelf fungi. Outside the study area, it occurs throughout much of Europe and in central Asia (Mongolia). It has often been found in nests of ants of the genus *Formica*.

Liposcelis mendax Pearman

Liposcelis mendax Pearman, 1946:243.

Recognition features. Known from both sexes. Specimens on hand (only females) pale brown with yellowish tinge (perhaps somewhat faded in preservative seven years). Sculpture of vertex (Fig. 524): areoles transverse in middle, arched forward on sides, their surfaces finely granulate, separated by depressed lines; sclerotized regions of abdominal terga (Fig. 525) with similar sculpture but areoles all transversely elongate. Compound eye with five to seven ommatidia.

Relationships. It is not possible to speculate on these with the present data.

Distribution and habitat. Within the study area, I have seen only a single specimen, collected at Goldfield, Teller County, Colorado, in a house. Outside the study area, the species has been recorded from England, France, Yugoslavia, and Italy entirely from domestic situations.

Group D

Diagnosis. Prosternal setae including one pair of laterals posterior to middle of sternum.

North American species: *L. bostrychophila* Badonnel, *L. corrodens* Heymons, *L. paeta* Pearman, *L. prenolepidis* Enderlein.

Liposcelis bostrychophila Badonnel

Liposcelis bostrychophilus Badonnel, 1931:63.
Liposcelis divergens Badonnel, 1943:139.
Liposcelis granicola Broadhead and Hobby, 1944:47.

Recognition features. Known from females only (thelytokous). Sclerotized areas of body pale to medium brown with yellowish tinge. Head and thorax slightly darker than abdomen. Sculpture of vertex (Fig. 526): areoles arched forward separated by narrow lines of thin cuticle and bearing relatively large nodules; sclerotized regions of abdominal terga with similar areoles but somewhat more transversely oriented (Fig. 99). Humeral seta short, not much longer than other setae of edge of lateral lobe of pronotum. Compound eye with seven ommatidia.

Relationships. *L. bostrychophila* and the forms which have been named as subspecies of it may be a series of agamic populations derived from similar sexual species, such as *L. nuptialis* Badonnel (Chile).

Distribution and habitat. Within the study area, the species has been found in out-door situations (under bark of trees, in bird nests, on mangroves [Florida Keys]) in the states of Indiana, Idaho, Florida, and Texas, also in domestic situations (stored grain, litter in chicken coops) in some of these states and in southeastern Canada. It has also been found in the provinces of Alberta and Saskatchewan, but collection data were provided for only two Alberta records: Lac la Biche and Seebe, where it was taken on *Pinus contorta* in association with the rust fungi *Cronartium* and *Peridermium*. Outside the study area, it is widely distributed, occurring in domestic situations in much of Europe, and free-living in Africa, Madagascar, southeastern Asia, and temperate South America.

Liposcelis corrodens Heymons

Troctes corrodens Heymons, 1909:452.
Liposcelis subfuscus Broadhead, 1947a:48.
Liposcelis corrodens (Heymons). Günther, 1974a:101.

Recognition features. Known from both sexes. Color on well sclerotized areas medium grayish brown; antennae and palpi paler. Sculpture of vertex (Fig. 527): areoles arched forward, separated by depressed lines, bearing small nodules; sclerotized regions of abdominal terga (Fig. 528) with transverse areoles separates by depressed lines bearing one to three rows of small nodules. Humeral seta relatively long, ~ 1.5-2X length of other setae on lateral lobe of pronotum. Female compound eye with seven ommatidia.

Relationships. The species seems to stand apart from other members of Group D.

Distribution and habitat. Within the study area, this species has been found in the provinces of Alberta, Saskatchewan, and Manitoba, and in the states of New Hampshire, Michigan, Illinois, Iowa, Missouri, Idaho, Washington, and California, primarily in semidomestic situations (nests of *Passer domesticus* L., *Hirundo rustica*, and *Sayornis phoebe*, bat guano in a barn) but also in fully free-living situations (nest of an ant, *Formica* sp., in *Polyporus* fungus on fir tree [Latah County, Idaho]). Outside the study area, the species is known from domestic and outdoor situations throughout Europe, in Crete, Australia, New Zealand, Campbell Island, and Japan.

Liposcelis paeta Pearman

Liposcelis paetus Pearman, 1942:289.

Recognition features. Known from both sexes. Color white to pale yellow, head and thorax slightly darker than abdomen. Vertex sculptured (Fig. 529) with areoles arched forward bearing small nodules, the nodules continuous along anterior margins of most areoles; areoles separated by thin clear lines; sculpture of sclerotized regions of abdominal terga: areoles mostly transversely oriented bearing small nodules in one to three rows and separated by lines of continuous minute nodules or granules. Female compound eye with three ommatidia.

Relationships. The sculpture and chaetotaxy are somewhat similar to those of *L. corrodens*, which has seven ommatidia in the compound eye. *L. rugosa* Badonnel (Morocco) has only five ommatidia in the compound eye, but its sculpture is very different.

Distribution and habitat. Within the study area, the species is known (to me) from a single female taken at Allendale, South Carolina in stored oats. Outside the study area, it has been found in storage facilities, primarily for grain in the United Kingdom, central and southern Europe, India, South Africa, and Zaire.

Liposcelis prenolepidis (Enderlein)

Troctes prenolepidis Enderlein, 1909a:338.
Liposcelis prenolepidis (Enderlein). Broadhead, 1950:381.
Liposcelis prenolepidis (Enderlein) (?). Mockford, 1965b:173.

Note. My (1965b) tentative determination of this species in material from South African termite nests was based on all available information. From arrangement of setae on the prosternum, there can be no reasonable doubt that the species belongs to Group D. It remains reasonable, but not certain, that the South African material, agreeing with the original description in size, color, number of ommatidia in the compound eye, and association with subterranean social insects (though termites instead of ants), may represent this species. The following description is based on the assumption that my tentative determination is correct.

Recognition features. Known from female only. Head pale rust yellow, slightly darker on labrum; thorax and abdomen grayish white. Sculpture of vertex (Fig. 530): medially with evenly-spaced small, rounded nodules, some nodules in rows larger than others, weakly delimiting areoles, the areoles becoming more distinct on periphery; sclerotized portions of abdominal terga same as central region of vertex; areoles visible only on tg1 and 2. Seven ommatidia in compound eye.

Relationships. The species appears to be close to *L. bostrychophila*.

Distribution and habitat. The species was described from an unspecified California locality, where it was taken in a colony of the ant *Prenolepis imparis* Say. Outside the study area it has been tentatively identified from termite colonies in South Africa.

Family Group Amphientometae

Diagnosis. Adults generally of moderate size, three to five mm. in length. Wings generally present and fully developed, often covered with scales. M three-branched in forewing and II A usually present. Coxal organ represented by mirror and usually small rasp. Ovipositor of three valvulae with v2 and v3 separate.

Family Amphientomidae

Diagnosis. Wings and body covered with scales. Front femur usually with row of spines on anterior carina (Fig. 104). In forewing (Fig. 10) IA and IIA ending separately on wing margin. Subgenital plate with a wedge-shaped or T-shaped sclerite (Fig. 531). Third ovipositor valvula deeply bilobed (Fig. 532). Phallosome open posteriorly.

North American genera: *Stimulopalpus* Enderlein, *Lithoseopsis* new genus.

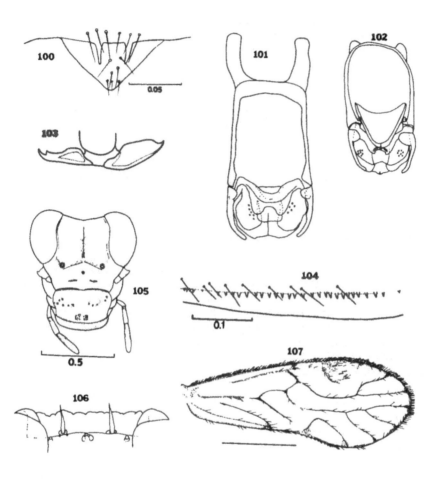

Figs. 100-107. 100 - distal margin of clunium and epiproct of *Tapinella maculata*; 101 - phallosome of *Pachytroctes aegyptius*; 102 - phallosome of *Pachytroctes neoleonensis*; 103 - pretarsal claws of *Nanopsocus oceanicus*; 104 - teeth in row of fore femur of *Lithoseopsis hystrix*; 105 - facial markings of *L. hystrix*; 106 - distal inner labral sensilla of *Bertkauia lepicidinaria*; 107 - forewing of *Epipsocus* sp.

Key to North American Genera and
Species of Family Amphientomidae

1.   Lateral ocelli close to compound eyes. Anterior carina of front
     femur with row of spines (Fig. 104) Genus *Lithoseopsis* 2
--   Lateral ocelli close to midline. Anterior carina of front femur
     lacking row of spines ....
     ....Genus *Stimulopalpus, S. japonicus* Enderlein

2.   Head unmarked, uniformly medium brown. Forewing relatively
     short, broad; length/greatest width ~ 2.83 ....
     ....*L. hellmani* (Mockford and Gurney)
--   Head yellowish white with light brown markings (Fig. 105).
     Forewing relatively long and slender; length/greatest width ~
     3.81 ....*L. hystrix* (Mockford)

Genus *Stimulopalpus* Enderlein

Stimulopalpus Enderlein, 1906f:65.

Diagnosis. Three ocelli present, close together in triangle in middle
of frontal region. Maxillary palpus with three to four spines on each of
segments 2, 3, and 4 distal to middle of segment. Pterostigma open
basally. In hindwing main R vein ending blindly in membrane; R1 and
basal piece of Rs absent; M simple. Front femur lacking row of spines
on anterior carina. Hind tibia with seven apical spurs of different sizes.
   Generotype. *S. japonicus* Enderlein.
   North American species. *S. japonicus* Enderlein.

Stimulopalpus japonicus Enderlein

Stimulopalpus japonicus Enderlein, 1906f:65.
Seopsis (Stimulopalpus) japonicus (Enderlein). Roesler, 1944:139.

   Recognition features. Body and wings dark brown except head
variegated: dark brown on labrum, clypeus, and two spots covering
most of frons, narrow border of median ecdysial line, and a pair of spots
bordering posterior margin of vertex (Fig. 533). Legs white except a
brown spot distally on each femur, each t1 brown in basal half, each
t2 and t3 completely brown.
   Relationships. The genus is small, including two described African species, an undescribed species from Hong Kong, and two
undescribed species from Sri Lanka. Interestingly, the character
triocellate vs. biocellate is variable in the Hong Kong and one of the Sri

Lankan species, some individuals having a well developed anterior ocellus, others with no trace. This species stands apart from the others in its striking head markings.

Distribution and habitat. Within the study area the species was introduced in the Washington, D. C. area about 1950, and has spread westward in a band from southern Ohio and southern Illinois south to northern Alabama. It occurs primarily on cement bridges, picnic benches, and occasionally on stone outcrops. Outside the study area, the species is known from Okayama, Japan (type locality) and Kohora, Assam, India (personal collection). Males are unknown, and it is presumably thelytokous.

Genus *Lithoseopsis* new genus

Diagnosis. Genus of Subfamily Amphientominae, tribe Amphientomini of Roesler (1944). Differing from all other genera of the tribe except Hemiseopsis Enderlein in having lateral ocelli very close to compound eyes. Differing from Hemiseopsis by (1) median ocellus being decidedly removed from clypeo-frontal suture, (2) 'sense spine' of mx2 present, (3) mx3 relatively long, nearly as long as mx4, (4) R1 absent in hindwing.

Other morphological features. Mx palpus with no spines other than sense spine of mx2, few to many scales among setae on mx2 and mx3. Fore femur with row of spines on anterior carina running much of its length (Fig. 104), their basal articulations relatively distinct. Pretarsal claw with single tooth before apex. Pterostigma closed or open basally. Rs fork stem in forewing one-half to about same length as R2+3. R1 absent in hindwing, the main R stem ending blindly in membrane. Hind tibia with seven apical spurs. Subgenital plate with median sclerite present, T-shaped or short and straight. A semicircular plate juxtaposed to clunium medio-dorsally on abdomen bearing numerous long setae (Fig. 534), field of long setae continuing on clunium posterior to plate. Genitalia of both sexes typical of the family. No spine on median margin of paraproct.

Generotype: '*Pseudoseopsis*' *hellmani* Mockford and Gurney.

North American species: *L. hellmani* (Mockford and Gurney), *L. hystrix* (Mockford).

Lithoseopsis hellmani (Mockford and Gurney)

Pseudoseopsis hellmani Mockford and Gurney, 1956:358.

Recognition features. Known from female only. Head unmarked, uniform medium brown. Forewing relatively short and broad (length/greatest width ~ 2.83). Subgenital plate with t-shaped median sclerite, the cross piece straight. Opening of spermathecal sac flanked by pair of sclerites bearing rows of short spines (Fig. 535). Spermathecal duct with wider portion leading from sac less than half as wide as length of small sclerite at opening of sac. Spermapore plate a simple sclerotized ring.

Relationships. This species is part of a complex of closely related species found from southern Texas and southern Arizona south to Central America.

Distribution and habitat. Within the study area the species has been found at Ezel's cave near San Marcos, Hayes County, Texas, where it was taken on limestone outcrops around the cave entrance. Outside the study area, it has been found at Cola de Caballo, El Diente, and Iturbide in Nuevo León State, Mexico, at each locality on rock outcrops.

Lithoseopsis hystrix (Mockford), new combination

Amphientomum hystrix Mockford, 1974a:117.

Recognition features. Known from female only. Head ground color yellowish white, marked with light brown longitudinal band through each parietal region of vertex, a band bordering median ecdysial line, a slender line along each eye medially terminating ventrally at lateral ocellus; postclypeus with slender striae, a band of spots through middle, some spotting along lower margin (Fig. 105). Forewing (Fig. 10) relatively elongate (length/greatest width ~ 3.81). Median sclerite of subgenital plate short and straight within a small triangular area (Fig. 531). Opening of spermathecal sac with no flanking sclerites. Spermapore plate a sclerotized ring thickened anteriorly and posteriorly, extended posteriorly as t-shaped sclerite (Fig. 536).

Relationships. This species and a similar one found on the Gulf Coast in Veracruz, Mexico, appear to form a distinctive section of the genus.

Distribution and habitat. Within the study area the species was taken at Arcadia, De Soto County, Florida, on branches of '*Prunus caroliniana*' (presumably *Laurocerasus caroliniana* [Mill.] Roem.). Outside the study area it was found at Mercedes, Matanzas Province, Cuba (type locality) where it was taken on mamoncillo.

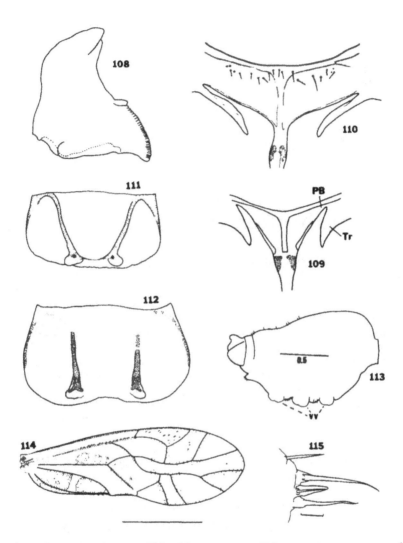

Figs. 108-115. 108 - mandible of *Epipsocus* sp.; 109 - mesosternum, precoxal bridges (PB), and Trochantins (Tr) of *Trichadenotecnum alexanderae* (typical of Family Group Psocetae); 110 - mesosternum, precoxal bridges, and trochantins of *Mesopsocus laticeps* (typical of Family Group Homilopsocidea); 111 - labrum of *Bertkauia lepicidinaria*; 112 - labrum of *Loneura* sp.; 113 - abdomen of *Xanthocaecilius* sp., lateral view showing ventral vesicles (vv); 114 - forewing of *Graphopsocus cruciatus*; 115 - spine of paraproctal margin of a typical dasydemellid (*Dasydemella* sp.).

Chapter 4
Suborder Psocomorpha

Diagnosis. Adults generally with 11 flagellomeres; flagellomeres never annulated but sometimes with other cuticular sculpture patterns (as Fig. 19). Hypopharyngeal filaments fused on midline most of their length, separate only near distal ends (Fig. 17). Labial palpus lacking basal segment (Fig. 537), consisting only of single rounded or triangular distal segment. Tarsi of adult two or three segmented. Wings generally present, sometimes reduced. Forewing generally with sclerotized pterostigma; Cu1a usually looping forward from marginal branching with Cu1b; Cu2 and 1A ending together on wing margin (Fig. 107); IIA sometimes present. Subgenital plate large, lacking median sclerite. Ovipositor with three valvulae on each side or reduced, occasionally absent. Paraproct usually without a stout spine on free margin except in some forms with much reduced wings.

North American taxa: Family Group Epipsocetae, Family Group Caecilietae, Family Group Homilopsocidea, Family Group Psocetae. See entry of each family group for included taxa.

Key to the Family Groups
and Families of Suborder Psocomorpha

1.   Pair of line-like sclerites running lengthwise through labrum
     (Fig. 111); usually an Rs-M crossvein in forewing (Fig. 107).
     Mandibles elongate, excavated posteriorly (Fig. 108). Pretarsal
     claws each with preapical denticle ....
     .... Family Group Epipsocetae 4

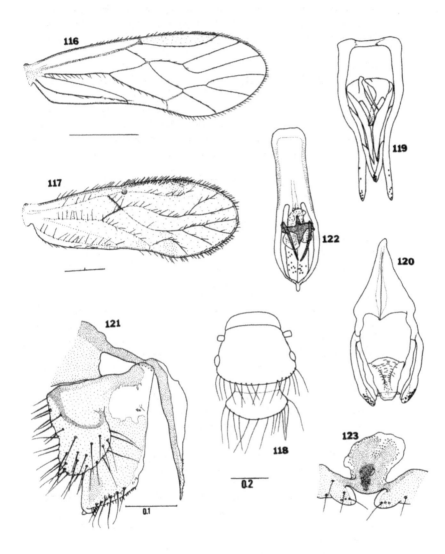

Figs. 116 - 123. 116 - forewing of *Hemipsocus chloroticus*; 117 - forewing of *Pseudocaecilius citricola*; 118 - head and prothorax of *Archipsocus nomas*, dorsal view; 119 - phallosome of *Heterocaecilius* sp.; 120 - phallosome of *Philotarsus kwakiutl*; 121 - ovipositor valvulae of *Kaestnerella fumosa*; 122 - phallosome of *Peripsocus madidus*; 123 - ovipositor valvulae and ninth sternum of *Ectopsocopsis cryptomeriae*.

-- 　Pair of longitudinal line-like sclerites never present in labrum but
　　sometimes pair of pigment lines in same positions. Rs-M junction in
　　forewing variable. If mandibles elongate and excavated posteriorly
　　then pretarsal claws lacking preapical denticle 2

2. 　Mandibles usually elongate and excavated posteriorly (exception:
　　Asiopsocids). Pretarsal claws lacking preapical denticle. Third ovi-
　　positor valvula either fused to body wall or to v2 ....
　　.. Family Group Caecilietae 5
-- 　Mandibles short, not excavated posteriorly. Pretarsal claws usually
　　with preapical denticle. Third valvula (rarely absent) not fused to
　　body wall or to v2 3

3. 　Meso-precoxal bridges narrow. Meso-trochantins wide basally (Fig.
　　109). Vein Cu1a in forewing usually joined in some manner to M
　　.. Family Group Psocetae 9
-- 　Meso-precoxal bridges wide (exceptions: some brachypterous forms).
　　Meso-trochantins narrow throughout (Fig. 110). Vein Cu1a in
　　forewing usually not joined to M ....
　　.. Family Group Homilopsocidea 11

4. 　Labral line-like sclerites running entire length of labrum with sclero-
　　tization continuing laterally and onto lateral labral margin (Fig. 111).
　　Tarsi two-segmented .... Family Epipsocidae
-- 　Labral line-like sclerites not reaching basal edge of labrum (Fig. 112).
　　Tarsi three-segmented ....
　　.............................. Family Ptiloneuridae, Genus *Loneura*, *L.* sp.

5. 　Mandibles short to moderate in length, not excavated posteriorly.
　　Ventral abdominal vesicles absent ....
　　...................... Superfamily Asiopsocoidea, Family Asiopsocidae
-- 　Mandibles elongate, excavated posteriorly. At least one, more com-
　　monly two or three ventral abdominal vesicles present (Fig. 113)
　　.. Superfamily Caecilioidea 6

6. 　In forewing pterostigma-R2+3 crossvein and M-Cu1a crossvein present
　　(Fig. 114). Hindwing margin without setae or with a few restricted to
　　cell R3 .... Family Stenopsocidae
-- 　In forewing pterostigma-R2+3 crossvein absent; Cu1a usually free
　　from M, occasionally fused to M 7

7. 　Ciliation of hindwing margin restricted to cell R3, or none. Mx4 longer
　　than mx2. Spine of free margin of paraproct relatively large (Fig. 115)
　　.. Family Dasydemellidae
　　.......................... Genus *Teliapsocus*, *T. conterminus* (Walsh)

--   Hindwing margin ciliated except basal two-thirds of front margin. Mx4 shorter than or equal to mx2. Spine of free margin of paraproct small or absent 8

8.   Setae of veins in distal half of forewing on both dorsal and ventral surfaces. Mx4 - mx2 in length .... Family Amphipsocidae .... Genus *Polypsocus, P. corruptus* (Hagen)

--   Setae of veins in distal half of forewing only on dorsal surface. Mx4 shorter than mx2 .... Family Caeciliidae

9.   Vein M in forewing two-branched, joined before branching to vein Cu1a by a long crossvein (Fig. 116). Dead-leaf inhabiting forms .... Family Hemipsocidae

--   Vein M in forewing three-branched, usually fused for a short distance to vein Cu1a before branching, rarely joined by a short crossvein. Bark-inhabiting forms 10

10.   Tarsi three-segmented. Forewings heavily blotched with brown, the margins with alternating brown and colorless banding .. .. Family Myopsocidae

--   Tarsi two-segmented. Forewing markings variable but margins never with alternating brown and colorless banding .... Family Psocidae

11.   Setae of median cell margins of forewing in two series forming crossing pairs (Fig. 117). Brachypterous and micropterous individuals with numerous long, backward- directed setae on vertex and thoracic dorsum 15

--   Crossing pairs of setae absent on wing margins; when wings are greatly reduced, body ciliation only moderate, never with numerous long, backward-directed setae on vertex and thoracic dorsum 12

12.   Tarsi two-segmented; forewing with vein Cu1a present and usually free from M; wings generally unciliated or lightly ciliated. Ovipositor reduced to one (v3), rarely two valvulae on each side. Male clunium usually without posterior comb or projection .... Family Lachesillidae

--   Either tarsi three-segmented and forewing with vein Cu1a or tarsi two-segmented and forewing without vein Cu1a, or, rarely, tarsi two-segmented but forewing with vein Cu1a; in latter case wings well ciliated. Usually all three ovipositor valvulae present on each side. Male clunium with or without posterior comb or projection 13

13.   Distal inner labral sensilla nine or eleven: median placoid and on
       each side three trichoids, one placoid (and one trichoid); if
       (rarely) sensilla reduced to five (as below) then tarsi three
       segmented. Tarsi otherwise either two- or three-segmented .
       14
--   Distal inner labral sensilla five: median placoid and on each side
       one trichoid and one placoid (Fig. 2). Tarsi two- segmented .
       17

14.   Distal inner labral sensilla nine, as described above. Tarsi three-
       segmented. Wings without ciliation, forewing (Fig. 787) with
       vein Cu 1a present and free from M; wings often much reduced
       in females .... Family Mesopsocidae
--   Distal inner labral sensilla either eleven, or five, as described
       above. Wings usually with some ciliation; vein Cu 1a present or
       absent; when present usually free from M but rarely fused to
       M a short distance. Wings sometimes greatly reduced ....
       .... Family Elipsocidae

15.   Venation in forewing vague; generally Rs in forewing not branched.
       Wings often greatly reduced; in both macropterous and short-
       winged forms, numerous long, backward-directed setae pres-
       ent on vertex and thoracic dorsum (Fig. 118). Colonial forms
       living under dense webbing .... Family Archipsocidae
--   Venation in forewing distinct; Rs in forewing with usual two
       branches. Wings seldom reduced. Vertex and thoracic dorsum
       without long, backward-directed setae. Not colonial, solitary or
       at most a few individuals living together, sometimes under
       webbing, sometimes in open 16

16.   Adults usually with tarsi two-segmented. External parameres
       generally much longer than aedeagal arch (Fig. 119). Leaf-
       inhabiting forms .... Family Pseudocaeciliidae
--   Adults usually with tarsi three-segmented. External parameres
       generally only slightly longer to shorter than aedeagal arch
       (Fig. 120). Bark inhabiting forms .... Family Philotarsidae

17.   Forewings fully developed, with vein Cu 1a. Third valvula broad,
       well ciliated; subgenital plate with two distal processes;
       endophallus not a complexly-folded structure of several scler-
       ites. Male clunium not ornamented ....
       .... Family Trichopsocidae
--   Forewing fully developed or reduced; when fully developed,
       forewing lacking vein Cu 1a. Third valvula broad, slender, or
       much reduced; when broad (Fig. 121), subgenital plate with

one distal process; when slender (Fig. 217) subgenital plate usually with two distal processes when much reduced (Fig. 123), subgenital plate broadly extruded distally. Endophallus variable. Male clunium with posterior comb or process 18

18. Third valvula broad. Subgenital plate with single distal process ('egg guide'). Endophallus a simple pair of sclerites or median fork-like sclerite (Fig. 122) .... Family Peripsocidae

-- Third valvula slender, sometimes short, thumb-like (Fig. 123b) Subgenital plate with two distal processes or a broad, extruded distal area. Endophallus containing several complex sclerites (Fig. 124) .... Family Ectopsocidae

Family Group Epipsocetae

Diagnosis. Labrum with slender sclerite running most or all of its length in each lateral half (Fig. 111). Five distal inner labral sensilla: three placoids alternating with two tricholds. Tarsi two or three segmented. Pretarsal claws with preapical denticle. Coxal organ with well developed rasp, generally with well developed mirror. Generally both fore- and hindwing with marginal setae except anterior margin of hindwing in basal one-third. Forewing (Fig. 107) generally with Rs-M crossvein. Ovipositor valvulae: v2 generally present and drawn into a long, acuminate point distally; v3 a separate lobe or a bulge on side of v2; v1 present or absent.

North American taxa: Families Epipsocidae, Ptiloneuridae.

Family Epipsocidae

Diagnosis. Labral sclerites running entire length of labrum, continuous with sclerotization on each side of labrum (Fig. 111). Distal inner labral sensilla in a straight line or tricholds slightly distad of placoids (Fig. 106). Adult tarsi two segmented; pulvillus with acuminate tip. Wing veins without spurs; pterostigma without internal crossveins; only one anal vein in forewing. Ovipositor valvulae: v1 present or not; v3 a bulge or finger-like lobe on side of v2. Phallosome closed posteriorly.

North American genera: *Epipsocus* Hagen, *Bertkauia* Kolbe.

Key to North American Genera and
Species of Family Epipsocidae

1. Both sexes macropterous. Base of v1 attached to sclerotized
    region of eighth abdominal tergum (Fig. 539). Female epiproct
    with three long setae in basal one-third ....
    .... Genus *Epipsocus*, E. sp.
-- Males (very scarce) macropterous, females apterous. Base of v1
    attached in membrane (Fig. 540). Female epiproct with no long
    setae in basal one-third .... Genus *Bertkauia* 2

2. Vertex, thoracic terga in middle, and preclunial abdominal
    segments dorsally shades of white. Rs-M crossvein in male
    forewing longer than preceding Rs segment ....
    .... *B. lepicidinaria* Chapman
-- Body in general shades of brown. Rs-M crossvein in male
    forewing about half length of preceding Rs segment ....
    .... *B. crosbyana* Chapman

Genus *Epipsocus* Hagen

Epipsocus Hagen, 1866:203.

Diagnosis. Lacinial tip broad, multidenticulate (Fig. 538). No
hyaline cones on first and third femora. Both sexes macropterous; M
in forewing three-branched. V1 present, its base attached to sclerotized
region of tg8; v3 a wide area on side of v2 (Fig. 539). Female epiproct
with three long setae in basal one-third.
   Generotype. *E. ciliatus* (Hagen).
   North American species. A single undescribed species occurs in
South-central Florida, where it was probably introduced.

Genus *Bertkauia* Kolbe

Bertkauia Kolbe, 1882b:208.
Lapithes Bertkau, 1883:180.

Diagnosis. Lacinial tip broad, multidenticulate. No hyaline cones
on first and third femora. Females apterous; M in male forewing three-
branched. V1 present, its base attached in membrane; v3 a decided
bulge on side of v2. Female epiproct with no long setae in basal one-
third.
   Generotype. *B. lucifuga* (Rambur).

North American species: *B. crosbyana* Chapman, *B. lepicidinaria* (Chapman).

Note. Pearman (1935) synonymized *Bertkauia* with *Epipsocus*, but Eertmoed (1973), on the basis of a phenetic analysis, found the two genera to be distinct. The distinction appears to be tenable, but a full understanding of relationships within this family awaits detailed morphological investigation of the many tropical American species.

Bertkauia crosbyana Chapman

Bertkauia crosbyana Chapman, 1930:364.
Epipsocus crosbyanus (Chapman). Mockford, 1950:195.

Recognition features. Male (exceedingly rare): body in general medium brown with dark purplish brown stripe along each side from antennal base to posterior end of precluntal region of abdomen. Wings unmarked, with tawny wash; Rs - M crossvein of forewing relatively short, about half length of Rs segment before crossvein. Female in general dark brown, paler on thoracic terga with dark brown longitudinal stripe through middle of meso- and metaterga.

Relationships. The species appears to be very close to *B. lucifuga* (Rambur) of Europe, differing in much smaller size and in that v3 forms a more prominent shoulder on the side of v2 (Fig. 540).

Distribution and habitat. Within the study area, the species occurs throughout eastern United States, southeastern Canada, the Turtle Mountains of Manitoba, the Black Hills of South Dakota, parts of the southern Rocky Mountains, northwestern Washington, and southwestern British Columbia. It generally occurs in woodland ground litter, but one male was taken on foliage of Virginia pine (*Pinus virginiana* Mill.). Outside the study area, the species has been recorded from the region of San Cristóbal, Chiapas State, Mexico. The species is presumably obligatorily thelytokous.

Bertkauia lepicidinaria Chapman

Bertkauia lepicidinaria Chapman. 1930:363.
Epipsocus lepicidinarius (Chapman). Mockford, 1950:195.

Recognition features. Male (exceeding rare): body coloration as described for female. Wings clear, unmarked. Rs-M crossvein of forewing relatively long, longer than preceding Rs segment. Female: head medium brown becoming paler on vertex dorsally. Thorax medium brown on sides and legs with dark purplish brown band above leg bases on each side; thoracic terga creamy white along broad median area. Preclunial abdominal segments grayish white, pale purplish brown on sides in basal half and along dorsal midline on terga 2-4, 7 and 8; terminal abdominal segments medium brown.

Relationships. The species appears to be close to *B. crosbyana* in shape of v2+3, but in color it is paler than either *B. crosbyana* or *B. lucifuga.* Male genitalic structures have not been compared, as these have not been described for the male of the European species.

Distribution and habitat. The species occurs throughout eastern United States north of the Gulf Coast and east of the Mississippi River, and in southern Ontario. It is found primarily on shaded rock outcrops. It is presumably obligatorily thelytokous.

Family Ptiloneuridae

Diagnosis. Labral sclerites incomplete, absent in proximal end of labrum (Fig. 112). Distal inner labral sensilla in approximately straight line. Lacinial tip broad, multi-denticulate (Fig. 541). Adult tarsi three segmented; pulvillus with acuminate tip. Wing veins without spurs; pterostigma without crossvein; two anal veins in forewing. Ovipositor valvulae: v1 present; v3 a finger-like lobe broadly attached to side of v2 (Fig. 542).

North American genus: *Loneura* Navás.

Genus *Loneura* Navás

Loneura Navás, 1927:49.

Diagnosis. M in hindwing more than two branched, in forewing usually five to six branched, occasionally seven branched.

Generotype. *L. crenata* Navás.

North American species: a single undescribed species occurs in Madera Canyon, Santa Rita Mountains, Arizona.

Family Group Caecilietae

Diagnosis. Labrum without a slender longitudinal sclerite in each lateral half, or, (rarely) these structures represented by pigment bands. Five to seven distal inner labral sensilla: alternating trichoids and placoids; when seven, trichoids outermost, when five placoids outermost. Tarsi two segmented; pretarsal claw lacking preapical denticle. Except in very neotenic forms, coxal organ well developed. Rs - M junction in forewing variable, usually not a crossvein. Non-neotenic forms with broad meso-precoxal bridge, narrow meso-trochantin (as in

Fig. 110). Ovipositor valvulae: three pairs; v3 often reduced to small bulge on side of v2, usually bearing one or two setae.

North American taxa: Superfamily Asiopsocoidea (Family Asiopsocidae), Superfamily Caecilioidea (Families Amphipsocidae, Dasydemellidae, Stenopsocidae, Caeciliidae).

Superfamily Asiopsocoidea

Diagnosis. See diagnosis of the family.

Family Asiopsocidae

Diagnosis. Lacinial tip broad with denticles of low relief. Mandibles of short to moderate length. Pulvillus variable: short and broad, slender and acuminate, or absent; coxal organ well developed, or represented by mirror, or absent. Winged forms with long Rs - M fusion in hindwing. Subgenital plate smoothly rounded on posterior margin; ventral abdominal vesicles absent. First and second valvulae, when present, with distal ends blunt, rarely pointed.

North American genera: *Asiopsocus* Günther, *Notiopsocus* Banks, *Pronotiopsocus* Mockford.

Key to the North American Genera and Species of Family Asiopsocidae

1. Five distal inner labral sensilla: median placoid and on each side one trichoid and one placoid. Male macropterous, female micropterous Genus *Asiopsocus* ....
................................*A. sonorensis* Mockford and Garcia Aldrete
-- Seven distal inner labral sensilla: as above plus one trichoid on each side. Male micropterous where known 2

2. Inner clypeal shelf narrow as in (Fig. 125). Female macropterous, lacking vein Cu1a in forewing ...... Genus *Notiopsocus*, *N.* sp.
-- Inner clypeal shelf broad as in (Fig. 126). Female usually micropterous, rarely macropterous, in latter case, vein Cu1a present in forewing .... Genus *Pronotiopsocus*, *P.* sp.

Genus *Asiopsocus* Günther

Asiopsocus Günther, 1968:128.

Diagnosis. Females apterous, males fully winged. Inner clypeal shelf (as in Fig. 125) narrow. Labrum with weakly developed longitudinal line from outer basal corner to distal margin in each lateral half (Fig. 543). Distal inner Labral sensilla five (three placoids alternating with two tricholds). Mandible of medium length, somewhat hollowed out postero-laterally (Fig. 544). Lacinia broadly bulging laterally before tip, the tip very wide with a few low, rounded denticles (Fig. 545). Pulvillus slender, acuminate-tipped; coxal organ present or absent. Mesepisternal sulcus absent. Wings unciliated; areola postica present; vein M three-branched. Third ovipositor valvula apparently fused with v2, the two structures forming a relatively large, mostly membranous flap (Fig. 546). Sheath of spermatheca short (Fig. 547).

Generotype. *A. mongolicus* Günther.

North American species: *A. sonorensis* Mockford and García Aldrete.

Asiopsocus sonorensis Mockford and García Aldrete

Asiopsocus sonorensis Mockford and García Aldrete, 1976:336.

Recognition features. The only North American species of its genus.

Relationships. Two other species of *Asiopsocus* are known: *A. mongolicus* Günther (Mongolia) and *A. meridionalis* Lienhard (Spain). *A. sonorensis* is smaller than either Old World species and lacks a coxal organ, which both Old World species possess.

Distribution and habitat. This species has been found in Cochise, Santa Cruz, and Graham Counties, Arizona, and in southwestern Catron County, New Mexico. It has been taken on evergreen oaks, yuccas, wild grape vines, and junipers usually near stream courses. Outside the study area, it occurs near the Arizona border in Sonora State, Mexico.

Genus *Notiopsocus* Banks

Notiopsocus Banks, 1913:84.
Lenkoella Machado-Allison and Papavero, 1964:312.

Diagnosis. Male apterous, female fully winged. Inner clypeal shelf narrow (as in Fig. 125). Labrum lacking longitudinal sclerites. Distal inner labral sensilla seven (four trichoids alternating with three placoids). Mandibles short, not hollowed out postero-laterally. Lacinia (Fig. 548) only moderately bulging laterally before apex; apex moderately broad with slight indication of rounded denticles, or none. Pulvillus and coxal organ absent. Mesepisternal sulcus present. Forewings (Fig. 549) ciliated on veins and margin; hindwings ciliated on margin except front margin in basal two-thirds; areola postica absent; vein M in forewing simple or once-branched. Ovipositor valvulae: v1 and v2 broad-tipped; v3 a distinct lobe relatively narrowly attached to v2 (Fig. 550).

Generotype. *N. simplex* Banks.

North American species: a single undescribed species occurs in southern Florida.

Genus *Pronotiopsocus* Mockford

Pronotiopsocus Mockford, 1983:243.

Diagnosis. Females apterous or macropterous; males unknown. Inner clypeal shelf broad (as in Fig. 126). Labrum without longitudinal sclerites. Distal inner labral sensilla seven (four trichoids alternating with three placoids). Mandibles short, not hollowed out postero-laterally. Lacinia as described for Notiopsocus. Pulvillus absent. Coxal organ represented by mirror only. Mesepisternal sulcus absent in apterous form, present in macropterous form. Wings ciliated as in Notiopsocus; areola postica present but small; M three-branched in forewing. Ovipositor valvulae: v1 long, blunt or pointed at tip; v2 same length as v1 or shorter, blunt-tipped; v3 represented only by a strong seta at base on side of v2 (Fig. 551).

Generotype. *P. amazonicus* Mockford.

North American species: a single undescribed species occurs in south-central Florida.

Superfamily Caecilioidea

Diagnosis. Lacinial tip variable. Mandibles elongate, decidedly excavated postero-laterally to accommodate the bulging galea (Fig. 552). Pulvillus broad, curved, somewhat flared distally (Fig. 553). Coxal organ generally well developed. Most species macropterous in both sexes but a few brachypterous or apterous, usually in the female. Subgenital plate rounded or bilobed or with pair of short apophyses on posterior margin. Ventral abdominal vesicles generally present (Fig.

127), one to three in number. First and second valvulae always present, with distal ends usually pointed, occasionally blunt.

Note. Great difficulty has been experienced in the attempt to determine relationships among the taxa of this superfamily. Here, a relatively split classification at the family level is adopted in the hope that it will help lead to a better understanding of relationships. The genera *Dasydemella* Enderlein, *Matsumuraiella* Enderlein, *Teliapsocus* Chapman, and *Ptenopsila* Enderlein are regarded as constituting a family, Dasydemellidae. The genera *Stenopsocus* Hagen and *Graphopsocus* Kolbe constitute the family Stenopsocidae. Amphipsocidae is as treated by Mockford (1978a) with the exclusion of the dasydemellids. The taxa of family Caeciliidae are as defined by Mockford (1989).

Family Amphipsocidae

Diagnosis. Inner clypeal shelf broad. Distal inner labral sensilla five. Lacinial tip relatively slender, usually bicuspid. Labral stylets present or absent. Maxillary palpus with mx4 > mx2 in length. Setae of forewing veins upright, relatively long, in more than one rank on at least some veins other than costa. Setae on veins in distal half of forewing on both dorsal and ventral surfaces. Hindwing margin mostly ciliated except basal two-thirds of front margin. Spine of free margin of paraproct small or absent. Ovipositor valvulae with v3 a sclerotic bulge from base of v2 or absent.

North American genus: *Polypsocus* Hagen.

Genus *Polypsocus* Hagen

Polypsocus Hagen, 1866:203.
Ptilopsocus Enderlein, 1900:140.

Diagnosis. Lacinial tip bicuspid. Labral stylets absent. No row of cones on front femur. In forewing (Fig. 554) Rs fork stem very short, Rs branches long and flexuous; Rs - M junction usually a short crossvein; M in forewing not more than two branched. Areola postica elongate, without internal crossveins. Basal brush of hindwing present (Fig. 555). Setae present on membrane bordering distal margin in both fore- and hindwing. Three ventral abdominal vesicles present (as in Fig. 127). Aedeagal arch somewhat constricted distally but not forming a slender distal process.

Generotype. *P. corruptus* (Hagen).
North American species. *P. corruptus* (Hagen).

Polypsocus corruptus (Hagen)

Psocus corruptus Hagen, 1861:13.
Psocus abruptus Hagen, 1861:13.
Polypsocus corruptus (Hagen). Hagen, 1866:211.
Polypsocus abruptus (Hagen). Hagen, 1866:211.
Ptilopsocus annulicornis Banks, 1903b:238.
Polypsocus corruptus var. pictilis Banks, 1938:72.
Polypsocus corruptus var. omissus Banks, 1938:72.
Polypsocus omissus Banks. Smithers, 1967:59.

Recognition features. The only species of its genus in the study area.

Relationships. The species shares the "corruptus" female forewing pattern -- mostly dark forewing interrupted by a colorless crossband immediately before the apex (Fig. 554) -- with several tropical American species.

Distribution and habitat. The species occurs throughout eastern United States west to Minnesota and Missouri, eastern Canada, (Nova Scotia, southern Quebec, southern Ontario, southern Manitoba) and the Pacific Coast from British Columbia south to northern California (Humboldt County). The Mexican species of Polypsocus have not yet been analyzed taxonomically, but one may be a geographic variant of P. corruptus.

Note. The forms described by Banks as 'varieties' pictilis and omissus appear to be local variant populations, perhaps not worthy of nomenclatural recognition. The form called pictilis appears to be a response to cool, damp climate, as it is found at higher elevations in the southern Appalachians and in the Pacific Northwest. I have not seen the form called omissus and cannot, therefore, judge its status. Some males from Florida agree rather closely with its description. Females associated with them are typical corruptus.

Family Dasydemellidae, **new status**

Diagnosis. Inner clypeal shelf broad. Distal inner labral sensilla five. Lacinial tip broad, flat or only slightly denticulate. Labral stylets absent. Maxillary palpus with mx4 > mx2 in length. Setae on forewing veins as described for Family Amphipsocidae except those in distal half of forewing only on dorsal surface. Hindwing margin with ciliation restricted to cell R3 or none. Spine of free margin of paraproct relatively large (Fig. 118). Ovipositor valvulae with v3 a sclerotic bulge or strap joined to v2 at base or distad of base of v2 (Fig. 556).

North American genus: Teliapsocus Chapman.

Genus *Tellapsocus* Chapman

Tellapsocus Chapman, 1930:334.

Note. Smithers (1970) proposed the synonymy of this genus with *Matsumuraiella* Enderlein. Mockford (1978a) argued against this proposed synonymy, but there is no question that the two genera are very close.

Diagnosis. As for the family plus following: outer surface of galea with field of only three or four setae; setae of forewing on membrane as well as on veins and margin; cell R5 of forewing parallel-sided immediately distad of its base (Fig. 557).

Generotype.. *T. conterminus* (Walsh).

North American species: *T. conterminus* (Walsh).

Tellapsocus conterminus (Walsh)

Psocus conterminus Walsh, 1863:185.
Elipsocus conterminus (Walsh). Hagen, 1866:207.
Psocus canadensis Provancher, 1876:177.
Caecilius definitus Aaron, 1883:38.
Tellapsocus conterminus (Walsh). Chapman, 1930:334.
Matsumuraiella contermina (Walsh). Smithers, 1970:81.

Recognition features. The only species of its genus in the study area.

Relationships. Another species assigned to this genus, *T. distinctus* Badonnel (Colombia) is probably a *Dasydemella.* The closest relatives are probably the species of *Matsumuraiella,* which occur in China and Japan.

Distribution and habitat. The species occurs throughout most of the United States and southern Canada (Quebec, southern Ontario, coastal British Columbia), but is apparently absent from the northern midwestern states (Ohio, Indiana, Illinois, probably Iowa, Kansas, Nebraska, and Dakotas) and probably the central provinces of Canada. It inhabits foliage of a great variety of trees, both broad-leaf and coniferous, and occasionally ground litter. It is not known outside the study area.

Family Stenopsocidae

Diagnosis. Inner clypeal shelf broad. Distal inner labral sensilla five. Labral stylets present or absent. Lacinial tip bicuspid. Maxillary palpus with mx4 - mx2 in length. Setae of forewing veins relatively short, slanting distad, generally in one rank; veins in distal half of forewing with setae only on dorsal surface. Hindwing margin without setae or with at most a few in cell R3. Forewing (Fig. 114) with pterostigma - r2+3 crossvein and m - cu1a crossvein. Two or three ventral abdominal vesicles. Spine of free margin of paraproct relatively small. Ovipositor valvulae: v3 a short sclerotic strap joined to v2 distad of base of latter (Fig. 558). Spermatheca with lateral pouch at junction of sac and duct (Fig. 559).

North American genus: *Graphopsocus* Kolbe.

Genus *Graphopsocus* Kolbe

Graphopsocus Kolbe, 1880:124.
Teratopsocus Reuter, 1894:43.

Diagnosis. As for the family plus the following: labral stylets present (Fig. 560); forewing (Fig. 114) with pattern involving several spots in basal half; Forewing margin with very sparse ciliation; no setae from base to pterostigma on front margin; two ventral abdominal vesicles; spermathecal pouch relatively large, the duct issuing lateral to it (Fig. 559).

Generotype. *G. cruciatus* (Linnaeus).
North American species: *G. cruciatus* (Linnaeus).

Graphopsocus cruciatus (Linnaeus)

Hemerobius cruciatus Linnaeus, 1768:225.
Hemerobius quadrifasciatus Fabricius, 1787:248.
Psocus subocellatus Stephens, 1836:126.
Psocus costalis Stephens, 1836:126.
Psocus nervosus Stephens, 1836:126.
Psocus cruciatus (Linnaeus). Brauer, 1857:219.
Stenopsocus cruciatus (Linnaeus). Hagen, 1866:219.
Graphopsocus cruciatus (Linnaeus). Kolbe, 1880:125.
Teratopsocus maculipennis Reuter, 1894:29.
Stenopsocus cruciatus var. *nervosus* (Stephens). McLachlan, 1881:211.
Graphopsocus cruciatus var. *brevipennis* Enderlein, 1903g:372.

Recognition features. The only species of its genus in the study area.

Relationships. The species is very close to *G. mexicanus* Enderlein (Mexico, South America). The two species differ consistently in the following: (1) head color, *G. mexicanus* having the head uniformly creamy yellow, *G. cruciatus* having the vertex and frons extensively marked with brown; (2) intensity of brown spots in basal half of male forewing, these being paler in *G. mexicanus*; (3) size of male compound eyes, these being relatively larger in *G. mexicanus*.

Distribution and habitat. The species was presumably introduced on both coasts of North America. Chapman (1930) reported no records on the East Coast prior to 1922 and none on the West Coast prior to 1927. It now occurs on the East Coast from Maine south to Georgia, inland to Cayuga Lake, New York and Macon, Georgia. On the West Coast it occurs from Vancouver, British Columbia south to Alameda County, California, but is rather closely restricted to areas bordering coastal waters and inlets. Outside the study area the species is widespread in Europe and has been reported from Japan.

Family Caeciliidae

Diagnosis. Inner clypeal shelf variable. Distal inner labral sensilla five to seven. Lacinial tip variable: broad and flat, broad and denticulate, slender and bicuspid, or slender and bilobed. Labral stylets present or absent. Maxillary palpus with mx4 < mx2 in length. Setae of forewing veins slanting distad, relatively short, usually in one rank except on costa (exceptions in *Maoripsocus* Tillyard and *Kodamaius* Okamoto). Setae on veins in distal half of forewing only on dorsal surface. Hindwing margin mostly ciliated except basal two-thirds of front margin. Spine of free margin of paraproct small or absent. Endophallus with median bulb, usually not divided on midline. ovipositor valvulae with v3 a sclerotic bulge from base of v2.

North American genera: *Caecilius* Curtis, *Xanthocaecilius* Mockford.

Key to the North American Genera and Species of Family Caeciliidae

1.   Five distal inner labral sensilla: median placoid and on each side one trichoid and one placoid. Two (rarely one) ventral abdominal vesicles (Fig. 127). Endophallus with median bulb not divided (Fig. 128) .... Genus *Caecilius* 3

--   Seven distal inner labral sensilla: as above plus an outer trichoid on each side. Three ventral abdominal vesicles. Endophallus with median bulb divided (Fig. 129) .... Genus *Xanthocaecilius* 2

2.   Forewing membrane dusky; pattern of dark and pale veins showing only obscurely. Eyes green in living and freshly-preserved individuals .... *X. quillayute* (Chapman)

--   Forewing membrane clear: pattern of dark and pale veins
       distinct (Fig. 130). Eyes yellow in living and freshly preserved
       individuals .... *X. sommermanae* (Mockford)

3.   Labral stylets present (Fig. 131). Lacinial tip flattened or shal-
       lowly bicuspid, usually with several low, rounded denticles
       (Figs. 132, 138). Forewings without a banding pattern ....
       .... Africanus Group 9
--   Labral stylets absent. Lacinial tip bicuspid or bilobed. Forewings
       with or without a banding pattern 4

4.   Forewing with extensive banding pattern, but pterostigma clear
       (Fig. 133). Micropterous forms (some females) with only one
       ventral abdominal vesicle ....
       .... Posticus Group *C. posticus* Banks
--   Forewing banded or not, but if banded, with a mark in pteros-
       tigma. Micropterous females with two ventral abdominal vesicles
       5

5.   Lacinial tip of two rounded lobes (Fig. 134). Male fore- and
       usually middle tibiae at least slightly swollen along part of their
       length (Fig. 135) .... Flavidus Group 12
--   Lacinial tip bicuspid or flat. Male fore- and middle tibiae not
       swollen 6

6.   Pterostigma decidedly angulate posteriorly (Fig. 136); males and
       some females with a small spur vein from apex of angle. Male
       vertex darker than surrounding head cuticle. Banding pattern
       absent in species from the study area ....
       .... Caligonus Group *C. indicator* Mockford
--   Pterostigma shallow, or, if angulate, without a spur vein from
       apex of angle and wings usually with a banding pattern. Male
       vertex not darker than surrounding cuticle 7

7.   Mesothoracic precoxal suture distinct, forming with pleural and
       mesepisternal sutures a K-shaped mark on side of mesothorax
       (Fig. 137). Wings banded. Pterostigma angulate ....
       .... Fasciatus Group 25
--   Mesothoracic precoxal suture absent or faint; suture system not
       forming a K-shaped mark on side of mesothorax. Wings
       banded or not. Pterostigma not markedly angulate 8

8.   Body color dull yellowish white (exception: *C. juniperorum* tawny
       brown). Pterostigma moderately angulate. Females without
       tendency towards wing reduction .... Subflavus Group 27

--   Body color pale to dusky brown. Pterostigma shallow. Females often brachypterous or micropterous .... Confluens Group 29

9.   Body dusky brown. Forewing relatively broad (length/greatest width ~ 2.8). Lacinial tip broad and flat with low denticles (Fig. 138) .... *C. africanus* Ribaga

--   Body yellow. Lacinial tip not so broad, shallowly bicuspid, denticulate or not. Forewings as above or relatively longer 10

10.   Head marked with brown stripe along midline from anteclypeus through vertex .... *C. casarum* Badonnel

--   No brown stripe on midline of head 11

11.   Forewings only moderately elongate; length/greatest width ~ 2.8 .... *C. insularum* Mockford

--   Forewings decidedly long and slender; length/greatest width ~ 3.7 .... *C. antillanus* Banks

12.   Head conspicuously marked with broad dark brown transverse band through frons and ocellar field back to eyes and hind head margin, on creamy yellow background .... *C. atricornis* McLachlan

--   Head not marked as above, generally without conspicuous markings, or these confined to postclypeus 13

13.   Forewings of normal shape, length/greatest width ~ 2.90 or less. Body without a lateral red-brown stripe 18

--   Forewings relatively long and slender, length/greatest width ~ 2.97 or greater. Body usually with a lateral red-brown stripe but exception as noted below 14

14.   With a lateral red-brown stripe at least partially developed. Antennae as long as or longer than forewings 15

--   Lacking a lateral red-brown stripe. Antennae shorter than forewings .... *C. tamiami* Mockford

15.   Larger species, found on palm foliage. Forewings > 2.25 mm. 17

--   Smaller species, found on grasses, sedges, and *Typha*. Forewings < 2.15 mm 16

16.   Lateral red-brown stripe nearly complete. Postclypeus with prominent brown lateral striations .. *C. manteri* Sommerman

--   Lateral red-brown stripe broken into series of spots bordering sutures on thorax. Postclypeus without lateral striations .... *C. lochloosae* Mockford

17.   Male fore tibiae greatly swollen (Fig. 139). Forewing length 2.3 -
      2.6 mm .... *C. micanopi* Mockford
--   Male fore tibiae not greatly swollen, but somewhat thicker than
      middle tibiae, with subtle bend and slight additional swelling
      at distal two-thirds (Fig. 140). Forewing length 2.7 - 3.3 mm
      .... *C. caloclypeus* Mockford and Gurney

18.   Body color predominantly yellow or ivory; ocellar field black ...
      19
--   Body color predominantly tawny brown; color of ocellar field
      variable 20

19.   Male middle tibiae of same shape as fore tibiae, with widest
      region near proximal end and narrowest region near distal end.
      Freshly preserved females without an irregular purplish-
      brown spot on tergum just mesad of each wing base. Females
      not brachypterous .... *C. flavidus* (Stephens)
--   Male middle tibiae cylindrical throughout. Freshly preserved
      females with an irregular purplish-brown spot on tergum just
      mesad of each wing base. Females frequently brachypterous
      .... *C. maritimus* Mockford

20.   Vein Cu2 in forewing usually without setae, occasionally with
      one or two setae, rarely (in some *C. perplexus*) with a row of
      setae. Ocellar field dark brown, contrasting with surrounding
      cuticle 21
--   Vein Cu2 in forewing regularly with a row of setae. Ocellar field
      concolorous with surrounding cuticle or very slightly darker
      23

21.   Veins in forewing bordered with cloudy brown, especially obvi-
      ous in basal half of wing (Fig. 141). Male fore and middle tibiae
      shaped as in *C. flavidus* .... *C. hyperboreus* Mockford
--   Veins in forewing not or scarcely bordered with cloudy brown.
      Male fore and middle tibiae nearly cylindrical, only very slightly
      swollen in basal half to two-thirds 22

22.   Head in lateral view with postclypeus greatly bulging (Fig. 142).
      Freshly preserved specimens lacking irregular red mark bor-
      dering eye in region of antennal base ....
      .... *C. perplexus* Chapman
--   Head in lateral view with postclypeus only moderately bulging
      (Fig. 143). Freshly preserved specimens with irregular red
      mark bordering eye in region of antennal base ....
      .... *C. pinicola* Banks

23. Thoracic notal lobes somewhat darker than surrounding cuticle. Male fore and middle tibiae decidedly stout for most of their length (Fig. 144) 24

-- Thoracic notal lobes scarcely darker than surrounding cuticle. Male fore and middle tibiae slightly swollen in middle, tapering towards both ends (Fig. 145) .... *C. burmeisteri* Brauer

24. Male epiproct without papillar field; male paraproct with field of small, irregularly scattered papillae (Fig. 146) ....
.... *C. boreus* Mockford

-- Male epiproct with moderate sized papillar field; male paraproct with field of moderate sized, evenly distributed papillae (Fig. 147) .... *C. croesus* Chapman

25. All adults macropterous. Forewings with distinct banding pattern involving anal, central, and marginal bands (Fig. 148) .
.... *C. distinctus* Mockford

-- Many females micropterous. Forewing banding vague, with no anal band, cloudy central and marginal bands (Fig. 149) 26

26. Males with narrow tergal sclerite on each of abdominal segments 3-8 (Fig. 150). Seta of v3 in membrane off the valvula, not attached to edge of sclerotized area ....
.... *C. totonacus* Mockford

-- Males lacking narrow tergal sclerites on abdominal segments. Seta of v3 arising from edge of sclerotized area ....
.... *C. nadleri* Mockford

27. Body dull yellowish white; wings clear with lemon-yellow wash
.... *C. incoloratus* Mockford

-- Either body tawny brown, or wings with spotting pattern, or both
.... 28

28. Body tawny brown; forewings either uniformly tawny brown or slightly indicated darker spot posterior to hind angle of pterostigma and another in base of cell R5 (Fig. 151) ....
.... *C. juniperorum* Mockford

-- Body ivory yellow except brown band from antennal base through eye to hind margin of head; thoracic notal lobes brown. Forewing with distinct spotting pattern (Fig. 152) ....
.... *C. subflavus* Aaron

29. Forewing uniform in color. Male epiproctal papillar field developed as a warty tubercle (Fig. 153) 30

--   Forewing with pigment spot over hind angle of pterostigma and another in base of cell M3 (Fig. 574). Male epiproctal papillar field developed as distinct papillae (Fig. 154) ....
....*C. gonostigma* Enderlein

30.   Adults sexually dimorphic in antennal length, those of males longer than forewing or body, those of females shorter than (macropterous) forewing or body. Body dusky reddish brown. Species primarily of conifer foliage .... *C. confluens* (Walsh)

--   Sexual dimorphism of antennal length not so striking; adults of both sexes with antennae longer than body, those of male longer than forewing, those of female not quite so long. Body mostly pale brown. Species primarily of tall grass ....
.... *C. graminis* Mockford

Genus *Caecilius* Curtis

Caecilius Curtis, 1837:648.

Diagnosis. As for the family, plus following: Distal inner labral sensilla five; generally two ventral abdominal vesicles (Fig. 127); endophallus with median bulb not divided (Fig. 128).

Generotype. *C. fuscopterus* (Latreille).

Mockford (1965a, 1966) proposed species groups for this large and confusing array of species. The groups are probably not adequate for the world fauna, but they apparently remain so for the Western Hemisphere and Palearctic species, and so are used here.

Africanus Group

Diagnosis. Labral stylets present (Fig. 131). Lacinial tip (Fig. 132) shallowly bicuspid or flattened, with several small rounded denticles. Forewing without banding pattern. Ciliation present or absent on vein Cu2 in forewing. Cell R3 in forewing elongate; approaching, even with, or extending basal to origin of vein M3 (Figs. 561, 562). Sheath of spermathecal duct of medium to great length (Figs. 563, 564).

North American species: *C. africanus* Ribaga, *C. antillanus* Banks, *C. casarum* Badonnel, *C. insularum* Mockford.

Caecilius africanus Ribaga

Caecilius africanus Ribaga, 1911:169.
Mepleres angolensis Badonnel, 1955:141.

Recognition features. Body dusky brown. Wings pale brown with colorless areas at base of pterostigma, nodulus, and base of areola postica. Lacinial tip (Fig. 138) broad, relatively flat, with low, rounded denticles. Forewing relatively broad with deep, angulate pterostigma (Fig. 561).

Relationships. An undescribed South African species appears to be closely related to this one.

Distribution and habitat. Within the study area the species has been found only in peninsular Florida probably as an introduction. Adults have been found resting on tree trunks, but the breeding habitat remains unknown. Outside the study area it is known from South Africa, Zimbabwe, and Angola.

Caecilius antillanus Banks

Caecilius antillanus Banks, 1938:288.

Recognition features. Lacinial tip (Fig. 565) relatively slender, shallowly bicuspid with a few very low denticles or none. Body yellow except pale to medium brown on thoracic notal lobes, on a spot in middle of frons, and in some specimens a very narrow border of median ecdysial line. Antennae black beyond fl. Wings clear, elongate and slender (length/greatest width ~ 3.7). Sheath of spermathecal duct relatively long (Fig. 563).

Relationships. The species is very close to, possibly identical with, *C. varians* Badonnel (Angola). Both are known only from females.

Distribution and habitat. Within the study area, the species has been found only in Alachua and Highlands Counties, Florida, where it was taken by sweeping grasses, primarily *Andropogon*, and on palm foliage. Outside the study area, the species is known from Cuba, Guatemala, the Mexican states of Chiapas, Hidalgo, Nuevo León, Tamaulipas, and Veracruz, and from Suriname. In these regions, it has been found on foliage of palms, cycads, *Typha*, pines, junipers, sedges, and bunch grasses.

Caecilius casarum Badonnel

Caecilius casarum Badonnel, 1931:234.
Caecilius palmarum Mockford and Gurney, 1956:361.

Recognition features. Lacinial tip (Fig. 132) of moderate width, shallowly bicuspid with a few low denticles on outer cusp. Body yellow except for medium brown areas on thoracic notal lobes and medium brown longitudinal stripe through midline of head from anteclypeus through vertex. Wings clear, elongate and slender (Fig. 562) (length/ greatest width ~ 3.35). Sheath of spermathecal duct relatively long (as in Fig. 563).

Relationships. An undescribed close relative, differing in head markings, occurs on palms in Sri Lanka. Another apparent close relative occurs on Leyte, Philippines.

Distribution and habitat. Within the study area the species is known only from Dade and Monroe Counties in far southern Florida, and from the lower Rio Grande Valley in Cameron County, Texas. In both areas it occurs only on foliage of palms. Outside the study area, the species has a wide range in the Tropics, including most of Mexico, Guatemala, Venezuela and the Guianas in the New World, and Mozambique in the Old World. Most populations appear to consist of only females. A male from New Guinea (Mockford, 1966) may represent a distinct species. A single male was taken in Guyana. The species appears to live exclusively on living or dead palm foliage.

Caecilius insularum Mockford

Caecilius insularum Mockford, 1966:157.

Recognition features. Lacinial tip (Fig. 566) of medium width, shallowly bicuspid with several small denticles on outer cusp and a denticle between cusps. Body pale yellow except thoracic notal lobes well pigmented with medium brown. Wing membranes clear except anal cell of forewing brown, membranes with slight tawny wash; wings not elongate and slender (length/greatest width ~ 2.78). Male epiproct and paraproct each with field of small papillae. Sheath of spermathecal duct of moderate length (Fig. 564).

Relationships. The species appears to be close to *C. analis* Banks (central Pacific Islands), differing in much smaller size, relatively larger papillar fields of the male epiproct and paraproct, and relatively shorter sheath of the spermathecal duct.

Distribution and habitat. Within the study area the species is known only from Key West and Loggerhead Key (Dry Tortugas), Florida. Outside the study area it has been found in the American Virgin Islands, Puerto Rico, Isle of Pines (Cuba), southern coast of Hispaniola, Panama Canal Zone, coastal Veracruz in Mexico, coastal Suriname, and coastal Venezuela. It has been taken on green leaves of coconut palm (*Cocos nucifera* L.), on foliage of *Citrus* sp., Australian Pine (*Casuarina equisetifolia* Forst.), and sea grape (*Coccolobis uvifera* (L.) Jacq.), and on persistent dead leaves of an unknown plant.

Caligonus Group

Diagnosis. Labral stylets absent. Lacinial tip slender, bicuspid, usually with high lateral cusp (Fig. 567). Forewings banded or not; if banded, anal band not extending to wing base. Pterostigma moderately angulate (Fig. 136); in males (occasionally females) a small spur vein usually present from vertex of posterior angle of pterostigma; vein Cu2 of forewing with or without ciliation. Males with vertex of head at least slightly darker than surrounding cuticle. Mesothoracic precoxal suture absent. Papillae of male epiproctal and paraproctal fields relatively large (Fig. 568). Sheath of spermathecal duct short to moderate in length.

North American species: *Caecilius indicator* Mockford.

Caecilius indicator Mockford

Caecilius indicator Mockford, 1969a:103.

Recognition features. Body yellow except male vertex dark brown. Wings clear, unmarked, with lemon-yellow wash seen against black background. Pterostigma decidedly angulate in both sexes, with small spur vein from posterior angle in male. Cu2 in forewing without ciliation. Sheath of spermathecal duct slightly more than twice as long as its greatest width (Fig. 569).

Relationships. Most of the species of the group occur in the areas around the Caribbean. *C. indicator* is most closely related to *C. dificilis* Mockford (Bahamas, Hispaniola, Puerto Rico).

Distribution and habitat. The species occurs throughout Florida and north to Baker County, Georgia. Outside the study area, it has been recorded from Gran Piedra, Oriente Province, Cuba. It is found on foliage of various trees, especially cabbage palms (*Sabal* spp.), several species of oaks, and in Spanish Moss (*Dendropogon usnioides* (L.) Raf.).

Confluens Group

Diagnosis. Labral stylets absent. Lacinial tip slender, shallowly bicuspid with a few low denticles (Fig. 571a). Body pigmentation usually dark; wings not banded, their membranes slightly tawny to dusky. Pterostigma relatively shallow; vein Cu2 in forewing with or without ciliation. Females usually with at least some size reduction in wings. Mesothoracic precoxal suture absent or represented by a pigment line. Papillae of male epiproctal and paraproctal fields relatively small (Fig. 578). Sheath of spermathecal duct of moderate length (Fig. 571b).

North American species: *C. confluens* Walsh, *C. gonostigma* Enderlein, *C. graminis* Mockford.

Caecilius confluens (Walsh)

Psocus confluens Walsh, 1863:185.
Caecilius confluens (Walsh). Hagen, 1866:205.
Caecilius umbrosus Banks, 1914:612.

Recognition features. Body in general dusky reddish brown, somewhat darker on thoracic notal lobes than elsewhere. Wings pale brown, the veins somewhat emphasized in distal half of forewing (Fig. 572) by cloudy brown borders. Males macropterous; most females brachypterous but some macropterous. Vein Cu2 in forewing without ciliation. Position of mesepisternal suture indicated by dorsal or ventral spur, or both (Fig. 573). Male epiproctal papillar field formed as a single, somewhat warty tubercle (Fig. 153); the paraproctal field of only two to three minute papillae.

Relationships. This species appears to be close to *C. kolbei* Tetens and *C. rhenanus* Tetens, both European species.

Distribution and habitat. In eastern North America the species has been taken in the states of New York, Michigan, Minnesota, Indiana, and Illinois and in southern Ontario. It is also known from Vancouver, British Columbia. It is found primarily on foliage of coniferous trees. At Vancouver, it was taken on broom (*Cytisus scoparius* (L.) Link).

Caecilius gonostigma Enderlein

Caecilius gonostigma Enderlein, 1906b:253.

Recognition features. Head, appendages, and preclunial abdominal segments pale brown, somewhat darker spottily on vertex and in middle of frons. Thorax reddish brown on mesopleura and notal lobes; remainder pale brown. Forewings (Fig. 574) obscurely marked with pale reddish brown, most pronounced in spot covering posterior angle of pterostigma and another spot immediately distad of areola postica in cell M3; a paler spot in distal end of cell R and another covering most of cells Cu2 and 1A, the latter visible only in some specimens. Vein Cu2 in forewing without ciliation. Position of mesepisternal suture marked by dorsal and ventral spurs (as in Fig. 573). Male epiproctal and paraproctal papillar fields of readily discernible individual papillae; those of epiproctal field of moderate size; those of paraproctal field relatively small (Figs. 154, 575).

Relationships. These remain uncertain.

Distribution and habitat. Within the study area, this species has been taken only as an introduction on reed fencing (*Phragmites* sp.) from Japan at Portland, Oregon. Outside the study area it is known from Japan and Taiwan.

Caecilius graminis Mockford

Caecilius graminis Mockford, 1966:148.

Recognition features. Antennae of both sexes longer than body, those of male longer than forewing. Males macropterous; females almost entirely brachypterous (rarely macropterous), the forewings extending back to second abdominal tergum. Body in general pale brown except dark brown on ocellar field, large spot on frons before ocellar field, mesopleura, and thoracic notal lobes; abdomen with reddish brown ring of subcuticular pigment on each preclunial segment; terminal segments and legs medium brown; antennae dark brown beyond paler scape and pedicel. Wings with tawny wash. Vein Cu2 in forewing without ciliation or at most with one or two setae. Mesepisternal suture indicated, if at all, by slight dorsal spur. Male epiproctal papillar field developed as large tubercle bearing numerous small papillae (Fig 570); paraproctal field of a few (four to eight) small papillae.

Relationships. The species is probably closest to *C. gynapterus* Tetens (central Europe).

Distribution and habitat. The species occurs across northern United States and southern Canada, from Ontario, Manitoba, and Saskatchewan west to British Columbia, also north into Yukon Territory (near Dawson) and southern Alaska. At the southern end of its range it occurs from Michigan and northern Indiana west to Colorado. It inhabits tall prairie grasses.

Fasciatus Group

Diagnosis. Labral stylets absent. Lacinial tip slender; bicuspid, or flat with a few rounded denticles, or a simple peg. Body well pigmented with brown or reddish brown. Wings extensively marked with brown or reddish brown, always with distal spot in pterostigma. Pterostigma angulate. Ciliation absent or sparse on vein Cu2 in forewing. Mesothoracic precoxal suture distinct (Fig. 137). Papillae of male epiproctal and paraproctal fields usually small. Sheath of spermathecal duct short to moderate in length (Fig. 576).

North American species: *C. distinctus* Mockford, *C. nadleri* Mockford, *C. totonacus* Mockford.

Caecilius distinctus Mockford

Caecilius distinctus Mockford, 1966:137.

Recognition features. Lacinial tip bicuspid or nearly flat with slightly raised edge around shallow depression. Forewing marked with distinctive banding pattern (Fig. 148): anal band over most of cell Cu2 extending into cell Cu1b and through middle of cell 1A to margin; central band crossing middle of wing basal to areola postica and pterostigma; marginal band along distal wing margin from distal end of cell R3 through distal end of areola postica, continuing forward through pterostigma with branch along Rs and R2+3 to wing margin. Ciliation absent on vein Cu2 of forewing. Spermathecal duct sheath of moderate length (as in Fig. 576).

Relationships. *C. distinctus* appears to be closest to *C. albiceps* Pearman (Central and West Africa). Lack of knowledge of the male terminal abdominal segment characters prevents more precise determination.

Distribution and habitat. Within the study area, this species has been taken on three occasions at Miami, Florida, as an introduction on orchid and *Dieffenbachia* plants from Peru. Outside the study area, it has been found in coastal Veracruz, Mexico (type locality), Trinidad (W.I.), Suriname, and the Galapagos Islands. It appears to be a forest-edge species, occurring on a variety of broad-leaf vegetation.

Caecilius nadleri Mockford

Caecilius nadleri Mockford, 1966:141.

Recognition features. Lacinial tip shallowly bicuspid (Fig. 577). Males macropterous, females dimorphic: macropterous or micropterous. Micropterous females with wings developed as tiny sacs, thoracic notal lobes and ocelli much reduced. Forewings (Fig. 149) with vague banding pattern interrupted by clear area across middle of wing and clear spot around nodulus. Central band most obvious in distal end of cell R and base of cell M3; banding in distal half of wing most obvious as spot covering distal half of pterostigma and extending across cell R1, another spot distally in areola postica extending into cell M3. Vein Cu2 in forewing without ciliation. Male epiproctal papillar field of crowded, small papillae; paraproctal field with papillae reduced to tiny spines

(Figs. 578, 579). Spermathecal duct sheath of moderate length (Fig. 576).

Relationships. The species is very close to *C. totonacus* Mockford (Mexico, southwestern United States), the two species probably having separated by a vicariance event of perhaps Pleistocene age.

Distribution and habitat. The species occurs throughout eastern United States north of the Gulf States, west to Wisconsin and Missouri, also in southeastern Canada. It is found in ground litter, especially in open woodland and hillsides in deeper forest.

Caecilius totonacus Mockford

Caecilius totonacus Mockford, 1966:143.

Recognition features. Very similar to *C. nadleri*, differing in following features: (1) lacinial tip with higher median cusp (Fig. 580), (2) forewing of macropterous form nearly uniform pale brown, colorless around nodulus, base of pterostigma, and in middle of cell R5, slightly darker in base of cell R5 and along veins Rs and Rs + M in cell R1, (3) male paraproctal papillar field with small rounded papillae instead of spines (Fig. 581), (4) male epiproctal papillar field smaller and more compact (Fig. 582), (5) seta of female v3 in membrane off the valvula rather than on edge of pigmented, sclerotized area, (6) male abdominal terga 3-8 each with a transverse brown sclerite (Fig. 150).

Relationships. These are as noted for *C. nadleri*.

Distribution and habitat. Within the study area the species is known from southern Colorado, New Mexico, and Yavapai County Arizona. It is usually found in ground litter of oak groves. Outside the study area, it is known from the Mexican states of Tamaulipas, Nuevo León, Hidalgo, Oaxaca, and Chiapas, also from Guatemala and Panama. Males and macropterous females frequently occur on foliage of low trees and shrubs in the Mexican mountains.

Flavidus Group

Diagnosis. Labral stylets absent. Lacinial tip somewhat expanded and rounded, usually as two lobes (Fig. 134). Body pigmentation variable: most species predominantly pale yellow in color, others tawny or brown; a few with banded wings. Fore and sometimes middle tibiae exhibiting sexual dimorphism: at least somewhat swollen in males, cylindrical throughout in females. Mesoprecoxal suture absent. Spermathecal duct sheath of short to medium length.

North American species: *C. atricornis* McLachlan, *C. boreus* Mockford, *C. burmeisteri* Brauer, *C. caloclypeus* Mockford and Gurney, *C. croesus* Chapman, *C. flavidus* Stephens, *C. hyperboreus* Mockford, *C. lochloosae* Mockford, *C. manteri* Sommerman, *C. maritimus* Mockford, *C. micanopi* Mockford, *C. perplexus* Chapman, *C. pinicola* Banks, *C. tamiami* Mockford.

Caecilius atricornis McLachlan

Caecilius atricornis McLachlan, 1869:197.

Recognition features. Both sexes with antennae longer than forewings, those of male longer than those of female. Body creamy yellow on lower parts of head and thorax, vertex between eyes, legs, all of preclunial abdominal segments. Antennae black beyond brown scape, pedicel, and base of f1; a broad dark brown band across head through frons and ocellar field, continuing backward on each side from antennal base, around eye, to hind margin of head, then continuing as paler band over thoracic pleura in their upper halves. Thoracic notal lobes dark brown. Wings tawny brown, paler along fore margin including pterostigma and in narrow band across middle. Male fore tibiae lengthened, slightly swollen near base; middle tibiae not lengthened but also slightly swollen near base.

Relationships. These are not known at present. The species appears to have no close relatives among the holarctic flavidus - group species.

Distribution and habitat. Within the study area, the species was taken as an introduction at Port Everglades, Florida, on reed fencing from Hungary. Outside the study area, the species is widespread in Europe. It occurs in low, wet vegetation, such as stands of reeds and other grasses in marshy areas.

Caecilius boreus Mockford

Caecilius boreus Mockford, 1965a:138.

Recognition features. Antennae of moderate length, about equal to that of forewing. Body in general pale brown except slightly darker on thoracic notal lobes, ocellar field, and male fore and middle tibiae. Paler on preclunial abdominal segments. Wing membranes tawny brown, pterostigma paler; wing setae somewhat darker than veins. Vein Cu2 in forewing ciliated. Male fore and middle tibiae decidedly stout most of their length, tapering at both ends (Fig. 144). Male epiproct consistently lacking papillar field. Male paraproct with field of scattered, relatively small papillae (Fig. 146).

Relationships. *C. boreus* is very close to *C. croesus* Chapman, the two probably having separated in the fairly recent geologic past by a vicariance event. Both are closely related to *C. burmeisteri* Brauer.

Distribution and habitat. The species is known from southern Alaska, southern Ontario, Nova Scotia, and the Appalachian Mountains in Tennessee. It has been taken on the foliage of spruce and larch. A record from Oregon (Mockford, 1965a) is a misdetermination of *C. burmeisteri* Brauer.

Caecilius burmeisteri Brauer

Psocus pedicularius Burmeister, 1839:776.
Caecilius burmeisteri Brauer, 1876:293.
Caecilius rufus Tetens, 1891:372.
Caecilius minutus Reuter, 1894:15.
Caecilius burmeisteri ab. *lipsiensis* Enderlein, 1901:541.

Recognition features. Antennae of moderate length, about equal to that of forewing. Body in general pale brown, slightly darker on ocellar field and male fore- and middle tibiae, but not or scarcely darker on thoracic notal lobes. Wing membranes pale tawny brown; wing setae somewhat darker than veins. Vein Cu2 in forewing ciliated. Male fore and middle tibiae slightly broadened in their mid-regions and slightly tapering towards each end (Fig. 145). Male epiproct usually with a field of two to twenty small papillae but these sometimes lacking. Male paraproct with well developed field of evenly distributed papillae (Fig. 583).

Relationships. These are as discussed for *C. boreus*.

Distribution and habitat. Within the study area, the species occurs from Vancouver, British Columbia, south on the coast to Humboldt County, California, primarily on foliage of various coniferous trees. Outside the study area, it occurs throughout most of Europe.

Caecilius caloclypeus Mockford and Gurney

Caecilius caloclypeus Mockford and Gurney, 1956:361.

Recognition features. Wings elongate and slender. Both sexes with antennae longer than forewings. Body and appendages in general creamy yellow; postclypeus marked with purplish brown striations on sides; a faint purplish brown band on each side from antennal base to compound eye, continuing on neck and upper half of propleuron in well pigmented specimens. Wings clear, unmarked. Male fore tibia with subtle bend and slight swelling at distal two-thirds (Fig. 140).·

Relationships. This species appears to be closest to *C. mexcalensis* Mockford (southern Mexico) and *C. micanopi* Mockford (Florida, Bimini), in morphological characters and habitat.

Distribution and habitat. Within the study area, the species has been found at several localities along and near the southern Rio Grande River from Mission, Texas southeastward. Outside the study area, it occurs along the Gulf Coastal Plain to southern Veracruz State, Mexico. It is found on foliage of palms and yuccas.

Caecilius croesus Chapman

Caecilius croesus Chapman, 1930:326.

Recognition features. With features describes for *C. boreus* (above) except following: male epiproct with moderate-sized papillar field; male paraproct with papillar field well developed, the papillae of moderate size (Figs. 147).

Relationships. These are as noted for *C. boreus.*

Distribution and habitat. Revision of material assigned to this species would be useful. At present it includes specimens from the following localities in the study area: Lee County, Texas, northwestern Arkansas, northern Mississippi, coastal plain in North Carolina (Lake Waccamaw, type locality), and Long Island, New York. Outside the study area, the species is widely distributed in the mountains of Mexico and extends south into Guatemala. It is found primarily on the foliage of *Juniperus* and *Cupressus.*

Caecilius flavidus (Stephens)

Psocus flavidus Stephens, 1836:122.
Psocus ochropterus Stephens, 1836:122.
Psocus flavicans Stephens, 1836:123.
Psocus subpunctatus Stephens, 1836:126.
Caecilius strigosus Curtis, 1837:648.
Psocus boreelus Zetterstedt, 1840:1053.
Psocus striatus Zetterstedt, 1840:1053.
Psocus aurantiacus Hagen, 1861:14.
Caecilius aurantiacus (Hagen). Hagen, 1866:205.
Caecilius flavidus (Stephens). Hagen, 1866:205.

Recognition features. Both sexes with antennae about equal to forewings in length (female) to slightly longer (male) than forewings. Male with fore and middle tibiae slightly swollen before middle, tapering towards both ends (Fig. 135). Body and legs in general bright to creamy yellow in color except thoracic notal lobes medium brown, ocellar field medium to dark brown, a pale brown band bordering ecdysial line through vertex; antennae with f1 pale brown contrasting somewhat with darker sc, p, and remainder of flagellum. Forewings (Fig. 584) pale tawny brown, emphasized along veins, especially in distal half of wing, and throughout cell IA.

Relationships. The species is in a small complex including *C. hyperboreus*, *C. maritimus*, and an undescribed species from the southern Appalachian Mountains.

Distribution and habitat. Within the study area the species occurs throughout eastern United States west to Arkansas and Missouri, across southern Canada from Nova Scotia to British Columbia, along the Pacific Coast from southern Alaska to southern Oregon (Coos County). Outside the study area, the species occurs throughout Europe, in the Canary Islands, Bermuda, and in the Mexican states of Nuevo León, Guerrero, and Chiapas. It is found on foliage of broad-leaf trees, rarely on conifers. The first generation in the deciduous forests is in the ground litter.

Note. Males tend to be scarce throughout most of the range of this species and females are known to reproduce by thelytoky. The populations from the Tennessee - North Carolina state line area designated (Mockford, 1971c) as including males represent an undescribed sibling species (pers. obs.)

Caecilius hyperboreus Mockford

Caecilius hyperboreus Mockford, 1965a:141.

Recognition features. Similar to *C. flavidus* in most morphological features. Differing in following features: (1) male compound eyes smaller and spaced farther apart (IO/d range for males from southern Alaska - 1.30 - 1.71, n = 9; for syntopic males of C. flavidus, range - 0.83 - 0.91, n- 10); (2) body in general pale brown with darker (medium) brown over much of vertex and thoracic terga; (3) forewings (Fig. 141) with brown wash over entire surface with darker brown clouding along R stem, M - Cu stem, and Cu1 its entire length except for interruption at base of areola postica.

Relationships. The species appears to be closely related to *C. flavidus*.

Distribution and habitat. The type locality is near Anchorage, Alaska. Female specimens assigned to this species on the basis of color are known from Labrador, Nova Scotia, Quebec, and Ontario. It is found primarily on foliage of spruce trees, but also occurs on birch foliage.

Caecilius lochloosae Mockford

Caecilius lochloosae Mockford, 1965a:143.

Recognition features. Male antennae longer than forewings; those of female about same length as forewings. Forewings elongate, slender (length/greatest width ~ 3.65). Female compound eyes dorsoventrally somewhat compressed (Fig. 585). Body color ivory yellow except ocellar field, antennae, and thoracic notal lobes pale brown; a reddish brown line from antennal base to compound eye and from compound eye to wing bases. Male fore tibia (Fig. 586) slightly swollen throughout middle, tapering slightly towards each end; middle tibiae very slightly less swollen.

Relationships. The species is in a complex including *C. bermudensis* Mockford (Bermuda) and *C. manteri* Sommerman (eastern United States).

Distribution and habitat. The species occurs throughout Florida and in southeastern Alabama. It is found on dead persistent stems of grasses, especially *Andropogon* spp.

Caecilius manteri Sommerman

Caecilius manteri Sommerman, 1943a:29.

Recognition features. As described for *C. lochloosae*, but differing in following features: (1) male fore and middle tibiae not differing in form; (2) reddish brown lateral stripe of thorax distinct and broad, reaching base of abdomen; (3) forewing not so long and slender, length/greatest width ~ 3.18; (4) spermathecal sheath longer, 76.8 - 78.4 µm vs. 69.0 µm; (5) female paraproctal sense cushion with 14 trichobothria vs. 9 - 12 trichobothria.

Relationships. These are as noted for *C. lochloosae*.

Distribution and habitat. The species has been found in most of the U.S. states east of the Mississippi River and north of Kentucky and Virginia, also in Minnesota, Missouri, and Nebraska. It inhabits cattails (*Typha* spp.), especially clumps of dead stalks, also dead and dying stalks of corn (*Zea mays* L.). Males are extremely scarce and appear to be confined to salt marsh areas on the northeastern coast. Brachypterous females of either this species or a close relative have been found in a salt marsh at Cedar Key, Levy County, Florida.

Caecilius maritimus Mockford

Caecilius maritimus Mockford, 1965a:145.

Recognition features. As described for *C. flavidus*, differing in following: (1) male middle tibia cylindrical throughout rather than swollen before middle; (2) larger number of papillae in male epiproctal papillar field (30-32 versus 0-26); (3) females frequently showing brachyptery with forewings reaching tip, or only middle, of abdomen; (4) female with darker spot on frons before ocelli and darker clypeal striations.

Note. Recently preserved females usually show a purplish brown spot below each wing base. This spot does not survive long preservation in alcohol. Macropterous females long in alcohol are not distinguishable from females of *C. flavidus*.

Relationships. These are as noted for *C. flavidus*.

Distribution and habitat. The species is known from coastal California, where it has been found from the San Francisco region north to Arcata. It inhabits ground litter and foliage of various trees and shrubs in woodland.

Caecilius micanopi Mockford

Caecilius micanopi Mockford, 1965a:147.

Recognition features. Antennae of both sexes longer than forewings. Wings relatively elongate and slender (Fig. 587). Female compound eyes somewhat elongate and dorso-ventrally compressed (Fig. 588). Body in general ivory yellow except medium brown: thoracic notal lobes, ocellar field, spot before and spot behind ocellar field, lateral postclypeal striations, band from antennal base to compound eye, flagellum beyond f2. Wings clear, unmarked. Male fore tibiae greatly swollen, especially immediately distad of middle (Fig. 139); middle tibiae normal.

Relationships. This species appears to be closely related to *C. mexcalensis* Mockford (southern Mexico) and *C. caloclypeus* (south Texas, northern Mexico). The three species share color and markings, size (decidedly longer than the species of the manteri complex), and habitat (palm foliage). A very similar undescribed species occurs in coastal Veracruz State, Mexico.

Distribution and habitat. The species is found throughout Florida, in southeastern Alabama, and on Bimini Island and Cat Cays in the Bahamas. It inhabits primarily foliage of palms, especially of the genera *Sabal* and *Coccothrinax*, and occurs more rarely on foliage and persistent dead leaves of dicotyledonous trees.

Caecilius perplexus Chapman

Caecilius perplexus Chapman, 1930:326.

Recognition features. Male antennae slightly shorter than forewings; female antennae little more than half length of forewings. Postclypeus in both sexes greatly protruding (Fig. 142), giving front of head a bulbous appearance in dorsal or lateral view (in dorsal view, postclypeus about one-third length of rest of head). Vein Cu2 in forewing without setae or with two to three setae along its length, or (rarely) with more complete row of setae. Body and appendages pale brown except ocellar field darker brown. Wings unmarked, with very faint brown wash. Male fore and middle tibiae slightly swollen before middle, tapering towards both ends.

Relationships. The species is probably closest to *C. pinicola* Banks.

Distribution and habitat. The species occurs from the desert mountain ranges of southern New Mexico and Arizona north through the Rocky Mountains to Alberta, also spottily in the Coastal States and British Columbia: Trinity County, California, Lake and Lincoln Counties, Oregon, Cluculz Lake and Houston in central British Columbia. It is usually found on foliage of pines, Douglas firs, and spruces, but occurs occasionally on foliage of broad-leaf trees.

Caecilius pinicola Banks

Caecilius pinicola Banks, 1903b:238.

Recognition features. Antennal proportions as described for *C. perplexus*. Postclypeus not so strongly bulging as in *C. perplexus*, forming about one-fourth length of head in dorsal view of male, somewhat more than that in female. Vein Cu2 in forewing usually without setae, at most with one or two setae along its length. Body and appendages pale brown except ocellar field darker brown; live and freshly preserved specimens with triangular red mark each side on vertico-frontal region mesad of antennal base (Fig. 589). Wings and male fore and middle tibiae as described for C. perplexus.

Relationships. The species is probably closest to *C. perplexus*.

Distribution and habitat. The species is found throughout eastern United States south to Highlands County, Florida, and Nacogdoches County, Texas, also in southern Ontario and southern Quebec. In western North America, the species is rather common in the Anchorage region of Alaska, and a population was found near Olympia, Lewis County, Washington. In eastern United States it inhabits primarily foliage of pines and junipers. In southeastern Canada it has been found also on foliage of spruce and *Thuja occidentalis*. The Lewis County, Washington specimens were from Douglas Fir.

Caecilius tamiami Mockford

Caecilius tamiami Mockford. 1965a:150.

Recognition features. Forewings elongate, slender. Antennae of both sexes somewhat shorter than forewings; male antennae stouter than those of female. Male fore and middle tibiae decidedly broadened throughout most of length, tapering at ends (Fig. 590). Body and appendages yellow (without lateral reddish brown marks) except ocellar field and thoracic notal lobes pale brown, male antennae dark brown, paler on sc and p. Female antennae somewhat paler throughout.

Relationships. These are uncertain, but the species appears to be a lowland tropical-subtropical representative of the conifer-inhabiting species set represented farther north by *C. burmeisteri, C. boreus, C. croesus, C. perplexus,* and *C. pinicola.*

Distribution and habitat. Within the study area, the species is known from two localities in southern Florida: Big Pine Key (Monroe County) and Archbold Biological Station (Highlands County). It inhabits the foliage of Caribbean pine (*Pinus caribaea* Morelet). Outside the study area, the species is known from Oriente Province, Cuba.

Posticus Group

Diagnosis. Labral stylets absent. Lacinial tip slender, bicuspid. Body well pigmented with brown or reddish brown. Wings extensively marked with brown or reddish brown but with pterostigma remaining completely clear (Fig. 133). Pterostigma relatively shallow. Ciliation absent from vein Cu2 in forewing. Mesothoracic precoxal suture weakly developed or absent. Sheath of spermathecal duct moderately long (Fig. 591).

North American species: *C. posticus* Banks.

Caecilius posticus Banks

Caecilius posticus Banks, 1914:612.

Recognition features. The only species of its group in the study area. Although some females are fully winged, others are micropterous with minute winglets not reaching the base of the abdomen. These are readily distinguishable from similar females of *C. nadleri* (fasciatus group) by absence of pigmentation in the distal segments of the flagellum.

Relationships. The other members of this group are found from the Gulf Coast in Veracruz, Mexico south through tropical South America, and include the named species *C. claripennis* New and Thornton (Brazil).

Distribution and habitat. The species is probably widely distributed in eastern United States, but it is rarely collected, occurring primarily in well drained leaf litter in open woodlands. There has been confusion in the literature of micropterous females of this species with those of *C. nadleri* Authentic records of *C. posticus* are from the states of Connecticut, Florida (Highlands County), Illinois, Michigan, and New York.

Subflavus Group

Diagnosis. Relatively small caeciliids with forewing always less than 2.5 mm in length. Labral stylets absent. Lacinial tip slender, bicuspid, with one or two low denticles associated with median cusp (Fig. 592). Body moderately to weakly pigmented. Wings with slight development of marking pattern or none. Pterostigma moderately angulate. Ciliation usually absent on vein Cu2 in forewing, but occasionally present. Mesothoracic precoxal suture absent. Papillae of male epiproctal and paraproctal fields of moderate to large size (Figs. 593, 594). Sheath of spermathecal duct short (Fig. 595).

North American species: *C. incoloratus* Mockford, *C. juniperorum* Mockford, *C. subflavus* Aaron.

Caecilius incoloratus Mockford

Caecilius incoloratus Mockford, 1969a:88.

Recognition features. Body and appendages uniformly dull yellowish white throughout except for faint brown tinge on lateral mesonotal lobes and faint brown subcuticular band across vertex between eyes. Ocellar field concolorous with surrounding cuticle. Freshly preserved females with compound eyes bicolored; a line running through middle separating eye into pale brown dorsal and deep violet ventral halves. Wings unmarked; forewings with lemon yellow wash. Vein Cu2 in forewing with at least two to three setae in distal half.

Relationships. This species is in a complex with four other species found around the Gulf of Mexico and on the Caribbean islands: *C. caribensis* Mockford (Jamaica, Martinique), *C. flavibrunneus* Mockford (Bahamas, Cuba), *C. olitorius* Banks (Puerto Rico), and *C. veracruzensis* Mockford (Veracruz, Chiapas, and Quintana Roo, Mexico).

Distribution and habitat. Within the study area this species is known from three of the Florida Keys: Big Pine, Key Largo, and Sugarloaf. It has been taken on foliage of mangrove, sea grape (*Coccolobis uvifera* [L.]), and thatch palm (*Thrinax* sp.). Outside the study area, it is known from South Bimini and New Providence Islands in the Bahamas.

Caecilius juniperorum Mockford

Caecilius juniperorum Mockford, 1969a:82.

Recognition features. Head, thorax, and abdomen tawny brown; abdomen somewhat paler than head and thorax. Sides of head, thorax, and basal half of abdomen with reddish brown cast. Well colored males with slightly indicated pattern in forewing (Fig. 151): brown spot below hind angle of pterostigma, brown spot in base of cell R5, colorless spot below base of pterostigma, pale area distad of spot in base of cell R5, another pale area around base of areola postica; rest of wing uniformly tawny brown. Female forewing uniformly tawny brown. Vein Cu2 in forewing without setae.

Relationships. This species is probably most closely related to *C. pallidobrunneus* Mockford, which is found on foliage of *Pinus caribaea*, *Podocarpus urbanii*, and *Juniperus lucayana* at moderate elevations (900-1200 m) in Jamaica (Turner, 1975). The two species differ in size and details of coloration.

Distribution and habitat. The species is known from three localities in Florida: Inglis (Levy County), Juniper Springs (Marion County), and West Melbourne (Brevard County). It has been taken on foliage of *Juniperus lucayana*, undetermined juniper, and *Thuja orientalis*.

Caecilius subflavus Aaron

Caecilius subflavus Aaron, 1886:13.

Recognition features. Body largely ivory yellow with medium brown band from antennal base to compound eye and medium brown spot behind compound eye to hind margin of head; thoracic notal lobes medium brown (darker on male than female). Forewing with distinct marking pattern (Fig. 152): brown spot below hind angle of pterostigma (extending forward into pterostigma in male, not in female), brown spot over distal half of cell R separated only by veins from brown spot in base of cell M3 and brown border of vein M + Cu1 and Cu1 to areola postica; brown spot in base of cell R5 and brown border on both sides of Cu loop in its distal two-thirds; cells Cu2 and IA brown except in their distal one-fourth; otherwise, wing membrane clear except for brown clouding near distal end of cell R5 and in base of cell M2 and adjacent region of cell M3. Freshly preserved females with compound eyes yellow dorsally darkening to reddish brown medially to dark purplish brown ventrally. Vein Cu2 of forewing without setae.

Relationships. The species is most closely related to *C. biminiensis* Mockford (Bimini Islands).

Distribution and habitat. The species is found throughout Florida, north on the Atlantic Coast to Hatteras Island, North Carolina. It also occurs in southern Texas, probably on the coast near Corpus Christi (type locality "Southern Texas"). Whether it is continuous around the Gulf Coast from Florida remains to be shown. It occurs on foliage of primarily evergreen trees and shrubs such as live oak, American holly, and wax myrtle.

Genus *Xanthocaecilius* Mockford

Xanthocaecilius Mockford, 1989:268.

Diagnosis. As noted for the family, plus following: inner clypeal shelf wide; distal inner labral sensilla seven; lacinial tip bicuspid; labral stylets absent; three ventral abdominal vesicles; endophallus with median bulb divided (Fig. 129); compound eyes of living and freshly preserved specimens yellow or pale green.

Generotype. *X. quillayute* (Chapman).

North American species: *X. quillayute* (Chapman), *X. sommermanae* (Mockford).

Xanthocaecilius quillayute (Chapman)

Caecilius quillayute Chapman, 1930:330.
Xanthocaecilius quillayute (Chapman). Mockford, 1989:270.

Recognition features. This species is very close to *X. sommermanae*, and when variation in both forms is well understood, they may prove to be the same species. At present, the following differences separate the two: (1) *X. quillayute* with dusky wings with pattern of dark and pale veins showing only obscurely (Fig. 596) versus in *X. sommermanae* wings clear with distinct pattern of dark and pale veins (Fig. 130); (2) *X. quillayute* with green eyes in life, vs. *X. sommermanae* with yellow eyes in life; (3) negative regression of IO:d of males falling off more rapidly in *X. quillayute* than in *X. sommermanae* (Fig. 597).

Relationships. The two species, very close to each other, appear not to be very close to other members of the genus.

Distribution and habitat. On the Pacific Coast, the species occurs from Humboldt County, California north to Vancouver, British Columbia. In eastern North America, material apparently representing this species has been found in southern Quebec, Nova Scotia, and in the southern Appalachian Mountains. The species is found primarily on foliage of conifers but occurs occasionally on foliage of broad-leaf trees.

Xanthocaecilius sommermanae (Mockford)

Caecilius aurantiacus Chapman, 1930 (in part), not Hagen, 1861.
Caecilius sommermanae Mockford, 1955b:438.
Xanthocaecilius sommermanae (Mockford). Mockford, 1989:270.

Recognition features. As noted for *X. quillayute* (above).
Relationships. These are as noted for *X. quillayute* (above).
Distribution and habitat. The species occurs in eastern Unites States from Connecticut south to central Florida (Marion County) and west to Arkansas. On the Pacific Coast populations apparently representing this species have been found in Del Norte County, California and Pacific County, Washington. It usually inhabits foliage of broad-leaf trees but also occurs on foliage of hemlock.

Family Group Homilopsocidea

Diagnosis. Labrum without slender longitudinal sclerite or pigment band in each half. Inner surface of distal margin of labrum with five to eleven sensilla (three placoids alternating with two trichoids, or median placoid flanked on each side by three trichoids, flanked by one placoid, in some cases flanked by one trichoid). Mandibles short, not excavated posteriorly (Fig. 598). Meso-precoxal bridges broad; mesotrochantins narrow (Fig. 110). Tarsi two to three segmented in adults. Pretarsal claws usually with preapical denticle (absent in Archipsocidae). Coxal organ well developed except in very neotenic forms. Ovipositor valvulae: one to three pairs, rarely absent; v2 frequently with two apices, one rounded and one acuminate tipped; v3 generally present, independent except at base (occasionally v2 fused into v3), bearing numerous setae. Female paraprocts not greatly extended and narrowed distally.

North American taxa: Families Lachesillidae, Ectopsocidae, Peripsocidae, Archipsocidae, Pseudocaeciliidae, Trichopsocidae, Philotarsidae, Elipsocidae, Mesopsocidae.

Note. Pearman (1936a), in proposing this and the other family groups, expressed some doubt about the naturalness of this group. As used here, with the hemipsocids excluded, it seems to consist of a series of related families except that the peripsocids stand apart from the others, especially in structure of male genitalia.

Family Lachesillidae

Diagnosis. Relatively small (length from head to tip of closed wings 2 - 4 mm) forms with body shades of brown, wings usually clear. Lacinial tip (Fig. 599) slender, bicuspid, the two cusps of about equal size and pointed apically or outer longer, sometimes with denticles. Five distal inner labral sensilla. Vertex and frons bearing numerous upright or forward-directed curved setae of various lengths. Ciliation sparse on veins and margin of forewing, usually absent on hindwing except, rarely, a few minute setae on margin in cell R3. Vein Cu 1a present in forewing (cubital loop). Rs and M joined in hindwing by relatively long fusion, about equal to or greater than length of preceding Rs segment. Ovipositor valvulae: v2, when present, with simple, rounded apex; v3 with field of setae of various lengths on and near its upper, outer surface.

Mockford and Sullivan (1986) proposed a subfamily and tribal classification for this family which is followed here.

Key to North American Subfamilies, Genera, and Species of Family Lachesillidae

1. Forewing sparsely ciliated on veins and margins. Phallosome a closed frame with endophallus of several sclerites or many denticles (Fig. 155). Major ovipositor valvula of each side usually tapering, always spinose towards distal end ....
    .... Subfamily Eolachesillinae 2

—   Forewing without obvious ciliation. Phallosome open distally (Fig. 156). Major ovipositor valvulae of each side usually not tapering, never spinose towards distal end ....
    .... Subfamily Lachesillinae, Genus *Lachesilla* 5

2. Forewings usually well developed; vein Cu 1a joined to M for a distance (Fig. 157). Ocelli present ....
    .... Genus *Anomopsocus*, *A. amabilis* (Walsh)

--   Forewings well developed or not; when well developed, vein Cu1a
      free from M. Ocelli absent 3

3.   Forewings well developed. Phallosome with basal apodeme (Fig.
      155). Major ovipositor valvula of each side blunt apically, with
      a small lobe at base (Fig. 158) .... Genus *Nanolachesilla* 4
--   Forewings greatly reduced, reaching only base of abdomen.
      Phallosome flat basally (Fig. 162). Major ovipositor valvula of
      each side bilobed, the outer lobe longer, bluntly pointed
      apically (Fig. 163) ....
      .... Genus *Prolachesilla*, *P. terricola* Mockford and Sullivan

4.   Hindwing with a few minute setae on margin in cell R3. Two major
      sclerites of endophallus resembling a swallow in flight (Fig.
      155) .... *N. hirundo* Mockford and Sullivan
--   Hindwing with no setae on margin in cell R3. Two major sclerites
      of endophallus not as above, lacking lateral wing- like pro-
      cesses (Fig. 161) .... *N. chelata* Mockford And Sullivan

5.   Hypandrium three-segmented; distal processes (claspers ?) not
      articulated at bases (Fig. 162) 6
--   Hypandrium one- or two-segmented. Claspers articulated or not
      at bases, sometimes absent 10

6.   Male epiproct not bilobed, with single short, bifid process in
      middle (Fig. 163). Subgenital plate without a separate median
      lobe .... Texcocana Group, *L. texcocana* Garcia Aldrete
--   Male epiproct bilobed, lacking median (single) process, but with
      a minute process on each side. Subgenital plate usually with
      a distinct median lobe 7

7.   Hypandrium with a small triangular plate between bases of
      distal processes (Fig. 164). Subgenital plate with only a slightly
      indicated or very small median lobe .... Centralis Group 8
--   Hypandrium (where known) with region between bases of distal
      processes heavily sclerotized, but without a triangular plate.
      Subgenital plate with a distinct, large median lobe ....
      .... Patzunensis Group, *L. sulcata* Garcia Aldrete

8.   Median tongue of subgenital plate distinct but small; subgenital
      plate with lateral rugose areas. Arms of phallosome with group
      of spinules before apex (Fig. 165).......................................
      .... *L. cintalapa* Garcia Aldrete

--   Median tongue of subgenital plate indicated by slight bulge or markings on surface; lateral rugose areas present or absent on plate. Arms of phallosome without group of spinules before apex 9

9.   Triangular distal plate of hypandrium small, rounded posteriorly, without setae (Fig. 166). Lateral rugose areas absent on subgenital plate .... *L. perezi* Garcia Aldrete

--   Triangular distal plate of hypandrium elongate transversely, pointed posteriorly, bearing two long setae (Fig. 627). Lateral rugose areas present on subgenital plate (Fig. 167) ....
      .... *L. centralis* Garcia Aldrete

10.   Subgenital plate in middle with either (1) a slender process, or (2) a tongue with bifid apex, or (3) a wide lobe deeply cloven in middle. Claspers/distal processes of hypandrium separated from hypandrium proper by, at most, an indistinct line, usually no separation 11

--   Subgenital plate without a median tongue or process; a wide median lobe, when present, not cleft or only shallowly so. Claspers either distinct at their bases from hypandrium proper or (rarely) totally absent 18

11.   Subgenital plate with wide, deeply cleft median lobe (Fig. 168). Male epiproct bilobed posteriorly ....
      ....Fuscipalpis Group, *L. fuscipalpis* Badonnel

--   Subgenital plate with either slender process, broader process or process with bifid apex. Male epiproct rounded posteriorly ..
      12

12.   Process of subgenital plate with bifid apex. Claspers simple ...
      .... Rufa Group 14

--   Process of subgenital plate simple at apex. Claspers, where known, bifid .... Corona Group 13

13.   Process of subgenital plate relatively broad, spade-like ....
      ....*L. albertina* Garcia Aldrete

--   Process of subgenital plate slender, finger-like (Fig. 169) ....
      .... *L. corona* Chapman

14.   Forewing with marking pattern (Fig. 170): a small brown spot at the end of each vein on wing margin from R1 to nodulus and an obscure brown spot in each of cells R3, R5, M1, M2, and M3 .... *L. nita* Sommerman

--   Forewing clear, unmarked 15

15.   Subgenital plate indented in middle of hind margin 16
--   Subgenital plate flattened or rounded in middle of hind margin
       17

16.   Median tongue of subgenital plate with short cleft region, not
       more than one-fifth length of tongue .... *L. arida* Chapman
--   Median tongue of subgenital plate with deeper cleft region,
       occupying about one-fourth length of tongue (Fig. 171). Phallo-
       somal apodemes completely separate ....
       .... *L. jeanae* Sommerman

17.   Subgenital plate rounded distally; median tongue with distal
       cleavage occupying about half length of tongue (Fig. 172).
       Phallosome with single basal apodeme divided at about basal
       one-third of length of entire structure .... *L. rufa* (Walsh)
--   Subgenital plate flattened distally; median tongue with distal
       cleavage occupying about one-fifth length of tongue (Fig. 173)
       .... *L. yakima* Mockford and García Aldrete

18.   Ovipositor valvulae single, short, directed caudad (Fig. 174).
       Phallosome either a single rod once-branched distally with
       branches subdivided or two asymmetrical, unbranched rods
       .... Andra Group 19
--   Ovipositor valvulae larger, directed mesad or meso-caudad,
       rarely two on each side. Phallosome a single rod, usually
       branching once distally, the branches not subdivided 25

19.   Phallosome of two asymmetrical rods (Fig. 175). Claspers absent.
       Female ninth sternum well sclerotized and four- lobed anteri-
       orly (Fig. 176) .... *L. kola* Sommerman
--   Phallosome of one basal rod once-branched distally with branches
       subdivided. Claspers present. Female ninth sternum bilobed
       or simple anteriorly 20

20.   Wings usually of normal shape, (sometimes greatly reduced)
       with distinct spotting pattern when macropterous 21
--   Wings somewhat elongated and slender (length/greatest width ~
       3.04), without spotting pattern .... *L. andra* Sommerman

21.   Subgenital plate with a distinct posterior median lobe (Fig. 177).
       Male epiproct with either two long lateral processes or ending
       in a long, median, horn-like projection 22
--   Subgenital plate without a posterior median lobe. Male epiproct
       with two lateral projections of only moderate length 24

22. Male epiproct with a long, horn-like median posterior projection (Fig. 178). Subgenital plate with main body bearing two small lobes at base of posterior median lobe (Fig. 177) .... *L. punctata* (Banks)

-- Male epiproct with two long lateral processes (Fig. 179). Subgenital plate not as above 23

23. Lateral lobes of subgenital plate blunt-tipped behind main body of plate (Fig. 180). Females micropterous. Clasper shafts straight basally, curved out near distal end (Fig. 181) .... *L. nubiloides* Garcia Aldrete

-- Lateral lobes of subgenital plate bluntly pointed (Fig. 186). Females macropterous. Clasper shafts curved mesad (Fig. 183) .... *L. dona* Sommerman.

24. Claspers shorter than basal apodeme of phallosome (Fig. 184). Females brachypterous .... *L. arnae* Sommerman

-- Claspers longer than basal apodeme of phallosome (Fig. 185). Females macropterous .... *L. nubilis* (Aaron)

25. Phallosome (Fig. 186) terminating distally in pair of pointed processes (distinct from clasper shafts). Female ninth sternum with two short projections on anterior surface (Fig. 187) .... Riegeli Group 26

-- Phallosome terminating distally in pair of rounded or triangular pads (clasper shafts often near midline or touching on midline), or single broad pad; pair of pads sometimes pointed at tips. Female ninth sternum usually lacking pair of short projections on anterior surface 28

26. Forewing with brown spot on margin at end of each M and Cu1 vein. Terminal processes of phallosome much shorter than clasper shafts (Fig. 186). Subgenital plate with distal margin indented in middle .... *L. riegeli* Sommerman

-- Forewing clear, unmarked. Terminal processes of phallosome, where known, slightly longer than clasper shafts. Distal margin of subgenital plate not indented in middle 27

27. Transverse internal thickening of subgenital plate large, covering most of area of plate (Fig. 188) .. *L. tropica* Garcia Aldrete

-- Transverse internal thickening of subgenital plate relatively slender, lying across middle of plate (Fig. 189) .... *L. ultima* Garcia Aldrete

28.   Male paraproct with no process or extremely short one. Claspers moderately long, frequently recurved, not approaching midline (Fig. 190). Subgenital plate weakly sclerotized, with posterolateral shoulders or short processes. Ovipositor valvulae (Fig. 191) broad, round- or blunt-tipped, directed mesally ...
.... Forcepeta Group 29

--   Male paraproct with a well developed process. Claspers relatively short, frequently close to midline, sometimes touching on midline. Subgenital plate moderately sclerotized, its hind margin rounded or bilobed. Ovipositor valvulae of moderate width, directed mesad or meso-caudad (Fig. 192) ....
.... Pedicularia Group 39

29.   Clasper shafts pointing mesad at apex (Fig. 190) 30
--   Clasper shafts straight or pointing outward at apex (Fig. 193)
.... 34

30.   Claspers with microspines on inner surface near distal end (Fig. 194). Hind margin of subgenital plate with a short, truncate lobe on each side (Fig. 195) .... *L. alpejia* Garcia Aldrete
--   Claspers without microspines on inner surface near distal end. Hind margin of subgenital plate without pair of lobes 31

31.   Clasper shafts with a tooth before apex (Fig. 196). Female ninth sternum with a bilobed anterior sclerotized region (Fig. 197)
.... *L. denticulata* Garcia Aldrete
--   Clasper shafts without a tooth before apex. Female ninth sternum lacking bilobed anterior sclerotized region 32

32.   Clasper shafts directed outward immediately beyond base (Fig. 198). Subgenital plate without internal sclerotizations ....
.... *L. bottimeri* Mockford and Gurney
--   Clasper shafts straight immediately beyond base (Fig. 190). Subgenital plate with a well defined or diffuse internal thickening 33

33.   Abdomen with reddish brown subcuticular ring on each preclunial segment. Subgenital plate with well defined lozenge shaped, orange internal thickening .... *L. major* Chapman
--   Abdomen with no subcuticular pigment rings. Subgenital plate with diffuse, uncolored, ovoid internal thickening................
....*L. forcepeta* Chapman

34.   Clasper shafts straight at apex 35
--   Clasper shafts pointing outward at apex 36

35.   Clasper shafts with a tooth before apex (Fig. 199). Subgenital plate with two ovoid, yellow internal sclerites (Fig. 200) ....
.... *L. anna* Sommerman
--   Clasper shafts with serrate inner margin but no tooth before apex (Fig. 201). Subgenital plate with single bilobed internal sclerotization (Fig. 202) .... *L. penta* Sommerman

36.   Clasper shafts with several denticles on median surface near apex (Fig. 203). Subgenital plate with two small ovoid internal sclerites (Fig. 204) .... *L. gracilis* García Aldrete
--   Clasper shafts with or without denticles near apex. Subgenital plate without internal sclerites 37

37.   Male epiproct with a median horn-like process (Fig. 205). Reddish brown spots scattered on sides of thorax and some preclunial abdominal segments 38
--   Male epiproct with no horn-like process. Sides of thorax and abdomen lacking reddish brown spots ....
.... *L. contraforcepeta* Chapman

38.   Clasper shafts widened before apex (Fig. 206). Subgenital plate with a broad median projection truncated at apex (Fig. 207)
.... *L. kathrynae* Mockford and Gurney
--   Clasper shafts slender before apex (Fig. 208). Subgenital plate without a broad median projection ....
.... *L. chapmani* Sommerman

39.   Known from female only. Head and thorax dusky brown; preclunial abdominal segments with wide dusky brown rings of subcuticular pigment .... *L. tectorum* Badonnel
--   Either known from both sexes, or body color usually not so dark, or both 40

40.   Known from female only. Ovipositor valvulae with pointed beak at apex (Fig. 209) .... *L. aethiopica* (Enderlein)
--   Known from both sexes. Ovipositor valvulae blunted or rounded at apex 41

41.   Male epiproct and clunium both lacking processes. Clasper shaft with a large tooth on outer surface (Fig. 210) ....
.... *L. quercus* (Kolbe)
--   Male either with pair of processes, directed posteriorly, on clunium and single short process on epiproct or no process on clunium but pair of processes, sometimes directed forward, on epiproct 42

42.   Male with pair of processes, directed posteriorly, on clunium and
      single short process on epiproct (Fig. 211) 43
--   Male with no process on clunium and pair of processes on
      epiproct 45

43.   Male paraproct with a well developed process (Fig. 212). Forew-
      ings marked with a spot distally in pterostigma, a spot border-
      ing cell Cu1a, and a spot at nodulus (Fig. 213). Females
      micropterous .... *L. pallida* (Chapman)
--   Male paraproct without a process. Forewings unmarked, or at
      most with slight brown borders of veins M1 - M3 and Cu1a.
      Females macropterous or micropterous 44

44.   Male epiproctal process curved. Sclerotized branches of
      phallosomal apodeme straight (Fig. 211). Both sexes usually
      macropterous .... *L. pedicularia* (Linn.)
--   Male epiproctal process bent at right angle. Sclerotized branches
      of phallosomal apodeme curved s-shaped. Males usually
      brachypterous, females micropterous ..... *L. greeni* (Pearman)

45.   Male epiproct with a long forward-directed process (Fig. 214).
      Distal wing veins bordered with brown before reaching mar-
      gins .... *L. rena* Sommerman
--   Male epiproct with asymmetrical curved processes (Fig. 215).
      Forewings unmarked .... *L. pacifica* Chapman

Subfamily Eolachesillinae

Diagnosis. Forewings sparsely ciliated on veins and margins.
Phallosome a closed frame (Fig. 155); endophallic sclerites variable in
size, generally some sclerites chelate. Ovipositor valvulae: v3 or fused
v2+3 elongate, tapering towards distal end. Female paraprocts with
field of short, relatively stout setae along median margin in ventral half
(Fig. 600).

Tribe Graphocaeciliini

Diagnosis. Epistomal suture developed only laterally, never present
dorsally (i.e., postclypeus continuous with frons). Outer cusp of lacinial
tip bidentate. Tarsi two-segmented. In forewing (Fig. 157): Rs fork stem
somewhat flexuous; Rs - M junction relatively long. Ovipositor valvulae:
v2 never present as an entity separate from v3.

North American genera: *Anomopsocus* Roesler, *Nanolachesilla* Mockford and Sullivan, *Prolachesilla* Mockford and Sullivan.

Genus *Anomopsocus* Roesler

Pseudopsocus Chapman, 1930:287, not Kolbe, 1882:208.
Anomopsocus Roesler, 1940:239.
Antipsocus Roesler, 1940:241.
Amapsocus Sommerman, 1944:359.

Diagnosis. Ocelli present. Fronto-clypeal region with field of short, blunt-tipped sensilla based on small pores. In forewing, Rs fork stem strongly sigmoidally curved; vein Cu 1a fused with M for a distance (Fig. 157). Phallosome (Fig. 601) with base rounded, external parameres weakly sclerotized. V2+3 with distal end protracted as slender, spinose process (Fig. 602).
  Generotype. *A. amabilis* (Walsh).
  North American species: *A. amabilis* (Walsh).

Anomopsocus amabilis (Walsh)

Psocus amabilis Walsh, 1862:362.
Pseudopsocus amabilis (Walsh). Chapman, 1930:288.
Anomopsocus amabilis (Walsh). Roesler, 1940:239 [implied].
Amapsocus amabilis (Walsh). Sommerman, 1944:359.

Recognition features. The only species of its genus in the study area. Flagellomeres f1 through f5 banded pale basally, dark distally.
  Relationships. Two other species of Anomopsocus are known, both from the American Tropics.
  Distribution and habitat. The species occurs throughout eastern United States and southeastern Canada, south to central Florida (Clay County) and west to Missouri and eastern Texas. It inhabits primarily hanging dead leaves and is found occasionally on foliage of conifers. Females from the southern end of the range (Texas and Florida) are frequently brachypterous.

Genus *Nanolachesilla* Mockford and Sullivan

Nanolachesilla Mockford and Sullivan, 1986:52.

Diagnosis. Ocelli absent. Fronto-clypeal region without field of blunt-tipped sensilla. In forewing (Fig. 603), Rs fork stem only slightly to moderately curved; Vein Cu 1a free from M. Phallosome (Fig. 155) a closed frame with basal apodeme; aedeagal arch and external parameres present; frame enclosing endophallus of several large sclerites. Hypandrium without lateral processes. Ovipositor valvulae (Fig. 158): v3 rounded at apex, with one or two lobes at base presumably representing vestiges of other valvulae.

Generotype. *N. hirundo* Mockford and Sullivan.

North American species: *N. hirundo* Mockford and Sullivan, *N. chelata* Mockford and Sullivan.

Nanolachesilla chelata Mockford and Sullivan

Nanolachesilla chelata Mockford and Sullivan, 1986:57.

Recognition features. Body in general creamy yellow; preclunial abdominal segments each with broad ring of pale purplish brown, absent ventrally. Two major sclerites of endophallus very long, about half length of phallosome and separated their entire length at least by a line (Fig. 161). Male epiproct with most of dorsal surface covered with papillar field (Fig. 604). Single lobe at base of v3.

Relationships. These remain unknown.

Distribution and habitat. The species has been collected only in Highlands and Sarasota Counties, Florida, where it has been found in foliage of a small oak tree and in persistent dead leaves of hickory.

Nanolachesilla hirundo Mockford and Sullivan

Nanolachesilla hirundo Mockford and Sullivan, 1986:53.

Recognition features. Body creamy white with diffuse reddish brown band along side of thorax; each preclunial abdominal segment with broad vertical reddish brown band on each side. Two major sclerites of endophallus about one-third length of phallosome, fused along midline in mid-length (Fig. 155). Male epiproct with small area in middle of dorsal surface covered with papillae. Double lobe at base of v3 (Fig. 586).

Relationships. The species is very close to *N. hirundoides* Mockford and Sullivan (eastern Mexico).

Distribution and habitat. The species is known primarily from the southern Florida counties of Monroe, Dade, Hendry, Sarasota, Highlands, and Polk, but there is one record from the coast in Flagler County. It is always associated with dead leaves of palms, primarily *Sabal palmetto*, either hanging or on the ground.

Genus *Prolachesilla* Mockford and Sullivan

Prolachesilla Mockford and Sullivan, 1986:33.

Diagnosis. Ocelli absent. Fronto-clypeal region without field of blunt-tipped sensilla. In forewing, Rs fork stem only slightly to moderately curved; vein Cu 1a free from M. Phallosome a closed frame with flat base with two sclerotic connections from base to distal region each side: internal one joining external paramere and external one joining aedeagal arch (Fig. 159). Endophallus of only minute sclerites. Hypandrium without lateral processes. Ovipositor valvulae: v3 generally pointed at apex, with no lobes at base; possible vestige of v1 represented by sclerotic band from clunium to ninth sternum at base of v3 (Fig. 160).

Generotype. *P. mexicana* Mockford and Sullivan.
North American species: *P. terricola* Mockford and Sullivan.

Prolachesilla terricola Mockford and Sullivan

Prolachesilla terricola Mockford and Sullivan, 1986:42.

Recognition features. The only species of its genus in the study area. The species is unique in its genus in being micropterous (both sexes), the forewing being equal in length to the reduced meso- plus metanotum.

Relationships. This species appears to be close to *P. mexicana* Mockford and Sullivan (Mexico south to Panama). Most species of *Prolachesilla* are Mexican, Central American and northern South America.

Distribution and habitat. The species occurs in the Rocky Mountains of southern Colorado and New Mexico, where it inhabits ground litter.

Subfamily Lachesillinae

Diagnosis. Forewing ciliation very sparse and short. Phallosome open distally (Fig. 156). Endophallus apparently absent. Ovipositor valvulae: v3 variable in shape; setae, when present, generally distributed over its entire surface. Female paraprocts lacking field of short, stout setae along median margin in ventral half.

North American genus: *Lachesilla* Westwood.

Genus *Lachesilla* Westwood

Lachesilla Westwood, 1840:47.
Pterodela Kolbe, 1880:118.
Leptopsocus Reuter, 1899:5.
Terracaecilius Chapman, 1930:343.

Diagnosis. Ocelli present. In forewing, Rs fork stem only slightly curved; vein Cu 1a free from M. Phallosome with base an apodeme or two separate apodemes; aedeagal arch absent, external parameres absent as such. Hypandrium usually bearing pair of lateral processes ('claspers'). Ovipositor valvulae: v3 (v2+3?) usually rounded at apex, occasionally pointed but point never spinose.

Generotype. *L. pedicularia* (Linnaeus).

Garcia Aldrete (1974) proposed species groups for this large assemblage of species. These are adopted here. The following species groups are found in the study area: andra group, centralis group, corona group, forcepeta group, fuscipalpis group, patzunensis group, pedicularia group, riegeli group, rufa group, and texcocana group.

Andra Group

Diagnosis. Wings clear or forewing with spotting pattern, the spots mostly along veins in distal half of wing. Hypandrium of two separate segments or the two fused showing line of fusion or not. Lateral processes of hypandrium ('claspers') on distinct plates or joined to sides of hypandrium, sometimes greatly reduced or absent. Phallosome of single basal apodeme divided distally, the two branches usually re-branching near their apices, or two separate apodemes simple at their apices. Male epiproct with two lateral processes, or one or two (one behind other) median processes. Male paraproct with a short, blunt process. Subgenital plate variable. Ovipositor valvulae one pair, short, directed caudad, rounded apically. Ninth sternum membranous; spermapore plate circular or ellipsoidal with pigmented rim.

North American species: *L. andra* Sommerman, *L. arnae* Sommerman, *L. dona* Sommerman, *L. kola* Sommerman, *L. rubilis* Aaron, *L. rubiloides* Garcia Aldrete, *L. punctata* Banks.

Lachesilla andra Sommerman

Lachesilla andra Sommerman, 1946:635.

Recognition features. Wings unmarked; forewings elongate and relatively slender in males (length/greatest width index ~ 3.10), those of females variable, from equal to those of males down to reaching about half length of abdomen at rest. Head and thorax pale to medium brown; preclunial abdominal segments creamy white, terminal segments medium brown. Hypandrium (Fig. 605) consisting of membranous distal region bearing claspers laterally and sclerotized basal region consisting of a triangular lobe on each side and transverse bar connecting lobes; claspers elongate, recurved, pointed apically. Phallosome (Fig. 605) of single basal apodeme, branching at distal three-fifths, each branch rebranched distally, outer branchlet expanded and bearing distal pointed process; inner branchlet short, rounded, denticulate on outer edge. Male epiproct (Fig. 606) with rounded median basal process with warty surface, longer distal process rounded at end. Male paraproct with relatively long process bearing beaked tip. Subgenital plate (Fig. 607) rounded apically with two short, spinose processes on margin near middle. Ovipositor valvulae as described for group. Female ninth sternum mostly membranous with somewhat sclerotized strip anteriorly; spermapore plate circular.

Relationships. The species appears to be closest to *L. kerzhneri* Günther (Mongolia) with which it shares form of the hypandrium, nature of the attachment of the claspers, the crook at the base and hooked beak at the tip of the paraproctal process, and a distal epiproctal process with a simple median base.

Distribution and habitat. Within the study area, the species is known from Connecticut south to Florida and west to California. The northern edge of its range has not been well established, but the species is known from Ohio, Michigan, Illinois, and Missouri. It ordinarily inhabits dead, standing grass stems and leaves. Outside the study area, it has been found in the Mexican states of Coahuila and Tlaxcala.

Lachesilla arnae Sommerman

Lachesilla arnae Sommerman, 1946:638.

Recognition features. Wings (Fig. 608) with distinct spotting pattern: a spot around marginal end of each vein of forewing from R1 through nodulus; a band through cells R5, M1, M2, and M3 connecting the distal spots. Body creamy white and medium brown, the latter over antennae, a spot on frons before ocelli, a ring around antennal base, a spot mesad of each eye, a spot along hind margin of vertex on each side of midline, thoracic notal lobes, a dorso-ventral band on each side of each abdominal segment 1-7, and terminal abdominal segments. Male with wings elongate, extending about one-third their length beyond tip of abdomen at rest. Females dimorphic: macropterous females with wings as in male; brachypterous females with wings extending from about end of basal one-third of abdomen to slightly beyond basal one-half of abdomen at rest. Hypandrium (Fig. 184) with broad, well sclerotized basal region continuous posteriorly on each side with sclerotized area tapering to point, the latter area bearing incurved clasper of moderate length. Phallosome (Fig. 184) with single basal apodeme sending off disto-lateral branch each side at about half its length and another pair slightly more distally; the two branches of each side apparently fusing together and fused to membranous distal region of hypandrium. Male epiproct with pair of short posteriorly directed processes (Fig. 609). Male paraproct with short, blunt, cylindrical process (Fig. 610). Female genitalia as noted for group; spermapore elliptical.

Relationships. This species appears to occupy a position intermediate between an undescribed species from southern Texas and northern Mexico (with which it shares wing spotting, relatively short claspers, heavy base of the hypandrium, and a blunt, cylindrical male paraproctal process of moderate length) and a cluster of species including *L. kerzhneri* Roesler (U.S.S.R), *L. nubiloides* Garcia Aldrete (northwestern United States) and *L. nubilis* (Aaron) (eastern and central United States) (with which it shares the single basal apodeme of the phallosome, re-branching distal branches of the phallosome, and, with some, the wing spotting pattern).

Distribution and habitat. The species has been found at Saskatoon, Saskatchewan (type locality), and at Potlatch Pine Orchard, Nez Perce County, Idaho. At Saskatoon these insects were taken by sweeping in native grassland. At the Idaho locality they were found on persistent dead leaves of blackberry (*Rubus* sp.).

Lachesilla dona Sommerman

Lachesilla dona Sommerman, 1946:641.

Recognition features. Forewing (Fig. 611) marked with brown spot at end of each vein from R1 through nodulus; otherwise unmarked except cubital loop bordered with brown its entire length beyond clear base. Body pale brown, somewhat darker on "dotted areas" of vertex (behind ocelli and mediad of each eye); mx palpi, tarsi, and thoracic notal lobes medium brown; preclunial abdominal segments each with a reddish brown ring; terminal segments pale to medium brown. Hypandrium (Fig. 612) weakly sclerotized; middle region setose, rounded, closely joined with phallosome at its point of branching; distal region a four lobed flap; base of hypandrium broad, included in two segments. Claspers elongate, curved, acuminate, each sturdily joined to well sclerotized basal plate, the latter not fused with hypandrium proper.

Phallosome (Fig. 612) with single basal apodeme branching immediately distad of middle, then branching again immediately; lateral branchlet producing a flat lamella; median branchlet also producing a flattened lamella and longer process, the latter re-branching to produce elongate lateral recurved process and double pointed median process; median processes of the two sides touching on midline at their tips. Male epiproct (Fig. 179) with warty tubercle on each side near base and distal cylindrical process on each side with papillate surface. Male paraproct with elongate, curved, beak-tipped process. Subgenital plate (Fig. 182) terminating distally in four lobes: on each side a broad, blunt-pointed, rough-surfaced outer lobe and a smaller, rounded, smooth-surfaced inner lobe. Ovipositor valvulae short, mesally directed. Female ninth sternum mostly membranous but with basal, somewhat thickened lobe each side; spermapore plate ellipsoidal.

Relationships. The species is probably closest to *L. nubiloides* García Aldrete, with which it shares hypandrial structure, similar phallosome, similar male epiproct, and a several-lobed distal end of the subgenital plate.

Distribution and habitat. Within the study area, the species is restricted to southern California (Santa Barbara, Los Angeles, Orange, San Bernardino, Riverside and San Diego Counties), where it probably inhabits foliage of grasses, herbaceous plants, shrubs, and small trees. Outside the study area, it has been found in the Mexican states of Baja California, Sinaloa, and Coahuila.

Lachesilla kola Sommerman

Lachesilla kola Sommerman, 1946:644.

Recognition features. Forewing (Fig. 617) with spots on ends of veins from R1 through nodulus except Cu1b. Body medium brown, darker on postclypeal striations, dotted areas of vertex, thoracic pleura and notal lobes, a ring on each preciuntal abdominal segment and terminal abdominal structures. Hypandrium (Fig. 175) long and broad, heavily sclerotized, probably including three to four sterna: terminating posteriorly on each side by rounded, membranous projection covered ventrally by a sclerotized, longer, slenderer, blunt-tipped process. Phallosome (Fig. 175) of two apodemes, both pointed posteriorly, the left somewhat curved and bent, J-shaped at base, the right straight and longer. Male epiproct (Fig. 614) with two short lateral processes from hind margin. Male paraproctal process short, blunt (Fig. 615). Subgenital plate of two regions (Fig. 616), the basal region terminating in four lobes, all short but laterals projecting beyond medians; a deep cleft between medians; distal region bowl-shaped, indented posteriorly, fitting into indentation formed by lateral lobes of basal region. Ovipositor valvulae short, directed posteriorly; ninth sternum membranous posterior to line across middle, more sclerotized anterior to line, with two blunt forward-directed projections on each side of anterior margin (Fig. 176). Spermapore ellipsoidal.

Relationships. The species appears to be closest to an undescribed species from Brazil, which has paler wing spots and shows symmetrical phallosomal apodemes.

Distribution and habitat. Within the study area, the species has been found in southern California (Riverside and San Diego Counties), southern Arizona (Cochise County), and southern New Mexico (Doña Ana County). It has been found in foliage of alfalfa, cocklebur, and peach, and probably occurs in foliage of various other herbaceous plants, shrubs, and small trees. Outside the study area it is known from the Mexican states of Sonora and Nuevo León.

Lachesilla nubilis (Aaron)

Caecilius nubilis Aaron, 1886:13.
Lachesilla nubilis (Aaron), Chapman, 1930:351.

Recognition features. Forewings (Fig. 617) marked as in preceding species plus a spot just before vein Cu 1b; spots in cells R5 and M 1 of well-marked females tending to coalesce. Body in general medium brown; darker areas as described for preceding species. Hypandrium (Fig. 185) with terminal piece membranous, slightly bilobed on margin, closely joined at base with distal end of phallosome; basal piece trapezoidal, setose; clasper bases well sclerotized, entirely separate from hypandrium proper; claspers straight, stout, tapering to acuminate tips. Phallosome (Fig. 185) with single basal apodeme slightly forked anteriorly, branching at about one half its length; outer branch forming relatively long process pointed at tip; inner branch forming shorter, rounded process; the longer and shorter processes of a side apparently fused part of their length before their apices. Male epiproct (Fig. 618) with stout, setose posterior process on each side. Male paraproct (Fig. 619) with short, stout, beak-like process. Female subgenital plate (Fig. 620) with hind margin cleft in middle, vaguely bilobed on each side; an inner sclerotization occupying most of anterior half of plate, terminating postero-laterally as acute point. Ovipositor valvulae short, posteriorly directed, tapering, setose; ninth sternum mostly membranous, with two broad, rounded lobes on anterior margin more heavily sclerotized. Spermapore ellipsoidal.

Relationships. The species is probably closest to *L. dona* Sommerman and *L. nubiloides* Garcia Aldrete, with both of which it shares the wing marking pattern and some male genitalic features.

Distribution and habitat. Within the study area the species occurs throughout eastern United States, west to eastern New Mexico, central Colorado, and Utah. In Florida it occurs south to the Gainesville region. Its northern known limit is Connecticut in the East, southern Ontario and southern Minnesota in the Midwest, and Saskatoon farther west. It generally inhabits dead grasses and other dead vegetation. Outside the study area, it has been found in the Monterrey region, Nuevo León State, Mexico.

Lachesilla nubiloides Garcia Aldrete

Lachesilla nubiloides Garcia Aldrete, 1975:217.

Recognition features. Dimorphic: females micropterous with winglets not reaching abdomen; males long-winged with wings greatly exceeding tip of abdomen at rest. Male wings marked as in *L. kola*. Body in general dusky brown; preclunial abdominal segments paler but each with broad reddish brown semi-ring dorsally and laterally. Hypandrium (Fig. 181) with distal region membranous, four lobed on margin, fused to distal end of phallosome; basal region well sclerotized, setose, transverse-trapezoidal with hind margin depressed in middle. Claspers each on well sclerotized transverse plate completely separate from hypandrium proper; clasper base swollen outward; rest of shaft slender, straight, curving outward near distal end. Phallosome (Fig. 181) with single basal apodeme branching at distal three-fifths, then each branch re-branching, each branchlet producing an elongate phylloid, the two lying one over the other and possibly fused in middle; upper phylloid acuminately pointed at apex. Male epiproct (Fig. 621) with elongate, blunt-ended process directed backward on each side (conspicuous on whole specimen due to its being white); short, pointed process in middle. Male paraproctal process (Fig. 622) elongate, pointed. Female subgenital plate (Fig. 180) with three rounded posterior lobes, median much broader than laterals. Ovipositor valvulae short, rounded, directed posteriorly; ninth sternum weakly sclerotized, slightly bilobed anteriorly.

Relationships. The species is probably closest to *L. nubilis*, which it resembles in wing markings, phallosome and hypandrial structure.

Distribution and habitat. The species is found in eastern and central Washington, central Oregon, and western Montana. It inhabits foliage of shrubs and small trees, including *Chrysothamnus nauseosus*,

Artemisia tridentata, and *Crataegus* sp., also dead leaves of cattail (*Typha latifolia*).

Lachesilla punctata (Banks)

Elipsocus punctatus Banks, 1905:1.
Lachesilla punctata (Banks). Chapman, 1930:357.

Recognition features. Forewings marked as in *L. nubilis*. Head and thorax pale to medium brown: middle of vertex and frons, median eye borders, and postclypeus medium brown, remainder pale brown; thoracic notal lobes medium brown, surrounding cuticle pale; thoracic pleura variegated pale and medium brown; preclunial abdominal segments colorless; terminal segments mostly medium brown. Hypandrium (Fig. 624) with distal region a broad, membranous flap bilobed on margin, closely fused to distal end of phallosome; basal region mostly membranous but including two small sclerotized, setose plates: a larger distal ovoid one and a smaller, more basal rounded one. Claspers slender, curved sickle-like, each based on well sclerotized plate completely separate from hypandrium proper; distally clasper held in groove formed by slender arm from clunium joining phallosome. Phallosome (Fig. 624) with single basal apodeme bent near base; branching at distal three-fifths, then rebranching immediately; lateral branchlets meeting distally to form closed frame, median branchlets forming two processes within frame, widened distally. Male epiproct (Fig. 178) tapering from base to tip to form hook-like structure. Male paraproctal process (Fig. 625) straight, slender, blunt-tipped. Female subgenital plate essentially as described for *L. nubiloides* but lateral lobes somewhat slenderer (Fig. 177). Ovipositor valvulae short, rounded, directed postero-medially; ninth sternum as described for *L. nubiloides*.

Relationships. Among the andra group species with a single basal phallosomal apodeme, *L. punctata* stands apart from the others in having the outer branchlets of the phallosome forming a closed frame and two processes from the clunium to the phallosome forming a groove to hold the claspers. Female genitalic features suggest proximity to *L. nubiloides*.

Distribution and habitat. Within the study area, the species is found in southern California north to the San Francisco area, in central and eastern Arizona (southern Coconino County, central Apache County), and over New Mexico in general. It inhabits dead leaves of various herbaceous plants and dead persistent tree leaves. Outside the study area, it is known from the Mexican states of Nuevo León, México, Michoacán, and Veracruz.

Centralis Group

Diagnosis. Wings clear; veins R, M, and Cu 1a in distal three-fifths of forewing somewhat darker than those in rest of wing. Hypandrium (Fig. 164) of three separate segments clearly showing intersegmental divisions; hypandrium terminating distally as pair of blunt, curved processes; the processes possibly representing claspers, but no articulated claspers present; a small, triangular distal plate on hypandrium between processes. Phallosome (Fig. 164) of single basal apodeme branching near its base to form two simple, curved rami pointed at tips. Male epiproct with two short processes near midline. Male paraproct (Fig. 626) with relatively long hooked or truncated process. Subgenital plate rounded distally with slight indication of median lobe (Fig. 167). Ovipositor valvulae one pair, relatively short, tapering to blunt apex, directed mesad. Ninth sternum membranous; spermapore circular with narrow pigmented rim.

North American species: *L. centralis* García Aldrete, *L. cintalapa* García Aldrete, *L. perezi* García Aldrete.

Lachesilla centralis García Aldrete

Lachesilla centralis García Aldrete, 1983:14.

Recognition features. Head and thorax tawny medium brown; thorax with reddish brown band each side above leg bases; preclunial abdominal segments pale brown, each with reddish brown ring; terminal segments tawny medium brown. Hypandrium as described for species group; hind margin of middle segment extended posteriorly in middle; distal processes bearing tooth on median side before apex; triangular distal plate of hypandrium (Fig. 627) distinct, elongate transversely, its arms straight. Single basal apodeme of phallosome about one-fourth entire length of phallosome. Process of male paraproct (Fig. 626) truncated. Subgenital plate (Fig. 167) with lateral rugose areas on margin, distinct pigmented area with lateral arms.

Relationships. García Aldrete (1983) found this to be a member of a complex of three species (this species, *L. perezi*, and *L. cintalapa* García Aldrete) distributed throughout Mexico and southern United States. Most of the characters of the species of this complex mark them as probably belonging to the corona species group (García Aldrete, 1974).

Distribution and habitat. Within the study area, the species has been found only in southern California (Tulare, San Bernardino, Riverside, and San Diego Counties), where the only recorded habitat was citrus foliage. It was also taken as an introduction from Mexico at

El Paso, Texas, and at Hidalgo, Texas, the former on citrus leaves, the latter on orchid root mat. Outside the study area, the species has a north-south distribution throughout most of Mexico, apparently being replaced in Chiapas by *L. cintalapa*.

Lachesilla cintalapa Garcia Aldrete

Lachesilla cintalapa Garcia Aldrete, 1983:19.

Recognition features. Color as described for previous species. Hypandrium with hind margin of middle segment rounded; triangular distal plate of hypandrium (Fig. 628) pointed posteriorly, with two long setae near point, arms of plate curved. Arms of phallosome with group of spinules before apex (Fig. 165). Process of male paraproct hooked at apex (Fig. 629). Subgenital plate (Fig. 630) with slight development of median lobe on hind margin, lateral rugose areas present on margin, pigmented area vague, without distinct arms.

Relationships. As noted for the previous species.

Distribution and habitat. Within the study area, a single male was taken in a blacklight trap at Miami International Airport. Outside the study area, it is known from two localities in Chiapas State, Mexico, where it was taken on dead persistent leaves.

Lachesilla perezi Garcia Aldrete

Lachesilla perezi Garcia Aldrete, 1983:23.

Recognition features. Color as in previous two species. Hypandrium with hind margin of middle segment gently curved; triangular distal plate (Fig. 631) small, rounded posteriorly, two relatively long setae borne on the plate in previous two species located anterior to the plate. Arms of phallosome without distinct group of spinules before apex. Process of male paraproct truncate at apex. Subgenital plate without median lobe of hind margin or lateral rugose areas of margin; pigmented area vague, without distinct arms.

Relationships. As noted for the previous two species.

Distribution and habitat. Within the study area, the species has been taken at New Orleans and Baton Rouge, Louisiana, in the former locality on banana leaves and miscellaneous vegetation, and in the latter on dried grasses. Outside the study area, the species has a long north-south range through eastern Mexico from eastern Coahuila State south to southern Veracruz.

Corona Group

Diagnosis. Wings generally clear with veins somewhat darker distally than basally. Hypandrium of one or two segments without distinct intersegmental division or division partially developed; hypandrium terminating distally as pair of straight or curved processes, sometimes once-branched, rarely articulated basally. Phallosome of one or two basal apodemes; if single, branching near its middle to form two rami, these sometimes rebranching once. Male epiproct bilobed distally, occasionally with two short, blunt, distal or sub-distal processes. Male paraproct with relatively long hooked or truncated process. Subgenital plate usually with median tongue, occasionally with median lobe. Ovipositor valvulae one pair, relatively short, tapering to blunt or pointed apex, directed mesad, clearly joined to ninth sternum. Female ninth sternum membranous; spermapore circular with narrow, pigmented rim.

North American species: *L. albertina* Garcia Aldrete (in press), *L. corona* Chapman.

Lachesilla albertina Garcia Aldrete

Lachesilla albertina Garcia Aldrete, in press.

Recognition features. Known from female only, micropterous. Subgenital plate with broad, spade-shaped median tongue slightly indented at distal end.

Relationships. Without information on the male or the wing venation, it is impossible to determine the relationships of this species at present.

Distribution and habitat. The species is known from two individuals collected in the Lake Louise area, Alberta.

Lachesilla corona Chapman

Lachesilla corona Chapman, 1930:350.

Recognition features. Both sexes macropterous. Male with distal processes of hypandrium divided, each process forming two slender, curved rami (Fig. 632). Female with a slender median tongue on subgenital plate (Fig. 169).

Relationships. This species appears to be closest to *L. dividiforceps* Garcia Aldrete (central Mexico), which it resembles in characters of the hypandrium and subgenital plate, but not the phallosome or ninth sternum.

Distribution and habitat. The species occurs throughout northeastern United States south to central Georgia and southwest to central Arkansas, also in southeastern Canada (Marmora, Ontario, and Montreal region, Quebec). It inhabits persistent dead leaves of a great variety of plants, both woody and herbaceous.

Forcepeta Group

Diagnosis. Wings unmarked. Hypandrium single-segmented, transverse; distal end of phallosome closely associated, in some cases completely fused with distal end of hypandrium; claspers on distinct antero-posteriorly oriented plates separate, at least by a suture, from main body of hypandrium (Fig. 190). Phallosome of single basal apodeme divided distally, the two branches usually forming each a broad lamella hinged onto or fused to distal end of hypandrium (Fig. 190). Male epiproct rounded posteriorly or with two lateral conical processes. Male paraproct usually with no process or an extremely short one. Subgenital plate weakly sclerotized, usually with posterolateral 'shoulders', or short processes, sometimes with internal sclerotized areas (Fig. 200). Ovipositor valvulae one pair, broad, rounded or blunt-tipped, directed mesally. Female ninth sternum transverse, largely membranous or partially sclerotized anteriorly; spermapore with broad pigmented rim.

North American species. *L. alpejia* García Aldrete, *L. anna* Sommerman, *L. bottimeri* Mockford and Gurney, *L. chapmani* Sommerman, *L. contraforcepeta* Chapman, *L. forcepeta* Chapman, *L. gracilis* García Aldrete, *L. kathrynae* Mockford and Gurney, *L. major* Chapman, *L. penta* Sommerman.

Lachesilla alpejia García Aldrete

Lachesilla alpejia García Aldrete, 1988a:39.

Note. I have not examined this species. The following notes are from the original description.

Recognition features. Body in general pale tan; dark brown bands in following positions: along fronto-clypeal sulcus and from each compound eye to fronto-clypeal sulcus surrounding antennal fossa; on each side of thorax above coxae; thorax, legs, and antennae somewhat more darkly pigmented than rest of body. Claspers arising from rounded bases, constricted and outcurved immediately beyond bases, then widening and incurved distally with field of microspines on distal mesal surface (Fig. 194). Subgenital plate with two broad, short projections on hind margin (Fig. 195). Ninth sternum mostly membranous with small area of heavier sclerotization along anterior margin, slightly lobed on each side.

Relationships. The species is probably closest to *L. penta* Sommerman (southeastern United States) and *L. denticulata* Garcia Aldrete (widely distributed in Mexico, Central America, and the Caribbean).

Distribution and habitat. Within the study area, the species was taken as an introduction from Mazatlán, Mexico, at an unknown United States port of entry on the Mexican border with pods of *Caesalpinia crista*. Outside the study area, it is known from Chamela, Jalisco, Mexico (type locality), where it has been found on foliage in forest and mangrove swamp.

Lachesilla anna Sommerman

Lachesilla anna Sommerman, 1946:636.

Recognition features. Head and thorax pale tawny brown, somewhat darker on thorax than head. Preclunial abdominal segments creamy yellow, each segment with reddish brown ring, incomplete ventrally; terminal segments tawny brown. Well colored specimens with reddish brown bands in following positions: one above and one below antennal base to compound eye; one dorso-laterally on neck and pronotum; one length of thorax above leg bases. Hypandrium rounded distally; clasper bases well separated from hypandrium proper; clasper shafts straight (Fig. 199), terminating distally in pair of tines: longer, somewhat curved, blunt outer tine and shorter, straight, pointed inner tine. Phallosome (Fig. 199) divided distally forming pair of triangular flaps joined at bases to distal end of hypandrium. Male epiproct with short pair of lobes, one on each side posteriorly. Male paraproct with very short process subtended by heavy spine. Subgenital plate (Fig. 200) slightly cornered posteriorly, with two yellow, ovoid sclerotized areas internally. Female ninth sternum membranous.

Relationships. The species is very close to an undescribed species which occurs on cattail (*Typha latifolia*) throughout eastern United States, and another undescribed species in the Rocky Mountains of Wyoming and Colorado.

Distribution and habitat. The species occurs throughout eastern United States and southeastern Canada, south to Marion County, Florida, west to Bowie County, Texas and Texas County, Missouri. It inhabits dead persistent leaves of trees and shrubs.

Lachesilla bottimeri Mockford and Gurney

Lachesilla bottimeri Mockford and Gurney, 1956:365.

Recognition features. Body creamy yellow except slightly darker on thoracic terga; a reddish brown band across fronto-clypeal suture, above antennal bases to compound eyes, continuing behind each compound eye and running entire length of body (immediately below wing bases on thorax), broken into spots on abdomen in poorly pigmented specimens. Hypandrium rounded distally; claspers (Fig. 198) with relatively broad bases distinctly separate from hypandrium proper; shafts curved outward immediately beyond base, elongate distal region strongly curved inward. Phallosome (Fig. 198) terminating in pair of triangular flaps well separated in middle. Male epiproct (Fig. 633) with upward-directed double lobe on each side. Male paraproct lacking process. Subgenital plate round-shouldered postero-laterally, without internal thickenings. Female ninth sternum membranous.

Relationships. The species seems to stand apart from most other members of the group in having elongate, incurved claspers. Other species with incurved claspers (*L. forcepeta, L. major*) have shorter claspers, a different body color and different habitat.

Distribution and habitat. Within the study area, the species has been found only at Southmost Palm Grove, Cameron County, Texas, where it inhabits the leaves of the palm Sabal texana Becc. Outside the study area, it appears to have a long north-south distribution in eastern Mexico, having been recorded from the states of Nuevo León, San Luis Potosí, Veracruz, Tabasco, and Quintana Roo. At most of the Mexican localities it was found on palm foliage, but it was also found on carrizo grass, banana foliage, and in miscellaneous beating in rain forest.

Lachesilla chapmani Sommerman

Lachesilla chapmani Sommerman, 1946:639.

Recognition features. Body creamy yellow to pale brown with small dark brown marks along sides of thorax around leg bases and below wing bases, and a few scattered dark brown marks on sides of abdomen. Hypandrium with hind margin rounded; small denticles present on sides of margin; claspers (Fig. 208) with bases well separated from hypandrium proper; shafts slender, curved inward at base, outward distally. Phallosome (Fig. 208) producing two triangular flaps distally, well separated in middle. Male epiproct (Fig. 634) with short, median horn-like process. Male paraproct without process. Subgenital plate flat posteriorly, slightly cornered postero-laterally. Female ninth sternum membranous, somewhat thickened in two ovoid areas along anterior margin.

Relationships. This species appears to be closest to three undescribed species from southern Mexico and Central American: *L.* spp. F-9, F-15, and F-42 of García Aldrete (1974) also to *L. kathrynae* of southern Texas and eastern Mexico with which it shares the curvature of the claspers and the single, median process of the male epiproct.

Distribution and habitat. The species is known throughout most of peninsular Florida, from Collier and Dade Counties in the south through Levy and Alachua Counties in the north. It inhabits primarily dead leaves of palms and yuccas, but occurs occasionally on persistent dead oak leaves.

Lachesilla contraforcepeta Chapman

Lachesilla contraforcepeta Chapman, 1930:347.

Recognition features. Wings tawny brown washed. Head and thorax pale tawny brown; abdomen creamy white; subgenital plate brown; in male, seminal vesicles seen through abdominal cuticle laterally and ventrally as large pair of orange ovoidal bodies. Hypandrium with hind margin rounded; claspers (Fig. 193) arising from slender bases completely separate from hypandrium proper; clasper shafts directed mediad at base, then curving outward. Phallosome (Fig. 193) terminating distally as pair of triangular flaps separated in middle and hinged to distal end of hypandrium. Male epiproct simple, rounded distally. Male paraproct without process. Subgenital plate with hind margin straight, cornered at sides. Female ninth sternum membranous, its surface rugose.

Relationships. Although numerous species of the forcepeta group have outcurved claspers, it is likely that this species is very close to *L. forcepeta*, in which the claspers curve inward distally. Habitat, color, and most other male and female genitalic characters of the two species are similar.

Distribution and habitat. The species occurs throughout eastern United States and southeastern Canada, from the Montreal region in the north south to Marion County, Florida; west, more or less continuously to central Nebraska and Saskatoon, Saskatchewan. It also occurs in the Rocky Mountains in Catron County, New Mexico, with records of single individuals from Fremont County, Colorado, and Boundary County, Idaho. It inhabits primarily the foliage of pines, junipers, and other coniferous trees, but is occasionally found in dead leaves of broad-leaf trees and in dead grass.

Lachesilla denticulata Garcia Aldrete

Lachesilla denticulata Garcia Aldrete, 1988a:43.

Recognition features. Wings clear. Head, thorax, and appendages tawny yellow; two reddish brown stripes from eye to antennal base, one above, other below antennal base; abdomen creamy white, each preclunial segment with reddish brown vertical stripe on each side, stripes continuous dorsally on segments 8-10. Hind margin of hypandrium slightly extruded in middle; claspers (Fig. 196) with bases well separated from hypandrium proper; shafts curved inward, then abruptly outward from base, bearing strong, pointed, medially directed tooth before apex; apex curved medially, pointed. Phallosome (Fig. 196) producing single broad, triangular flap distally, the flap sclerotized basally, membranous distally. Male epiproct (Fig. 635) with short, broad, median lobe posteriorly. Male paraproct without process. Subgenital plate (Fig. 636) with distinct postero-lateral angles, pair of sclerotizations. Female ninth sternum with bilobed anterior sclerotized region; spermapore with wide sclerotic rim, positioned between lobes of sclerotized region of sternum (Fig. 197).

Relationships. The species appears to be close to *L. penta* Sommerman (southeastern United States).

Distribution and habitat. Within the study area the species was taken once as an introduction from Mexico at Brownsville, Texas, on orchid plants, and once as an introduction from Panama at Miami, Florida on an unknown plant. Outside the study area, it has been found in most of the southern and central Mexican states, also in most of Central America, and in Trinidad (West Indies). It inhabits dead persistent leaves of various sorts.

Lachesilla forcepeta Chapman

Lachesilla forcepeta Chapman, 1930:348.

Recognition features. Wings clear except pterostigma opaque white. Head, thorax, and appendages orangish yellow; abdomen colorless; underlying pale yellow tissues showing through clear cuticle; transverse ovoidal internal thickening of subgenital plate seen orangish-yellow through cuticle of plate surface. Hypandrium with hind margin slightly concave in middle; clasper bases well separated from hypandrium proper (Fig. 190b); clasper shafts broad and denticulate on outer surfaces basally, abruptly narrowing, incurved distally, pointed at apex. Phallosome producing two triangular flaps distally separated by median cleft (as in Fig. 190a). Male epiproct rounded distally; male paraproct with short process. Subgenital plate (Fig. 637) decidedly cornered postero-laterally, sides with sclerotic reinforced edges posteriorly; diffuse ovoid sclerotic region in middle. Female ninth sternum membranous.

Relationships. The species appears to be closest to *L. major* and *L. contraforcepeta* in structure and color, but differs from *L. major* in habitat.

Distribution and habitat. The species occurs throughout eastern United States and in southeastern Canada (Montreal area), south to Sarasota County, Florida, west to Oklahoma and Kansas, and south in Texas to the San Antonio area. It inhabits primarily foliage of junipers, occasionally pine, and rarely persistent dead leaves of various broadleaf plants.

Lachesilla gracilis Garcia Aldrete

Lachesilla gracilis Garcia Aldrete, 1988a:49.

Recognition features. Wings elongate, slender, clear except pterostigma cloudy. Body in general creamy yellow, except a few buffy spots on vertex; thoracic notal lobes pale brown, a slender reddish brown bar above antennal base; reddish brown irregular spots on neck and thoracic pleura forming broken band; first five to six preclunial abdominal segments each with vertical reddish brown stripe on each side. Hypandrium flat posteriorly; clasper bases large, long-subquadrate, well separated from hypandrium proper; clasper shafts (Fig. 203) curving inward from base then abruptly outward, with minute denticles distally on median surface. Phallosome (Fig. 207) distally forming pair of triangular flaps separated by median cleft. Male epiproct (Fig. 638) with central process bearing setae, pointed apically; two small papillar fields before base of process. Male paraproct without process. Subgenital plate (Fig. 204) trapezoidal, slightly cornered posterolaterally, with two small ovoid thickened internal areas appearing yellow through cuticle. Female ninth sternum membranous, thickened along anterior margin; spermapore rim rounded, well sclerotized.

Relationships. This species is very similar in external genitalia to an undescribed species from northern Brazil. Several other tropical American species are similar in external genitalia, but *L. gracilis* stands apart in its wings being relatively more elongate and slender.

Distribution and habitat. Within the study area, this species has been found only at Lake Whitney State Park, Johnson County, Texas, on dead leaves of *Sorghum halepense* (specimen determined by Dr. Garcia Aldrete). Outside the study area, it is known from the Mexican states of Chiapas, Guerrero, and Jalisco, also from Belize and Guatemala. It usually inhabits foliage of pines.

Lachesilla kathrynae Mockford and Gurney

Lachesilla kathrynae Mockford and Gurney, 1956:365.

Recognition features. Wings relatively elongate and slender, clear except pterostigma opaque white. Body and appendages creamy yellow with reddish brown pigment over most of thoracic sterna and as irregular spots on thoracic pleura and sides of first three to four abdominal segments. Hypandrium rounded on distal margin; clasper bases separated by line but closely joined to hypandrium proper; clasper shafts (Fig. 206) curved inward from bases, then outward, swollen before rounded tips; a seta on shaft near base. Phallosome (Fig. 206) distally producing pair of triangular flaps pointed at apices. Male epiproct (Fig. 205) with posteriorly directed median horn-like process. Male paraproct without process. Subgenital plate (Fig. 207) with broad middle region of hind margin developed as truncated posterior projection. Female ninth sternum membranous; spermapore plate a wide, rounded sclerotic rim around spermapore.

Relationships. See relationships section under *L. chapmani.*

Distribution and habitat. Within the study area, the species appears to be confined to the southernmost counties (Hidalgo and Cameron) of the Lower Rio Grande Valley of Texas, where it inhabits dead leaves of palms. Outside the study area, it is found south in Mexico to coastal Tabasco State, primarily on palm foliage but also on carrizo grass.

Lachesilla major Chapman

Lachesilla forcepeta var. *major* Chapman, 1930:349.
Lachesilla major Chapman. Sommerman, 1946:645.

Recognition features. Wings clear except pterostigma opaque. Head, thorax, and appendages pale tawny brown; abdomen white with reddish brown subcuticular ring, incomplete ventrally, on each preclunial segment. Hypandrium (Fig. 190) with hind margin straight; clasper bases large, long subquadrate, separated by distinct line from hypandrium proper; clasper shafts (Fig. 190) straight to about distal three-fifths, there curving subtly inward, then outward, then inward again to acuminate tip. Male epiproct (Fig. 639) with transverse ridge in middle. Phallosome (Fig. 190) with pair of rounded flaps distally, separated by cleft in middle. Male paraproct without process. Subgenital plate (Fig. 640) with slightly protruding corners postero-laterally, with slight thickenings laterally, and transverse lozenge-shaped orange internal thickening. Female ninth sternum membranous except for two transverse postero-lateral thickenings.

Relationships. The species appears to be very close to *L. forcepeta*, also of eastern United States.

Distribution and habitat. The species occurs throughout most of eastern United States, with northern known limits in Mt. Carmel, Connecticut and Lake County, Michigan, southern limits in Collier County, Florida, and Orange County, Texas, and known western limit in southwestern Arkansas (Howard County). Usually it is found inhabiting persistent dead leaves of broad-leaf plants, but occasionally it has been found on foliage of conifers.

Lachesilla penta Sommerman

Lachesilla penta Sommerman, 1946:652.

Recognition features. Wings clear except pterostigma translucent. Head and thorax pale tawny brown, two bands from each antennal base to eye slightly darker; preclunial abdominal segments tawny yellow, each with a subcuticular reddish brown ring incomplete ventrally; terminal abdominal segments pale tawny brown. Hypandrium (Fig. 201) distally protruding as tapering process; clasper bases relatively slender, separated from hypandrium proper by a line; clasper shafts flat, relatively broad, curved inward at base, then outward, their median margin bearing saw-like blade. Phallosome (Fig. 201) bearing distally a single broad, bilobed flap sclerotized medially, hinged to distal margin of hypandrium. Male epiproct quadrate with short, rounded process at each disto-lateral corner (Fig. 641). Male paraproct lacking process; sense cushion along dorsal margin. Subgenital plate (Fig. 202) broad, flat posteriorly, slightly cornered postero-laterally, with bilobed internal thickened area. Female ninth sternum (Fig. 642) with very large, somewhat shield-shaped spermapore plate; two narrow, transverse sclerotizations posteriorly.

Note. The female described by Sommerman (1946) seems so different from the one described here that it may represent another species. The original description was based on a single male and female.

Relationships. These remain obscure.

Distribution and habitat. Within the study area, the species has been found at four localities: Brownsville, Texas and adjacent 'Southmost' Palm Grove; New Orleans, Louisiana; Gainesville (Alachua County), Florida; and Citra (Marion County), Florida. At two of these localities it was taken on banana leaves, at one on orange foliage, and at one on foliage of Sabal palm. Outside the study area the species occurs seemingly continuously through eastern Mexico from the United States border in Tamaulipas south to Chiapas State, where it has been found on persistent dead leaves of various kinds of plants and palm foliage.

Fuscipalpis Group

Diagnosis. Wings unmarked. Hypandrium (Fig. 643) of two segments separated by intersegmental line. Clasper bases slender, separated by texture change from hypandrium proper (more heavily sclerotized). Claspers slender at base, broader in middle, slender and bent outward distally. Phallosome (Fig. 647) of two simple, slender apodemes joined at base. Male epiproct with hind margin bilobed. Male paraproct with moderately long, truncated process. Subgenital plate with median lobe, the lobe deeply cleft in middle (Fig. 168). Ovipositor valvulae one pair, short, rounded distally, directed postero-medially. Female ninth sternum largely membranous, thickened in broad band across posterior margin, narrow band across anterior margin.

North American species. *L. fuscipalpis* Badonnel.

Lachesilla fuscipalpis Badonnel

Lachesilla fuscipalpis Badonnel, 1971c:35.

Recognition features. The only known species of its group. Females dimorphic: most brachypterous with forewings reaching only to third or fourth abdominal segment. A few with forewings reaching or exceeding tip of abdomen. Mx4 dark.

Relationships. The species appears to be close to the rufa group, differing principally by the less elaborate hypandrium and much smaller ovipositor valvulae.

Distribution and habitat. Within the study area the species was taken once as an introduction at Nogales, Arizona on bromeliads and orchids from Mexico. Outside the study area, the species is known from northern Chile (type locality Chanarcillo, Atacama), Peru, the Mexican states of Guerrero, Jalisco, Michoacán, Nuevo León, Puebla, San Luis Potosí, Veracruz, and the Distrito Federal. According to García Aldrete (1974), the species has been found in Mexico on foliage of *Juniperus* sp., *Pinus* sp., *Prosopis* sp., *Larrea* sp., palmate palms, and on dead leaves of *Quercus* sp. and *Yucca* sp.

Patzunensis Group

Diagnosis. Wings clear. Hypandrium of three segments, all separated by intersegmental lines. Claspers (presumably) represented by pair of slender, pointed processes arising distally on third hypandrial segment and separated only by ridge or texture change from rest of segment. Phallosome of two basal apodemes once branched or simple distally. Male epiproct with hind margin bilobed, a pair of short, rounded processes near base. Male paraproct with moderately long, truncated process. Subgenital plate (Fig. 645) with well developed broad, median tongue posteriorly, arising from straight margin of basal piece. Ovipositor valvulae one pair, moderately long, tapering to pointed apex, directed postero-mesad, clearly joined to ninth sternum. Female ninth sternum partially membranous, but with thickenings before, behind, to sides, or all around spermapore. Spermapore with narrow pigmented rim.

North American species. *Lachesilla sulcata* Garcia Aldrete.

Lachesilla sulcata Garcia Aldrete

Lachesilla sulcata García Aldrete, 1986a:57.

Recognition features. The only known species of its group from the study area. Distinguishable from other described species of the group by presence of a median longitudinal furrow running length of distal tongue of subgenital plate. Male unknown.

Relationships. Using a numerical procedure, García Aldrete (1986) found this species to be closest to *L. tapanatepeca* García Aldrete from Oaxaca State, Mexico.

Distribution and habitat. Within the study area, the species has been found in Alachua, Taylor, and Leon Counties in north peninsular and northwestern Florida and in Attala County in central Mississippi. It has been taken on dead leaves of palms, miscellaneous vegetation, and in a Berlese sample of hardwood litter. Outside the study area, it is known from the Mexican states of Nuevo León, Tamaulipas, and Chiapas, where it has been taken on dead leaves of various sorts of plants.

Pedicularia Group

Diagnosis. Wings either clear, (occasionally tawny, rarely deep tawny) or with extensive spotting pattern: preapical spot on each vein in distal half of forewing and spot bordering basal segment of Rs, and M before junction with Rs, and bordering Cu 1 before its branching, and surrounding nodulus, or spots reduced to M veins, and Cula in descending portion of Cu loop, or more reduced. Hypandrium usually of one well sclerotized segment, rarely two-segmented. Claspers relatively short processes on lateral plates closely joined to body of hypandrium proper. Phallosome usually with simple apodeme anteriorly, branching once near posterior end to form two rounded or triangular or long-oval flaps (single flap in a few forms), rarely two apodemes fused at anterior end. Male epiproct with single or double process directed forward, or single hook-like process directed backward and flanked by pair of processes arising on clunium. Male paraproct with relatively long process. Subgenital plate with hind margin rounded or bilobed, or excavated in middle, or drawn out as median lobe. Gonapophyses usually one pair, occasionally two, of moderate length, directed medially or postero-medially, arising from ninth sternum. Female ninth sternum mostly membranous, sometimes thickened anteriorly, or around spermapore, or both. Sclerotization around spermapore varying from none to broad, rounded area.

North American species: *L. aethiopica* Enderlein, *L. greeni* (Pearman), *L. pacifica* Chapman, *L. pallida* (Chapman), *L. pedicularia* (Linnaeus), *L. quercus* (Kolbe), *L. rena* Sommerman, *L. tectorum* Badonnel.

Lachesilla aethiopica (Enderlein)

Pterodela pedicularia var. *aethiopica* Enderlein, 1902:11.
Lachesilla aethiopica Badonnel (sic), 1949b:53.
Lachesilla aethiopica (Enderlein). Smithers, 1967:59.

Recognition features. Known from female only. Wings clear with slight tawny wash; pterostigma opaque. Head, thorax, and appendages medium tawny brown. Preclunial abdominal segments tawny yellow with broad reddish brown ring, incomplete ventrally, on each segment; terminal segments pale tawny brown. Subgenital plate (Fig. 646) with hind margin slightly excavated in middle. Gonapophyses single, beaked apically, directed medially, outer surface straight in basal half. Female ninth sternum membranous. Female epiproct with distal pigmented area nearly divided into two by basal clear area.

Relationships. According to Badonnel and García Aldrete (1980) this is a parthenogenetic sister species of the sexual species *L. nuptialis* Badonnel and García Aldrete (southeastern Mexico, Belize, Guatemala).

Distribution and habitat. Within the study area, a single specimen has been taken at Gainesville (Alachua County), Florida, on dead, persistent banana leaves. Outside the study area, the species has been found in the Mexican states of Veracruz, Puebla, and Oaxaca, in the Central- and South American countries of Belize, Costa Rica, Guatemala, Brazil, and Peru, the West Indian islands of Cuba, Hispaniola, Jamaica, Puerto Rico, and Trinidad, and the African countries of Tanzaniya, Uganda, Angola, and Zaire. It inhabits dead, persistent leaves of a wide variety of plants.

Lachesilla greeni Pearman

Terracaecilius greeni Pearman, 1933:81.
Lachesilla (Terracaecilius) greeni Pearman. Roesler, 1939:169.
Lachesilla greeni (Pearman). Badonnel, 1943:105.

Recognition features. Adults mostly brachypterous, forewinglets not quite reaching base of abdomen in female, reaching nearly to tip of abdomen in male; females rarely fully winged; the wings clear. Head medium brown; thorax and appendages somewhat paler brown; preclunial abdominal segments pale yellowish brown, lacking rings; terminal segments same shade as thorax. Male genitalic structures (described by Roesler, 1939) very similar to those of *L. pedicularia*, differing in: (1) claspers being somewhat longer and more pointed, (2) distal ends of phallosomal apodeme branches being curved s-shaped, (3) process arising from epiproct bent at right angle. Subgenital plate with hind margin straight or slightly depressed in middle. Gonapophyses relatively short, broad, blunt-pointed apically. Female paraproct with short blade on median margin. Female ninth sternum mostly membranous, somewhat thickened anteriorly; a narrow sclerotized ring around spermapore.

Relationships. These are as noted for *L. pallida*.

Distribution and habitat. Within the study area the species has been taken once (six females) at New York City as an introduction from Italy on bay leaves. Outside the study area, it is known from England, France, Germany, Belgium, Switzerland, and Morocco (Günther, 1974a, Lienhard, 1977), where it has been found in woodsheds on cut wood, on old boards in a cellar, in ground litter of spruce needles, and on an elm trunk.

Lachesilla pacifica Chapman

Lachesilla pacifica Chapman, 1930:353.
Lachesilla silvicola Chapman, 1930:361.
Lachesilla telsa Sommerman, 1946:656, new synonym.

Recognition features. Wings clear except for slight tawny wash and slightly opaque pterostigma. Head, thorax, and appendages medium brown, paler on vertex between "dotted areas". Preclunial abdominal segments dull white with bright white subcuticular spots mostly arranged in rings, incomplete ventrally, alternating with diffuse rings of reddish brown, also incomplete ventrally; terminal segments medium brown. Clasper bases each extended posteriorly as broad, rounded process (Fig. 647); claspers joined on midline forming slightly curved, spear-head shaped structure. Phallosome branches (Fig. 647) bearing posteriorly a single broad u-shaped flap. Male epiproct (Fig. 215) with two strongly asymmetrical processes, that of right behind that of left side, both curving around to form semicircle and together nearly enclosing complete circle. Male clunium medio-dorsally with broad, flat process (Fig. 648) rounded distally overlying base of epiproct. Male paraproct with relatively stout process (Fig. 649) hooked at tip. Subgenital plate (Fig. 650) slightly bilobed on hind margin, with pair of internal thickenings, one to each side of midline before hind margin, or these meeting posteriorly on midline, or forming near circle, open posteriorly. Female ninth sternum thickened anteriorly; spermapore sclerite rounded, moderately broad in front and on sides.

Relationships. The species forms a complex with two undescribed species, one in southern California and the other in the Rocky Mountains of Wyoming and Colorado.

Distribution and habitat. Within the study area the species occurs from southern California north to Vancouver, British Columbia. Sexual forms have also been found at Red Canyon, Garfield County, southwestern Utah. Parthenogenetic populations which have been considered the same species have been found in southern New Mexico (Otero County), also at several localities in central Illinois and one in Kentucky. The species occupies many habitats including bark of trees and vines, foliage of conifers, dead persistent leaves, and occasionally ground litter. Outside the study area, the species occurs in Baja California State, Mexico, and was recorded from several localities around Geneva, Switzerland, and the neighboring region of Haute Savoie, France (Lienhard, 1989).

Note. Variation in females of this species includes the characters on which the species *L. telsa* Sommerman (1946) was based. The latter named species was represented by the holotype female from Hollywood, California, and a single paratype female from Vancouver, British Columbia. One male of *L. pacifica* was taken with the holotype and two males of *L. pacifica* were collected with the paratype (both male collections recorded by Sommerman, 1946, under the synonym *L. silvicola*). These collections support the view, expressed here, that *L. telsa* is a synonym of *L. pacifica*.

Lachesilla pallida (Chapman)

Terracaecilius pallidus Chapman, 1930:343.
Lachesilla pallida (Chapman), Sommerman, 1946:649.

Recognition features. A dimorphic species with macropterous males and brachypterous females; wings of females not extending beyond first abdominal tergum. Male forewings (Fig. 213) marked with cloudy brown spots: distally in pterostigma, distally along veins R4+5, M1, M2, M3, along Cu1a from top of areola postica to wing margin, at nodulus and in some specimens in cell Cu1b. Male head and thorax dark brown, appendages medium brown; preclunial abdominal segments pale tawny brown, each with dark reddish brown ring, incomplete ventrally; well sclerotized portions of terminal segments dark brown. Female head, thorax, and appendages pale brown; abdomen same as male except rings broader and terminal segments pale brown. Hypandrium (Fig. 651) two segmented; claspers short, meeting (or overlapping) at midline and curved outward at apex. Phallosome (Fig. 651) of two apodemes fused at anterior end, each terminating in transversely elongate flap, the two flaps apparently joined by membrane on midline. Male epiproct with single hook-like process (Fig. 652) directed backward and flanked by pair of slender, straight processes arising on clunium. Male paraproct with slender, slightly curved process (Fig. 212) truncated apically. Subgenital plate (Fig. 653) with hind margin slightly depressed in middle. Ovipositor valvulae single, broad, somewhat tapering towards tip. Female ninth sternum with small circular sclerotization behind spermapore and wide thickening along anterior margin. Sclerotization around spermapore small, rounded.

Relationships. The species finds its closest relatives in two very similar undescribed species from eastern North America. It is also close to *L. greeni* (Europe), *L. pedicularia* (cosmopolitan), and *L. dimorpha* Lienhard (Greece).

Distribution and habitat. The species is relatively uncommon and has been collected to date only in the states of Illinois, Indiana, Minnesota, New York, and Tennessee, also in southern Ontario and southern Quebec. It is found primarily in leaf litter but males are sometimes taken in hanging dead leaves.

Lachesilla pedicularia (Linnaeus)

Hemerobius pedicularius Linnaeus, 1758:551.
Termes fatidicum Linnaeus, 1758:610.
Hemerobius abdominalis Fabricius, 1775:310.

Hemerobius fatidicus (Linnaeus). Fabricius, 1775:311.
Hemerobius pusillus Müller, 1776:146.
Psocus pedicularius (Linnaeus). Latreille, 1794:85.
Psocus flavicans (Linnaeus). Fabricius, 1798:203.
Psocus abdominalis (Fabricius). Fabricius, 1798:204.
Psocus fatidicus (Linnaeus). Fabricius, 1798:204.
Atropos fatidicus (Linnaeus). Stephens, 1829:314.
Psocus nigricans Stephens, 1836:127.
Psocus dubius Stephens, 1836:127.
Psocus domesticus Burmeister, 1839:777.
Psocus pulsatorius (Linnaeus). Zetterstedt, 1840:1054.
Psocus fatidicus (Linnaeus). Westwood, 1840:19.
Psocus binotatus Rambur, 1842:324.
Lachesis fatidica (Linnaeus). Hagen, 1861:22.
Psocus salicis Hagen, 1861:13.
Psocus geologus Walsh, 1862:362.
Caecilius salicis (Hagen). Hagen, 1866:206.
Caecilius pedicularius (Linnaeus). Hagen, 1866:206.
Caecilius pusillus (Müller). Hagen, 1866:207.
Elipsocus flavicans (Linnaeus). Hagen, 1866:207.
Psocus pusillus Harris, 1869:331.
Caecilius (Pterodela) pedicularia (Linnaeus). Kolbe, 1880:118.
Hyperetes fatidicus (Linnaeus). Hagen, 1883:320.
Leptopsocus exiguus Reuter, 1899:5.
Pterodela pedicularia ab. *giardi* Enderlein, 1901:546.
Pterodela pedicularia var. *brevipennis* Enderlein, 1903g:381.
Lachesilla pedicularia (Linnaeus). Enderlein, 1915:16.
Lachesilla limbata Enderlein, 1924:36.
Caecilius nigrotuberculatus Curran, 1925:292.
Lachesilla stigmalis Navás, 1932:106.

Recognition features. Adults primarily macropterous, occasionally with wings somewhat reduced, rarely males with wings extremely reduced. Normally, wings clear, occasionally with brown border of descending portions of Cu1a in forewing; shorter-winged adults with most distal veins brown-bordered. Head and thorax medium brown; legs and preclunial abdominal segments pale brown with reddish brown ring, incomplete ventrally, on each preclunial segment; terminal segments slightly darker brown than preclunials. Hypandrium (Fig. 654a) single-segmented; claspers apposed at bases on midline, straight. Phallosome (Fig. 654a) of single basal apodeme once branched at about distal three-fourths, the two branches bearing transverse lonzenge-shaped flap. Male epiproct with single hook-like process (Fig. 211) directed backward, flanked by pair of slender processes arising from clunium, straight in dorsal view, downcurved in lateral view. Male paraproct without process but with stout spine on free edge (Fig. 654b). Subgenital plate (Fig. 655) gently rounded posteriorly. Ovipositor valvulae (Fig. 192) single, bent near base, tapering to blunt tip. Female ninth sternum largely membranous with some irregular thickening anterior to spermapore; spermapore sclerite relatively wide circular rim around spermapore.

Relationships. As noted for *L. pallida*.

Distribution and habitat. Within the study area the species has been recorded from most of the United States, also from the Canadian provinces of Ontario (Bellville region), Saskatchewan (Drinkwater, Indian Head), and British Columbia (Kaslo, Vancouver). It occurs on dead grasses, including drying grain crops in the field, dead persistent leaves of various sorts, occasionally on foliage of coniferous trees, and occasionally in houses and bins of stored grain. Outside the study area it is nearly cosmopolitan, being readily transported in human commerce.

Lachesilla quercus (Kolbe)

Caecilius (Pterodela) quercus Kolbe, 1880:120.
Lachesilla quercus (Kolbe), Enderlein, 1915:19.

Recognition features. Wings clear except pterostigma opaque white. Head, thorax, legs, and terminal abdominal segments pale yellowish brown; preclunial abdominal segments white. Hypandrium one-segmented. Claspers curved outward (Fig. 210), each with pointed spur directed outward at distal two-thirds, apex pointed. Phallosome (Fig. 210) of single apodeme expanded as broad plate distally; two triangular semimembranous bodies each attached laterally on clasper and distally on clasper base probably homologs of distal phallosomal flaps of other species. Male epiproct (Fig. 656) with two short, truncated processes. Male paraproct with curved, pointed, spinulose process (Fig. 657). Subgenital plate broad-bilobed distally (Fig. 658). Ovipositor valvulae single, short, rounded distally. Female ninth sternum thickened in anterior half before spermapore; spermapore plate a narrow, round rim.

Relationships. The species is probably close to *L. bernardi* Badonnel (central Europe) and *L. rossica* Roesler (southwestern Soviet Union). The broad distal plate of the phallosome of this species is probably the 'piratized' homolog of the median distal lobe of the hypandrium of *L. bernardi* and *L. rossica*.

Distribution and habitat. Within the study area the species was found in large numbers in a flour warehouse in Pensacola, Florida (Escambia County), in May, 1980, where it was probably introduced on jute bags from Pakistan. It was also taken once as an introduction at Seattle, Washington, on cut foliage of *Larix* sp. from Switzerland. Outside the study area, the species is found throughout Europe, east to central Asia, and south to Morocco (Günther, 1974a). I also have a record from Turkey. It usually inhabits persistent dead leaves.

Lachesilla rena Sommerman

Lachesilla rena Sommerman, 1946:653.

Recognition features. Forewing (Fig. 659) marked with brown spots on vein borders; R1 around pterostigma bordered in distal half of pterostigma; all other veins surrounded by brown spot immediately before junction with margin, but Cu1a bordered most of its length beyond departure from Cu1b. Head and thorax pale tawny brown, slightly darker on thoracic notal lobes; appendages nearly colorless; preclunial abdominal segments creamy white with continuous reddish brown band along each side and along dorsal midline in males; a vertical streak extending up from each lateral band in each segment; terminal segments medium tawny brown. Hypandrium (Fig. 660) one-segmented; claspers short, straight, each on end of long sclerotic strip separating clasper base from hypandrium proper. Male epiproct (Fig. 214) bearing long, slender, slightly recurved, anteriorly directed process. Male paraproct bearing stout process (Fig. 661), tapering to point and upcurved near tip. Subgenital plate (Fig. 662) bilobed distally. Ovipositor valvulae broad basally, somewhat upcurved and pointed distally. Female ninth sternum entirely membranous.

Relationships. Several undescribed species from parts of tropical America are close to this one.

Distribution and habitat. Within the study area, the species occurs in southern California north to Riverside County, in Maricopa County, Arizona, and in Hidalgo and Brown Counties, Texas. It inhabits dead and dying low vegetation of various sorts. Outside the study area it has been found throughout much of Mexico (states of Campeche, Chiapas, San Luis Potosí, Sonora, Tamaulipas, Veracruz), in Guatemala, and on the islands of Hispaniola and Puerto Rico.

Lachesilla tectorum Badonnel

Lachesilla tectorum Badonnel, 1931:238.

Recognition features. Known from female only. Very similar to *L. pedicularia* in general appearance, but larger, darker, rings of preclunial abdominal segments very wide and dusky brown in color, distal margin of subgenital plate flatter, and angle on outer surface of ovipositor valvula more distal (Fig. 663).

Relationships. The species is presumably very close to *L. pedicularia*.

Distribution and habitat. Within the study area, the species has been taken in field collections at Vero Beach (Indian River County) and Stock Island (Monroe County), Florida, and at Southmost Palm Grove, Cameron County, Texas. At these localities it has been found on dead palm leaves and dead leaves of tall grass. It has been taken as an interception from Mexico at several localities along the Texas-Mexico border, on such plants as *Cucurbita* sp., *Zea mays*, *Agave victoria*, *Chamaedorea* sp., *Sorghum vulgare*, and orchids. It has been taken twice at Miami, Florida, as an interception, on cactus plants from Honduras, and on dried pepper (*Capsicum frutescens*) from Grand Cayman. Outside the study area, it has been taken in Nuevo León and Tamaulipas States, Mexico, where it was found in pine foliage and ground litter. It was originally described from Mozambique.

Riegeli Group

Diagnosis. Wings clear or forewing with spot on distal end of each M and Cu1 vein. Hypandrium one- or two-segmented. Clasper bases relatively slender, separated by a line from hypandrium proper. Claspers (Fig. 186) single-, double-, or triple-shafted, the shafts outcurved and pointed distally. Phallosome of single basal apodeme dividing either before its middle, immediately beyond its middle, or very near distal end; each branch producing small expanded area at distal end of hypandrium and beyond that an outcurved, pointed process. Male epiproct transverse, with single median, posteriorly directed process (Fig. 664). Male paraproct with stout, hook-tipped process. Subgenital plate rounded or bilobed distally with transverse internal thickening near distal end. Ovipositor valvulae (Fig. 187) single, relatively to extremely broad, rounded distally. Female ninth sternum mostly membranous, thickened along fore and hind margins; fore margin with pair of small processes directed antero-laterally (Fig. 187); spermapore sclerite a small, round rim.

North American species: *L. riegeli* Sommerman, *L. tropica* García Aldrete, *L. ultima* García Aldrete.

Lachesilla riegeli Sommerman

Lachesilla riegeli Sommerman, 1946:654.

Recognition features. Forewing (Fig. 665) with brown spot at distal end of each M and Cu1 vein. Head, thorax, and appendages tawny yellow with complex brown pattern on head including post-clypeal striations; narrow reddish brown line along side of thorax above leg bases; preclunial abdominal segments creamy white, each with reddish brown ring incomplete ventrally; terminal segments tawny yellow to brown. Hypandrium two-segmented. Clasper shafts (Fig. 667) single, relatively long, bent out near tip. Phallosomal apodeme (Fig. 667) dividing near distal end. Male epiproctal process (Fig. 664) with median marginal depression, resulting two lobes each bilobed. Subgenital plate (Fig. 666) tapering distally, with relatively narrow bilobed distal region. Ovipositor valvulae (Fig. 187) very broad, rounded. Processes of front margin of female ninth sternum close to midline (Fig. 187).

Relationships. The riegeli group is a small one, restricted to the American Tropics. Thus, all of the species are closely related.

Distribution and habitat. Within the study area the species is restricted to southern Florida north to Indian River and Polk Counties. There it has been found on a trunk and foliage of Serenoa repens, on foliage of live oak, and on black-eyed pea plants. Outside the study area, it has been found in the Mexican states of Campeche, Guerrero, Jalisco, Nayarit, Nuevo León, and Oaxaca, in Guatemala, Honduras, and Colombia (Cali), and on the islands of Puerto Rico, Hispaniola, and Cuba.

Lachesilla tropica Garcia Aldrete

Lachesilla tropica García Aldrete, 1982:204.

Recognition features. Wings clear. Color otherwise as described for *L. riegeli*. Hypandrium single-segmented. Claspers (Fig. 667) single-shafted, relatively short, curved, tapering gradually to acuminate tip. Phallosome (Fig. 667) divided only at extreme distal end, there producing two elongate, outcurved processes, longer than claspers, each tapering to acuminate tip. Male epiproctal process with distal margin entire. Subgenital plate (Fig. 188) tapering to rounded distal margin; transverse internal thickening large, covering most of area of plate. Ovipositor valvulae (Fig. 668) not as broad as in L. riegeli. Processes of front margin of female ninth sternum each about twice as long as distance of process from midline.

Relationships. See discussion for *L. riegeli.*

Distribution and habitat. Within the study area the species has only been found on Big Pine Key, Monroe County, Florida, where a single male was collected in August, 1951, by sweeping shrubs. Outside the study area, the species occurs throughout most of Mexico (states of Campeche, Chiapas, Guerrero, Jalisco, Nuevo León, Oaxaca, Quintana Roo, San Luis Potosí, Tamaulipas, Veracruz, and Yucatán), also in Guatemala, Honduras, and the island of Jamaica. It is usually found on foliage and dead leaves of small trees and shrubs.

Lachesilla ultima García Aldrete

Lachesilla ultima García Aldrete, 1982:208.

Recognition features. Known only from female. Wings clear. Color otherwise approximately as described for *L. riegeli.* Subgenital plate (Fig. 189) broadly rounded distally; inner thickening relatively slender, lying across middle. Ovipositor valvulae and female ninth sternum as described for *L. tropica.*

Relationships. The species appears to be very close to *L. sola* García Aldrete (Jamaica).

Distribution and habitat. The species is known only from the holotype, taken by beating vegetation in a live oak-cabbage palm hammock at Terra Cela on the Gulf Coast in Manatee County, Florida.

Rufa group

Diagnosis. Wings generally clear with veins uniform in color, occasionally veins in distal half of forewing darker than those in basal half. Hypandrium of two segments generally without distinct intersegmental divisions; sternum of preceding segment at least somewhat sclerotized. Distal margin of hypandrium depressed or bilobed. Claspers distinct, slender, their bases relatively slender and elongate, separated from hypandrium proper by a weak line. Phallosome of two apodemes completely separate or fused at base, or single basal apodeme branching once near anterior end (Fig. 670). Male epiproct simple, rounded posteriorly. Male paraproct with short, slender, acuminate process (Fig. 669). Subgenital plate with median tongue on outer surface arising near base of plate, bifid distally (Fig. 172). Ovipositor valvulae one pair, relatively long, broad basally, tapering distally, directed mesad or postero-mesally. Female ninth sternum membranous with rounded ring surrounding spermapore but well out from it, or broad, rounded sclerotized plate enclosing spermapore.

North American species: *L. arida* Chapman, *L. chiricahua* García Aldrete, *L. jeanae* Sommerman, *L. nita* Sommerman, *L. rufa* (Walsh). *L. yakima* Mockford and García Aldrete.

Lachesilla arida Chapman

Lachesilla arida Chapman, 1930:346.

Recognition features. (Information from Chapman, 1930, Sommerman, 1946, García Aldrete, 1972, 1974, 1990b). Known from female only. Wings clear; veins in proximal half of forewing pale yellow, in distal half brown. Body apparently pale brown to creamy white; preclunial abdominal segments with vertical grayish brown stripes. Subgenital plate with hind margin indented in middle; median tongue with very short cleft region, not occupying more than one-fifth of length of tongue. Ovipositor valvulae very broad basally, relatively short. Female ninth sternum with complex ring around spermapore, pair of very short processes on anterior margin (García Aldrete, 1974, Fig. 241).

Relationships. García Aldrete (1990b) places this species in his 'Group' I with *L. chiricahua* and six Mexican species.

Distribution and habitat. It has been recorded from Cochise and Apache Counties, Arizona (García Aldrete, 1972). The recorded habitats are dead oak leaves, *Pseudotsuga menziesii,* and *Juniperus* sp.

Lachesilla chiricahua García Aldrete

Lachesilla chiricahua García Aldrete, 1990b:38.

Note. I have not examined individuals of this species. The recognition features are taken from the original description.

Recognition features. Wings clear, their veins light brown. Body pale amber with band of ochre on each side of head from compound eye, around antennal fossa, to epistomal suture. Hypandrium with distal margin slightly bilobed between slender, forward-directed, outcurved claspers. Phallosome of two completely separate apodemes curved outward in their distal one-fourth; the outcurved portion with microspinose surface. Subgenital plate moderately indented in middle on posterior margin; cleft of median tongue of moderate depth, about one-fourth length of tongue. Ovipositor valvulae broad basally, tapering to blunt-pointed tips. Female ninth sternum membranous except thickened along anterior margin.

Relationships. García Aldrete (1990b) placed this species in his Group I, with *L. arida* and six Mexican species.

Distribution and habitat. The species has been found in the Chiricahua Mountains, Cochise County, Arizona, on dead oak leaves.

Lachesilla jeanae Sommerman

Lachesilla jeanae Sommerman, 1946:643.

Recognition features. Wing veins uniform brown in color. Head and thorax golden brown with two thin reddish brown bands from antennal base to compound eye, both continuing on neck and thorax as broken bands, dorsal one very weak or absent beyond prothorax, ventral band relatively distinct above leg bases for length of thorax. Appendages nearly colorless except tarsi and flagellum medium brown. Preclunial abdominal segments white with slender reddish brown rings, absent ventrally, on segments two through six. Terminal abdominal segments of male medium brown, those of female white. Hypandrium (Fig. 670b) with distal margin bilobed. Phallosome (Fig. 670b) of two completely separate apodemes curved outward distally. Subgenital plate with cleft of median tongue very deep, occupying nearly half length of tongue; hind margin of plate indented in middle (Fig. 171). Ovipositor valvulae broad basally tapering to blunt tip (Fig. 671b). Female ninth sternum membranous.

Relationships. Garcia Aldrete (1990b) placed this species in his group IIc, with *L. pinicola* Garcia Aldrete, *L. cupressicola* Garcia Aldrete, *L. broadheadi* Garcia Aldrete, *L. yuccalnicola* Garcia Aldrete, *L. abiesicola* Garcia Aldrete, and *L. turneri* Garcia Aldrete, all Mexican and Guatemalan species.

Distribution and habitat. The species occurs from the central Rocky Mountains in Colorado south to the Portal region in southeastern Arizona. It also occurs in the Black Hills in Lawrence County, South Dakota, and there is a single record from Burke County in northwestern North Dakota. Garcia Aldrete (in press) cites a record from "Mackenzie", Northwest Territory, W116.59, N61.43. It has been collected on dead oak leaves, foliage and bark of spruce, and on limber pine.

Lachesilla nita Sommerman

Lachesilla nita Sommerman, 1946:646.

Recognition features. Forewings marked (Fig. 170): a small brown spot at marginal end of each vein from R1 to nodulus; an obscure brown spot in each cell R3, R5, M1, M2, M3. Head, thorax, appendages, and abdomen tawny yellow, slightly darker on dotted areas of vertex and thoracic notal lobes: two reddish brown lines from antennal base to compound eye, continuing on thorax; dorsal line ending at forewing base, ventral line continuing length of thorax above leg bases. Preclunial abdominal segments each with broken vertical reddish brown band. Hypandrium (Fig. 672) depressed in middle on hind margin. Claspers short, closely attached to sides of hypandrium. Phallosome (Fig. 672) of two separate apodemes, each bowed out, meeting at base and meeting or overlapping at tip. Male epiproct transverse-quadrate. Male paraproct with short, pointed process (Fig. 673). Subgenital plate (Fig. 674) with median tongue represented by two pointed processes arising near distal margin of plate; distal margin rounded. Ovipositor valvulae (Fig. 675) broad basally, tapering distally, each with small process on outer distal margin. Female ninth sternum membranous, somewhat sclerotized along posterior margin. Spermapore sclerite rounded.

Relationships. Because of wing markings and the unique subgenital plate, this species stands apart from the other members of the group. García Aldrete (1990b) placed it alone in his Group III.

Distribution and habitat. Within the study area, this species is known from several counties (Alachua, Orange, Leon) in north-peninsular and northern Florida, from Savannah in southeastern Georgia, and from Panola County in eastern Texas. In these areas it has been found on foliage of cabbage palm, live oak, and willow oak. Outside the study area it has been found in the Mexican states of Chiapas, Nuevo León, and San Luis Potosí, also in Belize, Guatemala, and at El Volcán, Panamá.

Lachesilla rufa (Walsh)

Psocus rufus Walsh, 1863:185.
Caecilius rufus (Walsh). Hagen, 1866:206.
Caecilius impacatus Aaron, 1886:14.
Pterodela rufa (Walsh), Enderlein, 1906e:319.
Lachesilla rufa (Walsh), Chapman, 1930:358.

Recognition features. Wings clear except for opaque pterostigma. Head, thorax, and appendages yellowish brown. Preclunial abdominal segments creamy white, each with reddish brown ring absent ventrally. Terminal segments yellowish brown, darker on edges. Hypandrium (Fig. 670a) with hind margin incurved. Clasper bases relatively wide basally, tapering abruptly, partially folded under edges of hypandrium. Claspers curved out, slender, tapering. Phallosome (Fig. 670a) of single basal apodeme divided at distal one-third, the two branches slender, acuminately pointed apically. Male epiproct trapezoidal. Male paraproct with slender, pointed process (Fig. 669). Subgenital plate (Fig. 172) rounded distally; median tongue with distal cleavage reaching almost half length of tongue. Ovipositor valvulae (Fig. 671) of moderate width, tapering near distal end. Female ninth sternum (Fig. 671) with cresent-shaped sclerite to each side of spermapore, slender anterior extension before spermapore.

Relationships. García Aldrete (1990b) placed this species in his Group IIb, with seven Mexican and Guatemalan species and *L. yakima*.

Distribution and habitat. The species occurs in the Eastern States from Connecticut south to western North Carolina, west to eastern Missouri and western Illinois. It is usually found in persistent dead leaves.

Lachesilla yakima Mockford and Garcia Aldrete

Lachesilla silvicola Chapman (female), 1930:361.
Lachesilla yakima Mockford and Garcia Aldrete, 1974:236.

Recognition features. Known from female only. Wings clear. Head, thorax, appendages, and abdomen yellowish brown. Well colored specimens with two pale reddish brown lines from antennal base to compound eye. Upper line continuing on neck and propleuron to wing base; lower line running length of thorax above leg bases, darker on thorax than on head. First five preclunial abdominal segments each with diffuse vertical reddish brown band on each side. Subgenital plate (Fig. 173) flat on hind margin; median tongue broadening towards distal end, its cleavage about one-third length of tongue. Ovipositor valvulae (Fig. 676) broad basally tapering to point distally. Female ninth sternum (Fig. 676) mostly membranous, with small crescentic sclerite on front margin and narrow sclerotization along hind margin.

Relationships. These are as noted for *L. rufa*.

Distribution and habitat. The species occurs on the Pacific Coast from Vancouver Island, British Columbia south to Santa Barbara County, California and eastward to Jackson, Wyoming and Heber City, Utah. It has been found on foliage of spruce, hemlock, western red cedar, and Douglas fir.

Texcocana Group

Diagnosis. Differing from the centralis group (q.v.) in: 1) some species with forewing veins uniformly dark, 2) distal processes of hypandrium more slender, 3) small triangular distal plate absent between distal processes of hypandrium. Both groups with a heavy sclerotic strip along bases of gonapophyses.

North American species. *L. texcocana* Garcia Aldrete.

Lachesilla texcocana Garcia Aldrete

Lachesilla texcocana García Aldrete, 1972:123.

Recognition features. The only member of its species group in the study area. Differing from all known members of the group except *L. delta* García Aldrete by having a short articulated spine subapically on each distal process of hypandrium (Fig. 156). Differing from *L. delta* García Aldrete (Mexico) in 1) great reduction of subapical spinelets on phallosomal apodemes (Fig. 156), 2) single process with two points on male epiproct (Fig. 163) rather than two short, pointed processes.

Relationships. It is undoubtedly closest to *L. delta*, as indicated by García Aldrete (1985:61).

Distribution and habitat. Within the study area, the species is known from Washoe County, Nevada, and Grant County, New Mexico. At the Nevada locality it was taken in ground litter in a Jeffrey Pine forest, and at the New Mexico locality it was taken in alpine meadow. It has been recorded also from one locality in Alberta (Athabasca River near Brule), and one in British Columbia (Fort McLead near Lake McLead). Outside the study area, the species is widely distributed in Mexico, being known from the states of Chihuahua, Chiapas, Hidalgo, México, Morelos, Nuevo León, Puebla, Sinaloa, and the Distrito Federal (García Aldrete, 1985). It is usually found in persistent dead leaves of various sorts, and is fairly frequent in ground leaf litter.

Family Ectopsocidae

Diagnosis. Small forms (length from head to tip of closed wings 1.5 - 2.5 mm); body shades of brown; wings clear or variously marked. Adults usually macropterous but sometimes brachypterous or micropterous. Lacinial tip slender, bicuspid, lateral cusp usually somewhat broader than median. Five distal inner labral sensilla. Vertex bearing numerous setae, usually curved forward or backward. Fronto-clypeal suture present. Ciliation sparse to moderate on veins and margin of forewing, usually absent on hindwing but occasionally developed on margin in cell R3. Vein Cu1a absent in forewing. Rs and M joined by crossvein in hindwing. Adults with tarsi two-segmented. Ovipositor valvulae complete or reduced to v3, the latter in either case well ciliated.

North American genera: *Ectopsocopsis* Badonnel, *Ectopsocus* McLachlan.

Key to the North American Genera and
Species of Family Ectopsocidae

1. Male with sclerotized frame bearing several processes dorsally on abdomen, attached to clunium (Fig. 216). Female with ovipositor valvulae reduced to thumb-like v3 on each side (Fig. 123) .... Genus *Ectopsocopsis*, *E. cryptomeriae* (Enderlein)
-- Male with no sclerotized frame dorsally on abdomen but usually a comb of teeth on hind margin of clunium. Female with more complex ovipositor, usually with three valvulae on each side (Fig. 217) .... Genus *Ectopsocus* 2

2. Forewings marked at least with cloudy brown spots at ends of veins on wing margin, usually such a spot at Rs - M junction, and in some cases more elaborate markings. Both sexes usually macropterous, rarely brachypterous 3
-- Wings unmarked, or at most with cloudy brown border of vein M3. Both sexes of some species commonly brachypterous or micropterous 8

3. Forewing marked not only with spots at marginal ends of veins from R1 to Cu1 but also spots in basal cells R, Cu1, and Cu2 and distal cells R1, R3, and R5 (Fig. 218) ....
    ....*E. strauchi* Enderlein
-- Forewings marked only with spots at marginal ends of veins and at Rs - M junction 4

4. Subgenital plate truncated apically, slightly depresses in middle. Endophallus including several large, flat, overlapping sclerites each with a conspicuous distal prong (Fig. 219) ....
    ....*E. thibaudi* Badonnel
-- Subgenital plate with two apical processes. Endophallus with only one large, flat sclerite pointed at apex (Fig. 220) 5

5. Apical lobes of subgenital plate curved inward (Fig. 221). Spine of paraproct margin double, the two prongs nearly equal (Fig. 222). Phallosome with pair of thumb-like lobes distally ....
    ....*E. briggsi* McLachlan
-- Apical lobes of subgenital plate straight (Fig. 223). Spine of paraproctal margin either single or, if double, the two prongs very unequal (Fig. 224). Phallosome lacking pair of thumb-like lobes distally 6

6.   Apical lobes of subgenital plate < 1.5X as long as their basal width. Spine of paraproctal margin simple .... *E. californicus* (Banks)

--   Apical lobes of subgenital plate ~ 2.0X as long as their basal width or more. Spine of paraproctal margin double 7

7.   Known from both sexes. Females dimorphic: macropterous and brachypterous. Apical lobes of subgenital plate ~ 2.0X as long as their basal width .... *E. petersi* Smithers

--   Known from females only, all macropterous. Apical lobes of subgenital plate ~ 2.3X as long as their basal width .... *E. meridionalis* Ribaga

8.   Subgenital plate truncated at apex (Fig. 225). Adults of both sexes brachypterous, forewings not reaching beyond basal one-third of abdomen .... *E. richardsi* (Pearman)

--   Subgenital plate with two processes or lobes apically. Adults either macropterous or micropterous, not brachypterous .. 9

9.   A pair of rounded or truncated papillate lobes on aedeagal arch (Fig. 226). Processes of subgenital plate slender, each with an acuminate articulated spine at tip (Fig. 227)...................... ....*E. pumilis* (Banks)

--   Aedeagal arch without a pair of papillate lobes. Processes of subgenital plate variable in width but lacking an articulated spine at tip 10

10.   Adults of both sexes macropterous. Distal lobes of subgenital plate not wider at base than their length 11

--   Males micropterous; females dimorphic: macropterous and micropterous. Distal lobes of subgenital plate ~ 1.5X as wide at base as their length .... *E. vachoni* Badonnel

11.   Forward directed blade-like structure arising from base of phallosome (Fig. 228). Distal lobes of subgenital plate ~ as long as their basal width .... *E. salpinx* Thornton and Wong

--   No blade-like structure arising from base of phallosome. Distal lobes of subgenital plate longer than their basal width 12

12.   Longitudinal seam on midline below seminal vesicle capsule. Distal lobes of subgenital plate directed caudad with setae at tips (Fig. 229) .... *E. maindroni* Badonnel

--   No longitudinal seam below seminal vesicle capsule. Distal lobes of subgenital plate directed meso-caudad, with setae on sides as well as tip (Fig. 230) ....*E. titschacki* Jentsch

Genus *Ectopsocopsis* Badonnel

Ectopsocopsis Badonnel, 1955:185.

Diagnosis. Male of single North American species with elaborate sclerotic frame on dorsal surface of clunium bearing two curved processes anteriorly (Fig. 216). Female with ovipositor valvulae reduced to v3 (Fig. 123); subgenital plate slightly protruding in middle on hind margin.

Generotype. *E. balli* (Badonnel).

North American species. *E. cryptomeriae* (Enderlein).

Ectopsocopsis cryptomeriae (Enderlein)

Ectopsocus cryptomeriae Enderlein, 1907b:100.
Ectopsocus pumilis (Banks). Chapman, 1930:380 (misdet.).
Ectopsocus leprevoae Danks, 1955:180.
Ectopsocopsis pumilis (Banks). Badonnel, 1955:185 (misdet.).
Ectopsocopsis cryptomeriae (Enderlein). Badonnel, 1955:185.

Recognition features. The only species of its genus known from the study area. The elaborate dorsal sclerotization of the male terminal abdominal segments distinguishes it at once from other male ectopsocids of the area.

Relationships. The closest relative appears to be *E. annulata* Badonnel (Angola). There are no other New World species in the genus, and it seems likely that this one was introduced.

Distribution and habitat. Within the study area, the species is found from southern Florida and southern Texas north to southern New York, central Indiana, and central Illinois. It usually inhabits hanging dead leaves, including those of some crop plants such as *Zea mays*. Rarely it becomes a pest in warehouses. Outside the study area the species is widely distributed over the warmer parts of the world but has not yet been recorded from Africa or South America.

Genus *Ectopsocus* McLachlan

Ectopsocus McLachlan, 1899:277.
Micropsocus Enderlein, 1901:546.
Chaetopsocus Pearman, 1929:105.

Diagnosis. Males usually with row of denticles ('clunial comb') along hind margin of clunium before epiproct. Females usually with complete ovipositor valvulae (Fig. 217); subgenital plate usually at least slightly bilobed posteriorly.

Generotype. *E. briggsi* McLachlan.

North American species. *E. briggsi* McLachlan, *E. californicus* Banks, *E. maindroni* Badonnel, *E. meridionalis* Ribaga, *E. petersi* Smithers, *E. pumilis* Banks, *E. richardsi* (Pearman), *E. salpinx* Thornton and Wong, *E. strauchi* Enderlein, *E. thibaudi* Badonnel, *E. titschacki* Jentsch, *E. vachoni* Badonnel.

Ectopsocus briggsi McLachlan

Ectopsocus briggsi McLachlan, 1899:277.
Ectopsocus limbatus Navás, 1909:143.
Peripsocus parvulus Kolbe. Enderlein, 1927:224 (misdet.).

Recognition features. Forewing (Fig. 677) unmarked except for cloudy brown spot at marginal end of each vein R1 to nodulus and at Rs - M junction. Head, thorax, and appendages medium brown. Abdomen creamy yellow with ventral reddish brown band on each side of each preclunial segment. Clunial comb field separated anteriorly from rest of clunium by field of spinules (Fig. 678). Phallic apparatus bearing pair of thumb-like structures at its apex (Fig. 220). Spine of paraproctal margin (both sexes) double, the prongs nearly equal (Fig. 222). Apical lobes of subgenital plate decidedly curved inward, ending in a smooth-margined process between apical two setae (Fig. 221).

Relationships. The species is close to *E. californicus*, *E. meridionalis*, and *E. petersi*, q. v.

Distribution and habitat. Within the study area, this species is found along the coast of California from the San Francisco Bay region north to the Oregon state line, and at Vancouver, British Columbia. It has also been taken as an introduction from Europe at Boston and New York City. There is also a record of a single specimen taken on dead leaves of a jade plant at Plymouth, Orange County, Florida. It is found on both living and dead leaves of various broad-leaf trees, and occurs occasionally on conifer foliage. Outside the study area it is widely distributed, having been recorded from Mexico, Venezuela, Chile, much of Europe, much of Africa, Canary Islands, India, Australia, and New Zealand.

Ectopsocus californicus (Banks)

Peripsocus californicus Banks, 1903b:237.
Ectopsocus congener Tillyard, 1923:192.
Ectopsocus californicus (Banks, in part). Chapman, 1930:377.

Recognition features. Forewing (Fig. 679) spotted in same pattern as described for *E. briggsi*, but spots darker. Head creamy yellow except medium brown on postclypeus, dotted areas of vertex, and band from antennal base through compound eye to hind head margin. Thorax creamy yellow with reddish brown band running entire length above leg bases on well colored specimens. Legs and antennae creamy yellow, likewise abdomen; each preclunial segment with vertical reddish brown band on each side and dark reddish brown spot over each spiracle. Clunial comb field as described for E. briggsi. Phallic apparatus lacking pair of thumb-like structures at its apex. Spine of paraproctal margin simple (both sexes). Apical lobes of subgenital plate (Fig. 223) short, less than 1.5X their basal width.

Relationships. As noted for *E. briggsi*.

Distribution and habitat. Within the study area, the species occurs along the entire Pacific Coast of the United States and southern British Columbia. Outside the study area, the species is known from Baja California State, Mexico, the highlands of Guatemala, and New Zealand. It inhabits living and dead foliage of numerous kinds of trees and shrubs including both angiosperms and gymnosperms.

Ectopsocus maindroni Badonnel

Ectopsocus maindroni Badonnel, 1935a:76.

Recognition features. Small species, length from front of head to tip of closed forewing ~ 1.64 mm. Forewings clear except for opaque pterostigma and slight brown mark along distal ends of veins M3 and Cu1. Entire body creamy yellow. Sharp seam along mid-ventral surface under capsule of seminal vesicle clearly visible on male abdomen. Clunial comb field continuous anteriorly with rest of clunium. Endophallus complex (Fig. 680), including one dark helmet-shaped sclerite with spinulose surface as well as various smaller, paler sclerites. Apical lobes of subgenital plate (Fig. 229) straight, ~ 1.25X as long as their basal width.

Relationships. Superficially, the species resembles closely *E. titschacki*, but close relationship is doubtful. Thornton and Wong (1968) place it close to *E. marginatus* Thornton and Wong, *E. uncinatus* Thornton and Wong, and *E. intersitus* Thornton and Wong (Philippines and Pacific), but this grouping appears to be very tentative.

Distribution and habitat. Within the study area the species is established in the southern Florida counties of Dade, Monroe, and Indian River, where it is usually found on dead palm leaves. It has also been found in plant quarantine at several ports of entry, including New York, Miami, New Orleans, San Antonio, Nogales, and Seattle, on

various plants. Outside the study area, it is known from Cuba,
Jamaica, Puerto Rico, Antigua (West Indies), the Mexican states of
Veracruz, Tabasco, and San Luis Potosi, also, West Africa, Arabia,
southern India, Malaysia, and Hong Kong.

Ectopsocus meridionalis Ribaga

Ectopsocus briggsi var. *meridionalis* Ribaga, 1904:296.
Ectopsocus meridionalis Ribaga. Enderlein, 1907b:101.

Recognition features. Known from female only. Wings and body
color as described for *E. briggsi*, except preclunial abdominal bands
darker and continuous across dorsal surface. Spine of paraproctal
margin double, the two prongs very different in length (Fig. 224). Apical
lobes of subgenital plate (Fig. 681) straight, slender, ~ 2.3X as long as
their basal width.

Relationships. These are as noted for *E. briggsi*.

Distribution and habitat. Within the study area the species
occurs throughout eastern United States from Connecticut west to
central Wisconsin, south to Bexar County, Texas and Sarasota County,
Florida. It is found commonly on persistent dead leaves. Outside the
study area, it is known from central Mexico, Colombia, Chile, southern
Europe, Morocco, West Africa, East Africa, Japan, and Hong Kong.

Ectopsocus petersi Smithers

Ectopsocus petersi Smithers, 1978:144.

Recognition features. Females dimorphic, either macropterous
or brachypterous; brachypterous females with forewings extending at
rest from about half to about two-thirds length of abdomen. Males all
macropterous. Wings marked as described for *E. briggsi*, but spots on
forewings of brachypterous females more intense and relatively larger.
Males colored as in *E. briggsi* except vertical reddish brown bands of
preclunial abdominal segments continuous across dorsal surface.
Females with head and thorax somewhat darker than described for *E.
briggsi*, otherwise as males. Clunial comb field as described for *E.
briggsi*. Phallic apparatus without pair of thumb-like structures at its
apex. Spines of paraproctal margin double, the two prongs very
different in length (as in Fig. 224). Apical lobes of subgenital plate (Fig.
682) straight, of medium length, ~ 2.0X as long as basal width.

Relationships. These are as noted for *E. briggsi*.

Distribution and habitat. Within the study area, the species has been taken as an introduction at New York City and Boston on material of *Buxus*, *Chrysanthemum*, *Dianthus*, and other plants from England, Ireland, and Guernsey (Channel Islands). Outside the study area, it is known from a number of localities in Ireland.

Ectopsocus pumilis (Banks)

Peripsocus pumilis Banks, 1920:313.
Not *Ectopsocus pumilis* (Banks). Chapman, 1930:380.
Ectopsocus ghesquierei Ball, 1943:11.
Ectopsocus josephi Galil, 1984:3.

Recognition features. Wings unmarked. Body and appendages tawny yellow except terminal abdominal segments pale brown, processes of subgenital plate dark brown on base and tip, paler in middle. Clunial comb field (Fig. 683) completely separated from rest of clunium by lateral and anterior ridges. Male with acuminate-tipped external parameres (Fig. 236) and pair of papillate, truncated processes on aedeagal arch of phallosome. A dagger-like process arising from base of phallosome, directed forward. Apical processes of subgenital plate (Fig. 227) directed disto-mediad, tapering towards tip, length about three times basal width; a short, acuminate articulated spine at tip of each process.

Relationships. The species appears to be close to *E. vilhenai* Badonnel (Mexico, West Africa) in external genitalic structures of both sexes.

Distribution and habitat. Within the study area, the species has been found at Monticello (Jefferson County), and Gainesville (Alachua County), Florida, on dead banana leaves in an out-door situation, on walls of a building, and on moldy boots in a house. It has also been taken as an introduction at Miami, New York, and Boston. Outside the study area, the species is known from Hong Kong, Japan, India, West Africa, East Africa, and Azores. Males are known only from India and introduced material at Miami. The species has been found in cereals and stored grain in Hong Kong and Japan.

Ectopsocus richardsi Pearman

Chaetopsocus richardsi Pearman, 1929:105.
Ectopsocus richardsi (Pearman). Pearman, 1942:290.

Recognition features. Adults entirely brachypterous with forewings not reaching beyond basal one-third of abdomen. Fore-winglets clear except for clouding in distal R and M cells. Head medium brown; remainder of body and appendages pale brown. Clunial comb field (Fig. 684) continuous basally with rest of clunium. Phallosome (Fig. 685) with acuminate-tipped external parameres, endophallus with basal denticulate region (radula of Pearman, 1942). Subgenital plate (Fig. 225) truncate distally, lacking disto-lateral processes.

Relationships. Thornton and Wong (1968) placed the species in their hirsutus group of mostly Pacific species.

Distribution and habitat. Within the study area, this species has been taken at Houston (Harris County) and Beaumont (Orange County), Texas, in stored rice and in a rice mill. It has also been found in buildings at Gainesville (Alachua County) and Jacksonville (Duval County), Florida, and as an introduction at Miami, Florida, on plant material from Panama, Grand Cayman, and Argentina. It is known outside the study area also from Colombia, West Africa, England, Hawaii, Galapagos, and Hong Kong. It is generally associated with stored food products.

Ectopsocus salpinx Thornton and Wong

Ectopsocus salpinx Thornton and Wong, 1968:70.

Recognition features. Wings clear, with slight tawny wash. Head, thorax, and appendages medium reddish brown. Preclunial abdominal segments each with broad brown ring, incomplete ventrally; segments light reddish brown between rings; terminal segments medium reddish brown to dark brown. Male with knife-blade-like structure, readily visible through cuticle, arising at base of phallosome and lying between seminal vescles. Clunial comb field (Fig. 686) completely separated from rest of clunium by lateral breaks in cuticle. Phallosome (Fig. 228) with external parameres slender, directed mesad from broad base. Distal processes of subgenital plate (Fig. 687) short, blunt-tipped, about as long as their basal width.

Relationships. This species is close to *E. vilhenai* Badonnel (West Africa, Mexico), and *E. cinctus* Thornton (Southeast Asia).

Distribution and habitat. Within the study area, the species is known from a single introduction at Los Angeles on an orchid flower from Singapore. Outside the study area, it is known from Malaysia, Philippine Islands, and Guam. It has been found in coconut thatch, dead rattan, and banana leaves.

Ectopsocus strauchi Enderlein

Ectopsocus strauchi Enderlein, 1906b:315.
Peripsocus opulentus Navás, 1908:411.

Recognition features. In forewing (Fig. 218), veins R1 to Cu1 each with cloudy brown spot at distal end; cells R1, R3, R5 each with one cloudy brown spot running most of length of cell; spot sometimes divided into two in cell R1; cells R, Cu1, and Cu2 with basal and distal spot. Body and appendages pale brown. Clunial comb field (Fig. 688) broadly joined at base with rest of clunium. External parameres curving strongly inward from base, outward near tip; endophallus including several large, dentate sclerites. Apical lobes of subgenital plate (Fig. 689) short, broad based, tapering to narrow tips, about as broad at base as their length.

Relationships. Several features suggest proximity to *E. spilotus* Thornton and Wong (South Pacific, Bermuda).

Distribution and habitat. Within the study area, this species is known as an introduction at Miami and Boston on plant materials from the Azores, Italy and Peru. Outside the study area, it is known from the Azores, Canary Islands, Bermuda, Morocco, and Peru.

Ectopsocus thibaudi Badonnel

Ectopsocus thibaudi Badonnel, 1979:52.

Recognition features. Forewings (Fig. 690) with cloudy brown spot at marginal end of each vein and one at Rs - M junction. Body yellowish tan with slightly indicated reddish brown band running length of thorax along side. Clunial comb field (Fig. 691) continuous basally with rest of clunium. Phallosome (Fig. 219) with acuminate-tipped external parameres; endophallus including several large, flat, overlapping sclerites each with a conspicuous distal prong. Subgenital plate (Fig. 692) truncated apically, slightly depressed in middle.

Relationships. Badonnel (1979) noted that the species belongs to the hirsutus group of Thornton and Wong (1968). Most of the species are Pacific and Afro-Madagascan.

Distribution and habitat. Within the study area, the species occurs in Monroe and Indian River Counties, Florida, where it has been collected on mangrove and cabbage palm. It has also been taken twice as an introduction at Miami on orchid plants from the Bahamas and Suriname. Outside the study area, it is known from Suriname, Guyana, French Guiana, the Mexican states of Veracruz, Tabasco, and Campeche, and the islands of Guadelupe, Antigua, St. Kitts, and Puerto Rico.

Ectopsocus titschacki Jentsch

Ectopsocus titschacki Jentsch, 1939:120.
Ectopsocus gabelensis Badonnel, 1955:185.

Recognition features. Small: female body length ~ 1.48 mm from front of head to tips of closed wings. Wings clear except for cloudy brown mark along vein M3 of forewing. Body and appendages tawny yellow with reddish brown band along side of thorax in well colored specimens. Clunial comb field (Fig. 693) continuous basally with rest of clunium. Phallosome (Fig. 694) with external parameres broad in basal half, abruptly narrowing to slender, acuminate apical halves; aedeagal arch distally with pair of broad, truncated processes. Endophallus including long, strong, posteriorly directed processes. Distal processes of subgenital plate (Fig. 230) relatively short, tapering towards tip, ~ 1.5X as long as their basal width.

Relationships. Thornton and Wong (1968) placed the species near *E. amphithrix* Thornton and Wong, *E. argus* Thornton and Wong, and *E. innotatus* Thornton and Wong (Southeast Asia), but without information on males of any of these species.

Distribution and habitat. Within the study area, the species is known from an introduction at Mobile, Alabama from Panama on pineapple, and one at Miami from Suriname on orchid leaves. Outside the study area, it is known from the Mexican states of Veracruz, Tabasco, Campeche, and Chiapas, from Guatemala, Trinidad, Suriname, Guyana, French Guiana, Brazil (Belem), Venezuela, the islands of Cuba, Puerto Rico, Dominica, and Martinique, also from the Philippines, and West Africa.

Ectopsocus vachoni Badonnel

Ectopsocus vachoni Badonnel, 1945:44.
Ectopsocus dimorphus Mockford and Gurney, 1956:363.

Recognition features. Dimorphic: males entirely micropterous; females either micropterous or macropterous. Micropterous forms with ocelli vestigial (females) or absent (males). Wings clear, slightly tawny washed. Females (both forms) with head, thorax, and appendages medium reddish brown; preclunial abdominal segments pale reddish brown with faint, diffuse segmental rings of slightly darker reddish brown; terminal abdominal segments medium reddish brown. Males with head, thorax, and appendages pale reddish brown; preclunial abdominal segments reddish yellow, each with distinct brown ring, incomplete ventrally; terminal segments reddish yellow. Clunial comb field (Fig. 695) continuous anteriorly with rest of clunium. Phallosome (Fig. 696) with external parameres relatively elongate, blunt-tipped; endophallus of numerous small spines, one large and one smaller thorn-like sclerite. Distal processes of subgenital plate (Fig. 697) wide and short, ~ 1.5X as wide at base as long; their setae all apical or on median slope.

Relationships. Thornton and Wong (1968) suggested that the species may be close to *E. denervus* Thornton and Wong (Philippines, Micronesia, Samoa).

Distribution and habitat. Within the study area, the species is known from southern Georgia (Decatur County), central and western Texas (Bexar, Brown, and Jeff Davis Counties), and southern and central California (San Diego, Los Angeles, and Placer Counties). It usually is found in woodland ground litter (oak and madrone have been recorded), but may occur in dried cut grass and low foliage of trees. Outside the study area, it is known from the Mexican states of Nuevo León, San Luis Potosí, Hidalgo, Durango, and Oaxaca, also from Guatemala, Argentina, Chile, Morocco, Spain, France, and England.

Family Peripsocidae

Diagnosis. Forms of moderate size (length from head to tip of closed wings 2.0 - 4.0 mm). Body usually various shades of dusky brown. Wings occasionally clear, usually with diffuse bands, spots, or both. Adults usually macropterous, rarely brachypterous or micropterous. Lacinial tip as described for Family Ectopsocidae. Five distal inner labral sensilla. Vertex smooth or with few, minute setae. Fronto-clypeal suture absent (postclypeus extending, in effect, back to ocellar field). Ciliation usually absent, occasionally sparse to moderate on veins and margin of forewing, absent on hindwing. Vein Cu1a absent in forewing. Rs and M fused for a distance in hindwing. Adults with tarsi two segmented. Phallosome with 'external' (pore-bearing) parameres arising internal to 'internal' parameres (bases of aedeagal arch); aedeagal arch usually with an apical beak. Ovipositor valvulae complete; v3 broad, setose.

North American genera: *Kaestneriella* Roesler, *Peripsocus* Hagen.

Key to the North American Genera and Species of Family Peripsocidae

1. Forewing with at least a few setae visible at 70X or greater. Phallosome tapering only slightly from middle or widest region to base (Fig. 231); its distal process bent back on itself.

Ovipositor valvulae: v3 more than half length of v2 ....
.... Genus *Kaestneriella* 2

-- Forewing without visible setae. Phallosome usually tapering from broadest region to base (Fig. 232); its distal process straight, not bent back on itself. Ovipositor valvulae: v3 variable in length .... Genus *Peripsocus* 3

2. Female forewing length 1.99 - 2.52 mm. Median indentation of pigmented area of subgenital plate relatively deep (Fig. 233). Eyes of male small, IO/d ~ 1.48 ....
.... *K. tenebrosa* Mockford and Sullivan

-- Female forewing length 2.58 - 2.80 mm. Median indentation of pigmented area of subgenital plate shallow (Fig. 234). Eyes of male large, IO/d ~ 0.70 - 0.96 mm .... *K. fumosa* (Banks)

3. Forewing marked with distinct pattern: some colorless spots surrounded by pigmented membrane or colorless transverse bands in basal half of wing .... Group I 4

-- Forewing clear, or uniformly pale gray or brown, or, if spotted or banded, pattern consisting of darker marks surrounded or bordered by paler membrane. Colorless spots, if present, confined to cells of distal wing margin .... Group II 13

4. Forewing with colorless transverse bands in basal half but no colorless spots. Male clunial comb of one or more rows of denticles. Distal process of subgenital plate indented in middle
.... Subgroup A 5

-- Forewing with colorless spots, i. e., colorless regions surrounded by darker pigment. Male clunial 'comb' consisting of a single median process. Distal process of subgenital plate rounded at apex .... Subgroup B 8

5. Cell R5 of forewing with at least a lightly indicated pigment spot in its basal bulge. Adults entirely macropterous. Pigment spot of ocellar field not exceeding the field 6

-- Cell R5 of forewing without a pigment spot in its basal bulge. Most adults macropterous, some females brachypterous or micropterous. Pigment spot of ocellar field exceeding the field ....
.... *P. madidus* (Hagen)

6. Forewing with two transverse pigment bands -- one in middle and one at distal two-thirds plus a marginal pigment band in cells R3, R5, and M1, extending into bordering cell on each end (Fig. 235). Pigmented arms of subgenital plate broadly joined in middle (Fig. 236) .... *P. subfasciatus* (Rambur)

--   Forewing with at most an obscure transverse pigment band in middle of wing or pigmentation somewhat intensified along veins across middle of wing. Pigmented arms of subgenital plate joined, but relatively slender in middle 7

7.   Known only from females in the study area. Forewing with lightly indicated submarginal pigment band occupying cells M1 and M2, and extending into M3 (Fig. 237). A slender pigment spur extending forward from anterior angle of both pigmented arms of subgenital plate (Fig. 238) .... *P. paulianí* Badonnel

--   Known from both sexes in the study area. Forewing without a submarginal pigment band, but veins M2 and M3 bordered with nebulous pigment. Pigmented arms of subgenital plate without a forward-extending spur from their anterior angle . ... *P. reductus* Badonnel

8.   No Dark radial arcs of spots in cells R1, R3, R5, M1, and M2 of forewing but only continuous dusky tone except for two radial series of colorless spots in these cells (Fig. 239). Inner sclerite of subgenital plate absent .... *P. potosi* Mockford

--   Two radial series of dark spots present, forming inner and outer dark radial arcs in cells indicated above (Fig. 714). One or two inner sclerites of subgenital plate present 9

9.   Outer colorless radial series of spots in forewing all obscure; spots posterior to cell R3 frequently absent (fig. 717). Egg guide shorter than its basal width; two internal sclerites of subgenital plate (Fig 240). Arms of fork sclerite barely touching on midline (Fig. 241a) .... *P. minimus* Mockford

--   At least two anterior spots of outer colorless radial series distinct (i.e., those of cells R1 and R3), remainder usually present. Subgenital plate: egg guide either longer than its basal width and one or two internal sclerites of the plate, or egg guide shorter than its basal width and only one internal sclerite of the plate. Arms of fork sclerite, where known, capable of overlapping or lying side by side on midline 10

10.   Two internal sclerites of subgenital plate (Fig. 242). Arms of fork sclerite straight .... *P. alboguttatus* (Dalman)

--   One internal sclerite of subgenital plate, sometimes tending to be bilobed. Arms of fork sclerite curved (Fig. 243a) 11

11.   Forewing (Fig. 239) in general relatively pale; dark spots of radial series obscure and diffuse; those of transverse series reduced, with only one large dark spot (no. 4 of second row). Egg guide

shorter than its basal width; internal sclerite of subgenital
plate pentagonal (Fig. 245) .... *P. alachuae* Mockford
--   Forewing in general relatively dark; dark spots of most series
relatively well developed. Egg guide longer than its basal width;
internal sclerite of subgenital plate round or transversely
lengthened 12

12.   Known only from female. Internal sclerite of subgenital plate
transversely lengthened (Fig. 246). Little if any pigmentation
extending forward from region of junction of pigmented arms
of subgenital plate .... *P. maculosus* Mockford
--   Known from both sexes. Internal sclerite of subgenital plate
rounded. A tongue of pigment extending forward from region
of junction of pigmented arms of subgenital plate (Fig. 247)
.... *P. madescens* (Walsh)

13.   Forewings uniformly pale brown. Index ve/vd ~ 0.33 ....
.... Subgroup B, *P. phaeopterus* (Stephens)
--   Forewings at least with dusky areas in cells M1 and M2. Index
ve/vd ~ 0.65 .... Subgroup C, *P. stagnivagus* Chapman

Genus *Kaestneriella* Roesler

Kaestneriella Roesler, 1943:10.

Diagnosis. Forewing with at least a few setae visible at 70X or
greater; some forms with moderately setose forewings, the setae visible
at lower magnification. Cell R5 in male forewing relatively short and
broad, R5 index (Fig. 698, see also Mockford and Wong, 1969) range
1.11 - 1.38. Forewing largely gray or gray-brown with (at least) three
clear regions: one above, one below, one just distad of Rs - M junction.
Clunial comb (Fig. 699) broad, consisting of one or more rows of short,
irregular teeth, set on a quadrate posterior extension of clunial margin.
Phallosome (Fig. 231) tapering only slightly, if at all, from middle or
widest region of aedeagal arch to base; distal beak of phallosome with
tip bent at 30° to basal part. Third ovipositor valvula more than half
length of second.

Generotype. *K. pilosa* Roesler.

North American species: *K. fumosa* (Banks), *K. tenebrosa* Mockford
and Sullivan (1990).

Kaestneriella fumosa (Banks)

Peripsocus fumosus Banks, 1903a:237.
Kaestneriella fumosa (Banks). Mockford and Wong, 1969:245.

Recognition features. Marginal setae of forewing visible only above 70X. Forewing (Fig. 698) grayish brown except clear spots in middle (indicated in generic diagnosis) and clear spot at distal end of vein Cu1. Phallosome (Fig. 231) parallel-sided, its distal beak slender from base to tip. Major endophallic sclerites (Fig. 700) a pair of processes, each bearing a basal and distal prong, the two processes joined together anteriorly by a broad ribbon-like sclerite. Median indentation of female subgenital plate (Fig. 234) shallow, the arms broad basally.

Relationships. This is probably a sister species with *K. tenebrosa.*

Distribution and habitat. The species occurs in southeastern Arizona and south central New Mexico. The type locality (southwestern Colorado) is probably in error (Mockford and Sullivan, 1990). It has been found on a considerable variety of woody plants, both conifer and broad-leaf.

Kaestneriella tenebrosa Mockford and Sullivan

Kaestneriella tenebrosa Mockford and Sullivan, 1990:287.

Recognition features. Forewing characters as described for *K. fumosa.* Phallosome (Fig. 701) parallel-sided, its distal beak broad basally, tapering towards tip. Ribbon-like sclerite of endophallus long and slender, its greatest width not exceeding that of base of lateral process (Fig. 702). Median indentation of female subgenital plate (Fig. 233) deep, the arms relatively slender basally.

Relationships. As for *K. fumosa.*

Distribution and habitat. The species occurs from extreme southeastern Arizona north through west-central New Mexico into southern Colorado. It has been collected primarily on *Pinus ponderosa*, but may also be found on *Juniperus, Picea, Pseudotsuga,* and *Arctostaphylos.*

Genus *Peripsocus* Hagen

Peripsocus Hagen, 1866:203.

Diagnosis. Forewing without setae. Cell R5 in male forewing relatively long and slender, R5 index range 1.38-1.70. Forewing usually with cloudy brown bands, spots, or both, occasionally clear.

Clunial comb generally slender, consisting of single row (occasionally more than one) of regular teeth on a slender extension of clunial margin. Phallosome usually tapering from broadest region near distal end to base. Distal beak of phallosome straight, not bent back on itself. Third ovipositor valvula variable in length.

Generotype. *P. phaeopterus* Stephens.

Mockford (in press) proposed a classification of three groups and five subgroups for the Western Hemisphere species of this genus, which is adopted here.

Note. The index ve/vd (Badonnel, 1986c) is the length of v3 from its common base with v2 divided by the length of v2.

Group I

Diagnosis. Forewings marked with distinct pattern, including either some colorless spots surrounded by pigmented membrane or colorless bands across wing in basal half from margin to margin.

Subgroup A

Diagnosis. Forewing with colorless bands in basal half but no colorless spots (i.e., all colorless regions extensive and reaching wing margin). Male clunial comb of one or more rows of denticles. Distal process (egg guide) of female subgenital plate indented in middle.

North American species: *P. madidus* (Hagen), *P. pauliani* Badonnel, *P. reductus* Badonnel, *P. subfasciatus* (Rambur).

Peripsocus madidus (Hagen)

Psocus madidus Hagen, 1861:12.
Psocus permadidus Walsh, 1863:185.
Caecilius permadidus (Walsh). Hagen, 1866:206.
Peripsocus madidus (Hagen). Hagen, 1866:210.

Recognition features. Males and most females macropterous, some females brachypterous with wings extending to half length of abdomen or less at rest. All forewing markings obscure (Fig. 703); no pigment spot in base of cell R5. Head dull ivory with large dark brown spot including ocellar field; well colored specimens with pair of brown spots on vertex; reddish brown rings of preclunial abdominal segments broken on each side of dorsal midline. Phallosome (Fig. 122) slender, widest area (in aedeagal arch) less than twice basal width of phallosome; outer tines of endophallic fork straight or slightly curved, long, nearly half length of aedeagal arch (Fig. 122). Pigmented arms of subgenital plate (Fig. 704) almost completely separated in middle. Index ve/vd ~ 0.50.

Relationships. The species is probably close to the European species *P. didymus* Roesler and *P. phaeopterus* (Stephens), although the wing markings place it in a separate group.

Distribution and habitat. The species is found from Nova Scotia (Lockeport) west to Minnesota, south to Sarasota County, Florida and Bexar County, Texas. It inhabits bark of trunks and branches of many kinds of broad-leaf trees, and also may be found on juniper, spruce, hemlock, pine, and arbor vitae.

Peripsocus paulianl Badonnel

Peripsocus paulianl Badonnel, 1949a:42.

Recognition features. Known only from females (males reported, but not described, from Jamaica by Turner, 1975). Macropterous. All forewing markings obscure (Fig. 237): a lightly indicated pigment spot in basal bulge of cell R5; a pigment band bordering veins across middle of wing and another through cells M1 and M2, entering M3. Head pale brown except medium brown on dotted areas of vertex and postclypeal striations; ocellar field black in middle. Thorax medium brown except paler around edges of notal lobes. Antennae, legs, and preclunial abdominal segments pale brown; terminal abdominal segments slightly darker. Pigmented arms of subgenital plate (Fig. 238) joined in middle; a slender pigment spur extending forward from anterior angle of each arm. Index ve/vd ~ 0.70.

Relationships. These remain unknown.

Distribution and habitat. Within the study area, the species is known only from Monroe and Broward Counties, Florida, where it has been collected on foliage of shrubs, including mangrove, and on cabbage palm. Outside the study area, it is known from southwestern Mexico, the Guianas, northern Brazil, the islands of Cuba, Jamaica, Puerto Rico, and Antigua, West and Central Africa, and Sri Lanka.

Peripsocus reductus Badonnel

Peripsocus reductus Badonnel, 1943:98.

Recognition features. Macropterous. Forewing markings (Fig. 705) a nebulous crossband in middle of wing, a nebulous spot in base of cell R5, a nebulous band along each of veins M2 and M3. Head with dotted areas of vertex and postclypeal striations medium brown, ocellar field dark brown, remainder creamy yellow. Thorax dark brown;

antennae and legs pale brown. Preclunial abdominal segments dull white, each with a broad purplish brown ring, incomplete ventrally; well sclerotized regions of terminal segments dark brown, remainder paler. Phallosome (Fig. 232) slender, but widest area, in aedeagal arch, ~ 2.5X width of base. Outer tines of endophallic fork straight, less than one-fourth length of aedeagal arch. Terminal beak of phallosome long and slender. Clunial comb (Fig. 706) on decidedly protruding region, with ~ eight spaced out teeth. Pigmented arms of subgenital plate (Fig. 707) slender, angled at distal ends, their median junction relatively wide. Median lobe of subgenital plate ~ 1.5X as long as its basal width. Index ve/vd ~ 0.25.

Relationships. These remain uncertain.

Distribution and habitat. Within the study area, the species is known from the San Francisco Bay region of California, where it was probably introduced. It was taken on dead branches of oaks and other broad-leaf trees, and on shaded rocks in damp woodlands. Outside the study area, it is known from France, England, Chile, and eastern Australia. Males are known only from Chile and California.

Peripsocus subfasciatus (Rambur)

Psocus subfasciatus Rambur, 1842:322.
Peripsocus alboguttatus (Dalman). McLachlan, 1867:373 (part).
Psocus quadrifasciatus Harris, 1869:331, new synonym.
Peripsocus subpupillatus McLachlan, 1883:183.
Peripsocus subpupillatus ab. *quadriramosus* Enderlein, 1901:541.
Peripsocus quadrifasciatus (Harris). Chapman, 1930:372, new synonym.

Note: Parthenogenetic females from eastern North America and females in general from the Pacific Coast are identical with (parthenogenetic) European females in the characters studied. Sexual females from eastern North America are distinct. Males from the Pacific Coast differ consistently from eastern males in having smaller eyes. I regard all material of both sexes from the Pacific Coast and the eastern parthenogenetics as representing *P. subfasciatus*. I have looked at the types of *P. quadrifasciatus*, but owing to their condition, the necessary features, beyond confirming that they are females, cannot be seen. The type locality, around Cambridge, Massachusetts, is well within a large area of parthenogenetics, hence the synonymy. The sexual form in eastern North America is here regarded as a distinct species, at present un-named.

Recognition features. Macropterous. Forewing markings (Fig. 235) generally distinct, always with a pigment spot in basal bulge of cell R5, a pigment band across middle of wing, another across at distal two-thirds, and a marginal band in cells R3, R5, and M1, extending into R1 and M2. Head in general pale brown with a dark brown pigment spot covering ocellar field. Thorax and appendages medium brown except notal lobes dark brown. Preclunial abdominal segments pale brown, each with a vertical purplish brown stripe on each side; terminal segments medium brown. Phallosome (Fig. 708) short, broad, but tapering somewhat to base from broadest area in aedeagal arch (widest point ~ 2.1-2.7X basal width); external parameres weakly sclerotized on outer surfaces; outer tines of endophallic fork slightly over half length from base of (short) external parameres to distal end of beak of aedeagal arch; the tines blunt apically. Clunial comb (Fig. 709) of short, crowded teeth. Pigmented arms of subgenital plate (Fig. 236) broadly joined in middle. Index ve/vd ~ 0.40-0.42; v3 triangular.

Relationships. This species is part of a small complex including an un-named species in eastern United States, an un-named Mexican species (cited by Badonnel, 1986c, as *P. quadrifasciatus*), an un-named Central American species, and *P. nigrescens* Williner (Bolivia).

Distribution and habitat. Within the study area, the species occurs throughout eastern United States and southeastern Canada (southern Ontario, southern Quebec) south to the Gulf Coast in Louisiana, west to Voyageurs National Park, Minnesota, and western Arkansas. It inhabits branches of a great variety of trees, both broad-leaf and needle-leaf, and also occurs on shaded stone outcrops. On the Pacific Coast it occurs from San Francisco Bay north to southern British Columbia. Males are locally common in Pacific and Snohomish Counties, Washington, and in Vancouver, British Columbia. Elsewhere near the coast they are scarce and further inland virtually absent. Outside the study area, the species occurs throughout Europe.

Subgroup B

Diagnosis. Forewing with colorless spots, i. e. colorless regions enclosed by darker pigment. Male Clunial comb consisting of a single median process (Fig. 711). Distal lobe of subgenital plate rounded at apex (Fig. 240).

North American species. *P. alboguttatus* (Dalman), *P. alachuae* Mockford, *P. maculosus* Mockford, *P. madescens* (Walsh), *P. minimus* Mockford, *P. potosi* Mockford.

Note. The North American species of subgroup B are not distinguishable one from another on body color other than wing markings. In general they conform to the following description. Head medium brown except dull yellowish white on genae and anterior one-third of vertex, ocellar field dark brown to black. Thorax medium brown, somewhat darker on notal lobes. Legs and antennae pale to medium brown. Preclunial abdominal segments yellowish white ringed with broken bands of purplish brown. Terminal abdominal segments medium brown.

Peripsocus alachuae Mockford

Peripsocus alachuae Mockford, 1971b:92.

Recognition features. Known from both sexes. Forewing marked with complex pattern of dark and colorless spots (Fig. 244); colorless spots in basal half of wing tending to run together to form continuous bands: two transverse series of these separated partially by transverse series of dark spots. Distal half of forewing with two radial series of colorless spots separated by slightly indicated dark spots in cells R1, R3, R5, M1, M2, M3. Phallosome (Fig. 710) parallel-sided; its beak short, ~ 1.5X as long as its basal width; endophallic fork with outer tines curved. Male clunial process with moderate lateral flanges (Fig. 711). Subgenital plate broad in middle; inner sclerite pentagonal, lacking field of granules (Fig. 245). Index ve/vd ~ 0.45.

Relationships. Within the P. alboguttatus group (Mockford, 1971b) the species is probably closest to P. madescens (Walsh).

Distribution and habitat. The species is found throughout most of Florida from Santa Rosa County south to Highlands County. It has been collected in leaf litter, on *Andropogon* grass, and in oak foliage.

Peripsocus alboguttatus (Dalman)

Psocus alboguttatus Dalman, 1823:98.
Psocus quadrimaculatus Stephens, 1836:124.
Psocus pupillatus Walker, 1853:493.
Peripsocus pupillatus (Walker). Hagen, 1866:210.
Peripsocus alboguttatus (Dalman). Hagen, 1866:210.

Recognition features. Forewing as described for P. alachuae except outer colorless spots of cells R1 and M2 not reaching wing margin. Phallosome (Fig. 712) parallel sided. Beak of phallosome elongate, ~ 2.5X its basal width. Outer tines of endophallic fork straight. Male clunial process (Fig. 713) with moderate lateral flanges. Subgenital plate (Fig. 242) with pigmented arms separated in middle; pair of rounded inner sclerites. Index ve/vd ~ 0.54.

Relationships. The species seems to be intermediate between *P. minimus* and the complex *P. maculosus* - *P. madescens* - *P. alachuae* in several characters.

Distribution and habitat. Within the study area, this species has a divided range. Sexual forms occur over much of British Columbia and south near the coast in Washington. Parthenogenetic forms occur in eastern Unites States from Maine south in the mountains to Sevier County, Tennessee and west to western New York. They inhabit primarily foliage of conifers (hemlock, spruce, larch), but in the Pacific Northwest, they have also been found on maple, alder, and Scotch broom. Outside the study area, the species occurs throughout Europe and east to Mongolia.

Peripsocus maculosus Mockford

Peripsocus maculosus Mockford, 1971b:100.

Recognition features. Known from females only. Forewing marked much as in *P. alachuae*, but with distinct basal transverse row of dark spots (Fig. 714, d.t.r.#1); spot 1 of third transverse dark row (Fig. 772, spot 1, d.t.r.#3) elongate, continuous from behind Rs - M junction to vein M3. Subgenital plate (Fig. 246) with pigment arms broadly joined in middle; sometimes a small spot of more intense pigment in space before junction region of arms. Inner sclerite of subgenital plate single, broad, with some indication of doubling; occasionally double, the two in contact along midline. Index ve/vd ~ 0.50.

Relationships. The species is part of a complex including *P. madescens* (eastern United States, coastal Veracruz), and *P. alachuae* (Florida).

Distribution and habitat. The species occurs spottily throughout eastern North America from southern Quebec (Montreal region) and central Ontario (St. Joseph's Island) south to north-central Florida, west to northern Minnesota and southeastern Arkansas. It inhabits primarily foliage and branches of conifers, including larch, pine, juniper, and spruce, but also oak, *Vaccinium*, and sumac.

Peripsocus madescens (Walsh)

Psocus madescens Walsh, 1863:186.
Peripsocus madescens (Walsh). Hagen, 1866:210.

Recognition features. Known from both sexes. Forewings marked in general as described for *P. alachuae*, but with distinct basal transverse row of dark spots; spot number 2 of second pale transverse row curved (Fig. 715, spot 2, p.t.r.#2); spot 1 of third transverse dark row pale or non-existent.

Phallosome (Fig. 243) broadened and rounded in posterior half, its beak somewhat swollen beyond base; outer tines of endophallic fork curved. Male clunial process lacking lateral flanges (Fig. 716). Subgenital plate broadened in middle; inner sclerite single, rounded, with posterior transverse field of granules (Fig. 247). Index ve/vd ~ 0.46.

Relationships. These are as noted for *P. maculosus*.

Distribution and habitat. Within the study area the species occurs in southern Ontario (Marmora), New Brunswick (Kouchibouguac Provincial Park) and throughout eastern United States from Maine (Mount Desert Island) south to Dade County, Florida and west to Illinois and western Arkansas. It inhabits primarily foliage and branches of pines, but also occurs on broad-leaf trees and shrubs. Outside the study area, it is known from the Mexican states of Puebla and Veracruz.

Peripsocus minimus Mockford

Peripsocus minimus Mockford, 1971b:105.

Recognition features. Known from both sexes. Forewing not exceeding 2.2 mm in length, mostly dark (Fig. 717), the colorless spots greatly reduced and outer radial series represented only by minute spots in cells R1 and R3. Phallosome (Fig. 241) approximately parallel-sided but tapering slightly from base distad; beak of phallosome elongate, ~ 3X as long as its basal width; outer tines of endophallic fork straight. Male clunial process (Fig. 718) with broad lateral flanges. Subgenital plate (Fig. 240) with pigmented arms joined medially and a pigmented tongue extending forward from region of junction; internal sclerite two separate rounded areas. Index ve/vd ~ 0.55.

Relationships. The species appears to stand rather close to *P. alboguttatus*.

Distribution and habitat. The species occurs from southern Illinois (Hamilton County) southwestward through southeastern Missouri and Arkansas to Lee County, Texas. It has been found only on the foliage of *Juniperus virginiana*.

Peripsocus potosi Mockford

Peripsocus potosi Mockford, 1971b:110.

Recognition features. Known from females only over most of its range. Forewing (Fig. 24) marked in general as described for *P. alachuae* but second transverse dark row paler, dark radial arcs replaced by continuous dusky tone over most of distal one-third of wing except for spots of inner and outer pale radial arcs. Subgenital plate

(Fig. 719) with pigment interrupted at base of each arm; a tongue of light pigmentation extending forward in middle; inner sclerite absent; distal lobe parallel-sided, about as long as its basal width. Index ve/vd ~ 0.28.

Relationships. This species appears to be close to *P. teutonicus* Mockford (southern Brazil).

Distribution and habitat. Within the study area this species has been collected in three Florida counties: Citrus, Highlands, and Lee. It has been found on branches and foliage of oaks, sand pine, and juniper, also on imported orchids (*Cymbidium* sp.). The earliest Florida record is in 1983, and the species was probably introduced shortly before that time. Outside the study area, it is known from central Mexico southward to northern South America (Colombia) and on the islands of Jamaica, Puerto Rico, Guadeloupe, and Trinidad. The species is undoubtedly thelytokous, and only a single male has been recorded (Badonnel, 1986b).

Group III

Diagnosis. Forewing clear, or uniformly pale gray or brown, or, if spotted or banded, pattern consisting of darker marks surrounded or bordered by paler membrane. Colorless spots, if present, confined to cells of distal wing margin.

Subgroup B

Diagnosis. Forewing gray or brown.
North American species. *P. phaeopterus* (Stephens).

Peripsocus phaeopterus (Stephens)

Psocus nigricornis Stephens, 1836:126.
Psocus phaeopterus Stephens, 1836: 127.
Psocus obscurus Rambur, 1842:322.
Peripsocus phaeopterus (Stephens). Hagen, 1866:210.

Recognition features. Wings (Fig. 720) pale brown. Head medium brown except ocellar field black, region surrounding ocellar field pale brown. Thorax, legs, and antennae medium brown. Preclunial abdominal segments dull white to pale reddish brown, lacking segmental rings. Terminal segments medium brown. Phallosome (Fig. 721) elongate, slender, tapering slightly from widest region in aedeagal arch to near base, then slightly widened at base; aedeagal arch acute-angled in distal half; apical beak ~ 1.5X as long as its basal width; outer tines of endophallic fork elongate, slender, acuminate-tipped, about half length of pore-bearing parameres. Clunial comb (Fig. 722) of ~ 11 denticles on median protrusion. Subgenital plate (Fig. 723) with lateral arms narrowly meeting in middle; distal lobe slightly longer than its basal width, indented at tip. Index ve/vd ~ 0.33.

Relationships. The species appears to be close to *P. didymus* Roesler (central Europe).

Distribution and habitat. Within the study area the species is known only from Vancouver, British Columbia, where it was presumably introduced. Specimens have been collected on foliage and branches of *Thuja plicata* and Scotch broom. Outside the study area, the species occurs throughout Europe.

Subgroup C

Diagnosis. Forewings with darker spots or bands on paler background.

North American species. *P. stagnivagus* Chapman.

Peripsocus stagnivagus Chapman

Peripsocus stagnivagus Chapman, 1930:376.

Recognition features. Known from both sexes but male extremely local (see distribution, below). Forewing (Fig. 724) usually with dusky areas in cells M1 and M2, sometimes extending as continuous band through these cells, into cells R5, R3 and M3; a dusky spot slightly to moderately developed in base of cell R5; sometimes a dusky spot slightly to moderately developed in middle of cell Cu1. Head, body, and appendages pale brown except ocellar field black, postclypeal striations, dotted areas of vertex, and thoracic notal lobes medium brown; terminal abdominal segments medium to dark brown. Phallosome (Fig. 725) decidedly tapering from widest region of aedeagal arch to slender base; with pair of appendages arising at base of aedeagal arch, curving antero-medially; apical beak slender, about twice as long as its basal width; outer tines of endophallic fork very long, slender, curved, acuminate-tipped, about same length as a pore-bearing paramere; median tine slender, pointed, about half length of lateral tine. Clunial comb (Fig. 726) slightly protruding, with ~ 12 teeth. Subgenital plate (Fig. 727) with lateral arms slender, separated in middle; distal lobe broader than long, with slender baso-lateral extensions, tip indented. Index ve/vd ~ 0.65.

Relationships. The species is very close to an undescribed species in central Mexico and a second undescribed species in southern Mexico. The three are readily separated on male characters, but females appear to be indistinguishable morphologically.

Distribution and habitat. The species occurs from Maryland to central Indiana, south to eastern Texas (Liberty County) and throughout Florida. It inhabits trunks and branches of trees, Spanish moss, and occasionally stone outcrops. Due to the current impossibility of distinguishing females of this species from females of its Mexican close relatives, female populations south of the study area cannot be determined except where associated with males. Males of this species are known from central and southeastern Indiana (Marion and Ripley Counties).

Family Archipsocidae

Diagnosis. Small to moderate-sized forms (length from head to tips of closed wings 1.0-3.0 mm) usually found in colonies under dense webbing; body shades of red-brown or orange -brown, sometimes paler. Forewings somewhat coriaceous with veins obscure; hindwings membranous with veins distinct. Adults usually polymorphic: males micropterous (rarely apterous), females macropterous, micropterous, or some intermediate stage. Lacinial tip slender, bicuspid; lateral cusp broader than median and sometimes denticulate. Distal inner labral sensilla: three placoids; apparently no tricholds on inner surface. Short setae abundant on head and row of long setae directed backward on hind margin of vertex. Setae also abundant on rest of body; a row of long setae directed backward across pronotum. Wings setose with relatively short setae on membrane, longer setae on margins; setae of median margin of forewing in two series forming crossing pairs (Fig. 728). Cell Cu 1a (areola postica, delimited by "cubital loop") in forewing relatively flat, not looped forward. Hindwing venation reduced (Fig. 729): one R, one M, and one Cu vein. Adults with tarsi two segmented. Phallosome with external parameres membranous or absent; endophallic sclerotizations usually minute or absent. Subgenital plate rounded or flattened distally. Ovipositor valvulae lacking v1; v2 slender; v3 broad with long marginal setae; all valvulae absent in viviparous species.

North American genera: *Archipsocopsis* Badonnel, *Archipsocus* Hagen.

Key to the North American Genera and Species of Family Archipsocidae

1.   Females viviparous, lacking ovipositor valvulae, or with these reduced to rudiments. Subgenital plate broad and flat posteriorly (Fig. 248). Phallosome with aedeagal arch constricted at about half its length to form a median arched area (Fig. 249) .... Genus *Archipsocopsis* 2

--   Females oviparous, with ovipositor consisting of a broad valvula
       and a slender one on each side (Fig. 250). Phallosome rounded
       or gradually tapering towards distal end (Fig. 251) ....
       .... Genus *Archipsocus* 3

2.   Females macropterous and brachypterous, not micropterous;
       males micropterous. First flagellomere longer than pedicel.
       Phallosome closed at base (Fig. 249) ......*A. frater* (Mockford)
--   Females macropterous, brachypterous, and micropterous; males
       apterous. First flagellomere shorter than pedicel. Phallosome
       open at base (Fig. 252) .... *A. parvula* Mockford

3.   Micropterous female forewing length ~ 0.17 mm. Female para-
       proctal zone C (Fig. 253) with at least 23 setae. Aedeagal arch
       index (Fig. 254) not over 2.30 .... *A. floridanus* Mockford
--   Micropterous female forewing length ~ 0.28 or somewhat great-
       er. Female paraproctal zone C with not over 24 setae. Aedeagal
       arch index at least 2.22, usually > 2.50 4

4.   Body in general medium reddish brown. At least eight marginal
       setae on v3. Male epiproct with at least 16 setae. Leaf-
       inhabiting (usually palm leaves) ....*A. gurneyi* Mockford
--   Body in general dark brown except preclunial abdominal seg-
       ments dark reddish brown. Not over ten marginal setae on v3.
       Male epiproct with not over 14 setae. Bark-inhabiting ....
       ....*A. nomas* Gurney

Genus *Archipsocopsis* Badonnel

Subgenus *Archipsocopsis* Badonnel, 1948:294.
Genus *Archipsocopsis* Badonnel, 1966:413.

   Diagnosis. Females viviparous, without gonapophyses or with
these reduced to rudiments of ninth segment ones. Subgenital plate
broad and flat posteriorly (Fig. 248). Phallosome with aedeagal arch
constricted at about half its length to form a median arched area (Fig.
249). Coeloconic sensilla of flagellomeres 6 and 10 with only a short
central cone (Fig. 730).
   Generotype. *A. mendax* Badonnel.
   North American species: *A. frater* (Mockford), *A. parvulus*
(Mockford).

Archipsocopsis frater (Mockford)

Archipsocus frater Mockford, 1957b:33.
Archipsocopsis frater (Mockford). Badonnel, 1966:413.

Recognition features. Females macropterous and brachypterous, not micropterous; males micropterous. All body regions orange-brown; ocellar field concolorous with surrounding cuticle. Wings washed with pale orange-brown. Macropterous females ~ 1.4-1.5 mm in body length; brachypterous females and males ~ 1.1-1.2 mm. First flagellomere longer than pedicel, coeloconic sensillum at mid-length. Phallosome closed basally; endophallus rugose (Fig. 249). Females lacking any rudiments of ovipositor valvulae.

Relationships. The species is probably rather close to *A. inornata* New (South America).

Distribution and habitat. Within the study area, the species occurs in peninsular Florida from Gainesville south to Everglades National Park. It inhabits primarily living and dead leaves, and has been found on several species of oak (water oak, laurel oak, turkey oak), *Citrus limon*, *Myrica* sp., *Phragmites* sp., and *Thalia* sp. Outside the study area it has been found in coastal Veracruz State, Mexico, Jamaica, and the Sao Paulo region of Brazil.

Archipsocopsis parvula (Mockford)

Archipsocus (*Archipsocopsis*) *parvulus* Mockford, 1953:123.
Archipsocopsis parvulus (Mockford). Badonnel, 1966:413.

Recognition features. Females macropterous, brachypterous, and micropterous; males apterous. Head medium brown; thorax, abdomen, and appendages paler, orange-brown. Female ~ 1.1 mm in body length, male ~ 0.90 mm. First flagellomere somewhat shorter than pedicel, with coeloconic sensillum before mid-length. Struts of phallosome abutting at base but not fused; endophallus denticulate (Fig. 252). Females lacking any rudiments of ovipositor valvulae.

Relationships. The species is probably very close to *A. inornata* New (South America).

Distribution and habitat. Within the study area, the species is known from Alachua, Indian River, and Sarasota Counties, Florida, and from Southmost Palm Grove, Cameron County, Texas. Its webs are found primarily on trunks of cabbage palms and laurel oaks. Outside the study area, it is known from El Salto, San Luis Potosi State, Mexico.

Genus *Archipsocus* Hagen

Archipsocus Hagen, 1882:225.

Diagnosis. Females oviparous; ovipositor valvulae (Fig. 256) as described for the family. Subgenital plate (Fig. 250) rounded or somewhat angulate posteriorly. Aedeagal arch of phallosome (Fig. 251) rounded or gradually tapering towards distal end. Coeloconic sensilla of flagellomeres 6 and 10 with a long central hyaline filament (Fig. 731).

Generotype. *A. puber* Hagen.

North American species: *A. floridanus* Mockford, *A. gurneyi* Mockford, *A. nomas* Gurney. There are several undescribed species, including one misdetermined as *A. panama* Gurney by Mockford (1953). The described species all belong to Group IIBb-a (Mockford, in press), characterized by smooth flagellar diagram, male tenth abdominal tergum without fields of heavy setae or spines, t1 < t2, P > f1.

Archipsocus floridanus Mockford

Archipsocus (*Archipsocus*) *floridanus* Mockford, 1953:116.

Recognition features. Females macropterous, brachypterous, and micropterous. Micropterous females with forewings ~ 0.17 mm in length; forewings of males slightly shorter. Head and body dark brown in micropterous individuals, paler brown in macropterous individuals. Wings pale brown-washed. In antenna: f2 < f3 < f4 ~~ f5. Aedeagal arch index (Mockford, 1977 and Fig. 254) = 1.87 - 2.40. Male epiproct with 13 to 17 setae. Anterior indentation of subgenital plate moderate, tapering to rounded end (Fig. 250). Ovipositor valvulae (Fig. 250): v2 round- or blunt-tipped, occasionally acute-tipped; v3 with 14 to 19 marginal setae; 9 to 12 non-marginal setae. Female paraproct zone C (Badonnel, 1976 and Fig. 253) with 23 to 29 setae.

Relationships. This and the next two species form a complex within the genus.

Distribution and habitat. Within the study area, the species is found throughout Florida, west to coastal Louisiana and south to southern Texas (Cameron County). It makes its webs on trunks and branches of hardwood trees and on trunks of cabbage palms. Outside the study area, the species is recorded from the Sao Paulo area, Brazil (Badonnel, 1978). Cuban material tentatively assigned to this species (Mockford, 1974a) is *A. nomas* (pers. obs.).

Archipsocus gurneyi Mockford

Archipsocus (*Archipsocus*) *gurneyi* Mockford, 1953:120.

Recognition features. Females macropterous, micropterous, and in several states of brachyptery. Micropterous females with forewing ~ 0.28 mm; males with forewing ~ 0.16 mm. Head and body medium reddish brown, somewhat darker on heavily sclerotized basal and terminal regions of abdomen; legs and antennae pale reddish brown. Wings clear. In antenna: f2 < f3 < f4 < F5. Aedeagal arch index = 2.59 - 3.75. Male epiproct with 16 to 20 setae. Anterior indentation of subgenital plate (Fig. 791) moderate, tending to be broader in middle than at anterior end. Ovipositor valvulae: v2 tip varying from rounded to acuminately pointed; v3 with 10 to 18 marginal setae, 8 to 15 lateral setae. Female paraproct zone C with 16 to 21 setae.

Relationships. These are as noted for *A. floridanus.*

Distribution and habitat. Within the study area, the species occurs in peninsular Florida from Alachua County south to Collier County. Its webs are found on living and persistent dead leaves of cabbage palms and occasionally other palms. Outside the study area it is known from southeastern San Luis Potosi State in Mexico, Jamaica, and Roraima State, Brazil.

Archipsocus nomas Gurney

Archipsocus nomas Gurney, 1939:502.

Recognition features. Females macropterous, brachypterous, and micropterous. Micropterous females with forewing ~ 0.29 mm; males with forewing ~ 0.16 mm. Head and body dark brown except preclunial abdominal segments medium to dark reddish brown; legs and antennae pale brown; wings clear. In antenna: f2 < f3 < f4 > f5. Aedeagal arch index = 2.22 - 3.85. Male epiproct with 10 to 14 setae. Subgenital plate indentation approximately as in A. floridanus. Ovipositor valvulae: v2 usually acuminate-pointed; v3 with 10 to 14 marginal setae, 5 to 10 lateral setae. Female paraproct zone C with 17 to 24 setae.

Relationships. These are as noted for *A. floridanus.*

Distribution and habitat. Within the study area, the species occurs throughout Florida, west around the Gulf Coast to New Orleans, Louisiana, and south to Cameron County, Texas. Its webs are frequent on trunks and branches of hardwood trees. Outside the study area, it is known from Cuba and Bermuda.

Family Pseudocaeciliidae

Diagnosis. Medium sized forms (length from head to tip of closed wings 2.0 - 3.0 mm); body brown or yellow. Often found single or in small groups under sparse webbing on leaves. Adults macropterous. Wings usually membranous; forewings occasionally somewhat coriaceous but with distinct venation. Lacinial tip slender, bicuspid; lateral cusp longer than median and frequently denticulate at tip. Five distal inner labral sensilla. Setae abundant on head, dorsal surface of thorax, forewing margin and veins, hindwing margin but sparse on hindwing veins. Setae of median margin of forewing and most of distal and hind margin of hindwing in two series forming crossing pairs (Fig. 257). Cell Cu1a present in forewing, its limiting vein free from M. Adults of most species with tarsi two segmented. Phallosome with external parameres elongate, generally longer than aedeagal arch. Ovipositor valvulae complete; v1's of some species closely associated with subgenital plate; v2 spiculate at apex or with pointed spiculate process in addition to rounded, bare lobe; v3 setose (Fig. 256).

North American genera: *Heterocaecilius* Lee and Thornton, *Pseudocaecilius* Enderlein.

Key to the North American Genera and Species of Family Pseudocaeciliidae

1.   Known from both sexes in the study area. Ovipositor valvulae: v1 and v2 both with rounded apical lobe and pointed spiculate process (Fig. 255) .... Genus *Heterocaecilius, H.* sp.

--   Known only from females in the study area. Ovipositor valvulae: v1 and v2 both terminating in a single structure, a relatively broad process tapering and spiculate at tip ....
.... Genus *Pseudocaecilius* 2

2.   Faint reddish brown band from eye to antennal base. No spots at marginal ends of Rs and M branches in forewing ....
....*P. citricola* (Ashmead)

--   Strong reddish brown band from eye to antennal base. A pigment spot at marginal end of each of Rs and M branch in forewing (Fig. 257) .... *P. tahitiensis* (Karny)

Genus *Heterocaecilius* Lee and Thornton

Heterocaecilius Lee and Thornton, 1967:13.

Diagnosis. Venation approximately as in *Pseudocaecilius*; Rs fork stem of forewing at least as long as R4+5. Vein ends usually without cloudy pigment. Male forewing usually without sense papillae.

Hypandrium with pair of postero-lateral sclerotized projections. External parameres longer than aedeagal arch; endophallus with conspicuous rod-like sclerites. Subgenital plate bilobed (Fig. 733), each lobe approximately triangular, bearing a seta at or near apex and another near base mesally. First and second valvulae each with rounded apical lobe and pointed spiculate process (Fig. 261).

Generotype. *H. minutus* Lee and Thornton by present designation.

North American species. Within the study area, the genus is represented by a single undescribed species collected at Great Sippewissett Marsh, Barnstable County, Massachusetts. It probably represents an introduction, as most species are from the Pacific Basin and southeastern Asia. The specimens were collected in a single season, and it is not known if the species is permanently established.

Genus *Pseudocaecilius* Enderlein

Hageniella Enderlein, 1903b:258 (name preoccupied).
Pseudocaecilius Enderlein, 1903b:260.

Diagnosis. Rs fork stem in forewing as long as or longer than R4+5; areola postica relatively low (Fig. 257). Ovipositor valvulae (Fig. 256): v1 and v2 with spinules near tip but not bilobed; v3 narrowing somewhat towards apex. Ventral eversible vesicle on abdominal segments 5/6. Subgenital plate (Fig. 734) bilobed distally, each lobe with single apical seta. Hypandrium with disto-lateral pair of processes. External parameres only slightly exceeding tip of aedeagal arch; endophallus with denticles or no sclerotization.

Generotype. *P. citricola* (Ashmead).

North American species: *P. citricola* (Ashmead), *P. tahitiensis* Karny.

Pseudocaecilius citricola (Ashmead)

Psocus citricola Ashmead, 1879:228.
Elipsocus criniger Perkins, 1899:85.
Pseudocaecilius elutus Enderlein, 1903b:261.
Kilauella criniger (Perkins). Enderlein, 1913:357.
Caecilius pretiosus Banks, 1920:311.
Pseudocaecilius wolcotti Banks, 1924:423.
Pseudocaecilius pretiosus (Banks). Chapman, 1930:332.
Trichopsocus indicatus Navás, 1934:45.
Pseudocaecilius citricola (Ashmead). Mockford and Gurney, 1956:364.
Pseudocaecilius criniger (Perkins). Thornton et al., 1972:112.

Recognition features. Known only from female. Body and appendages yellow except dark reddish brown band entire length of thorax above leg bases, continuing forward on neck, faintly on head behind and before eye to antennal base, also continuing backward onto first abdominal segment. Lateral mesonotal lobes pale brown. Wings (Fig. 117) clear except faint brown spot through distal one-third of pterostigma and another through middle of areola postica; veins colorless in basal half of forewing except Rs and M brown before their junction; veins in distal half of forewing brown.

Relationships. The species is probably close to the following one.

Distribution and habitat. Within the study area, the species occurs throughout Florida, in the San Antonio region of Texas, and was found in a greenhouse in Washington, D.C. It inhabits living leaves of citrus, evergreen oaks, palms, and other trees. Outside the study area it is virtually pan-tropical. In the American Tropics it is recorded south to Colombia and French Guiana, also in the Galápagos Islands.

Pseudocaecilius tahitiensis (Karny)

Epipsocus tahitiensis Karny, 1926:288.
Pseudocaecilius tahitiensis (Karny). Lee and Thornton, 1967:79.

Recognition features. Known only from female. Similar to preceding species, differing in following characters: reddish brown band as strong on head before and behind eye as on thorax; lateral mesonotal lobes concolorous with surrounding cuticle; forewing (Fig. 257) with brown spots at distal ends of Rs and M branches; spot through areola postica at acute angle to wing axis rather than perpendicular.

Relationships. These are as noted for *P. citricola.*

Distribution and habitat. Within the study area the species is known from two adult specimens, one collected at Miami, Florida and the other on Long Key, Monroe County, Florida. Outside the study area, it is known from the Mexican states of Veracruz, Tabasco, Chiapas, Campeche, and Yucatán, also from Guatemala, Guyana, Suriname, French Guiana, northeastern Brazil, Puerto Rico, Galápagos, Tahiti, and the Southern Mariana Islands. It is found on foliage of various trees and shrubs.

Family Trichopsocidae

Diagnosis. Medium sized forms (length from head to tips of closed wings ~ 2.5 mm); body largely yellow with reddish brown band along each side of thorax. Found on living and dead leaves. Adults macropterous. Wings membranous with distinct venation. Lacinial tip as described for the Pseudocaecilidae. Five distal inner labral sensilla. Setae abundant on vertex, frons, postclypeus, and antennae, sparse on rest of body except abundant on wing margin, moderate on wing veins. Distal and hind margins of wings without crossing pairs of setae. Cell Cu1a present in forewing, relatively short and low (Fig. 735). Adults with two tarsomeres. Adults with a ventral eversible vesicle on abdominal segments 6/7; nymphs with two vesicles: one on abd. 6/7, one on 7/8. Phallosome (Fig. 742) with external parameres somewhat longer than aedeagal arch and usually closely adhering to arch. Subgenital plate (Fig. 737) bilobed posteriorly. Ovipositor valvulae (Fig. 738) complete; v1 terminating in slender, acuminate process denticulate on median surface; v2 terminating in rounded lobe and slender acuminate process similar to that of v1; v3 rounded, setose.

North American genus: *Trichopsocus* Kolbe.

Genus *Trichopsocus* Kolbe

Trichopsocus Kolbe, 1882a:25.

Diagnosis. As for the family.
Generotype. *T. dalii* McLachlan.
North American species: *T. acuminatus* Badonnel, *T. dalii* McLachlan.

Trichopsocus acuminatus Badonnel

Trichopsocus acuminatus Badonnel, 1943:89.
Caecilius clarus Banks, 1908:258, new synonym.
Pseudocaecilius (?) *clarus* (Banks). Chapman, 1930:334, new synonym.

Recognition features. Areola postica relatively long and low (Fig. 735). Apex of vein Cu1 in hindwing bordered with brown only on proximal edge (Fig. 739, scarcely visible in male). Phallosome with aedeagus absent or membranous (Fig. 736). Endophallus lacking central core. V2 with long, slender pointed process, about as long as an adjacent marginal seta of v3 (Fig. 738). Spermathecal duct slender throughout with almost imperceptible proximal sheath (Fig. 740).

Relationships. Most of the species of *Trichopsocus* are Atlanto-European and seem to form a compact group.

Distribution and habitat. Within the study area, the species is established in coastal California, where it has been collected in Alameda, Contra Costa, Humboldt, Marin, and San Diego Counties. It inhabits foliage of ferns, rhododendrons, *Quercus agrifolia*, jojoba, and persistent dead leaves. It has also been taken three times at Boston as an introduction from Europe and the Azores. Outside the study area, it is widely distributed in southern Europe and is also known from the Canary Islands, Madeira, Azores, and coastal Peru.

Trichopsocus dalii (McLachlan)

Caecilius dalii Mclachlan, 1867:272.
Caecilius hirtellus McLachlan, 1877:54.
Trichopsocus hirtellus var. *angulata* Navás, 1916:599.
Trichopsocus hirtellus (McLachlan). Badonnel, 1938:18.
Trichopsocus dalii (McLachlan). Badonnel, 1943:87.

Recognition features. Areola postica approximately semi-circular. Apex of vein Cu1 in hindwing bordered with brown on both edges (Fig. 741). Phallosome with sclerotized aedeagus (Fig. 742); endophallus with conspicuous central core decidedly broadened anteriorly. V2 with pointed process much shorter than adjacent marginal setae of v3 (Fig. 743). Spermathecal duct (Fig. 744) wide at departure from sac, with a conspicuous dark sheath around proximal end.

Relationships. As noted for *T. acuminatus*.

Distribution and habitat. Within the study area this species has been taken as an introduction at Miami (from Lima, Peru), New York City (from Italy), and Laredo, Texas (from Mexico). It was associated with bay leaves and citrus leaves. Outside the study area, the species is known from Atlantic and Mediterranean coastal regions of Europe and southern England, and from North Africa.

Family Philotarsidae

Diagnosis. Medium sized forms (length from head to tip of closed wings 2.0 - 3.5 mm); body shades of brown and gray. Inhabiting bark, either with little or no webbing, or single or in small groups under dense webbing. Adults usually macropterous, occasionally brachypterous, rarely micropterous. Lacinial tip slender, bicuspid; lateral cusp wider than median and denticulate on free edge. Nine distal inner labral sensilla. Setae moderate on vertex, frons, and mesonotum; sparse on rest of head and body; abundant on wing margins, moderate on forewing veins; moderate or absent on hindwing veins. Setae of most of median margin of forewing and part of that of hindwing in two series forming crossing pairs. Cell Cu1a present in forewing, its limiting vein free from M. Adults of most species with tarsi three segmented. Phallosome (Fig. 120) with external parameres relatively short, only slightly exceeding tip of aedeagal arch. Ovipositor valvulae complete (Fig. 751); v2 with short, blunt or rounded spiculate process; v3 setose.

North American genera: *Aaroniella* Mockford, *Philotarsus* Kolbe.

Key to the North American Genera and Species of Family Philotarsidae

1. In antenna, f11 partially or completely fused to f10, terminating in long, slender process bearing a seta at tip (Fig. 258). Phallosome not or scarcely widened at base. Sense cushions of male paraprocts round. Pigmented arms of subgenital plate not over 2.5X as long as their basal width .... Genus *Aaroniella* 2

-- In antenna, f11 completely separate from f10 and not terminating in a long, slender process (Fig. 259). Phallosome with a broad, triangular basal area (Fig. 120). Sense cushions of male paraprocts elongate (Fig. 260). Pigmented arms of subgenital plate at least 3.0X as long as their basal width .... Genus *Philotarsus* 4

2. Forewing length < 2.30 mm. Distal piece of subgenital plate bearing several setae .... *A. achrysa* (Banks)

-- Forewing length > 2.40 mm. Distal piece of subgenital plate without setae 3

3. Known from females only. Thoracic pleura marked with slender, pale longitudinal band between two dark bands (Fig. 261) ... . *A. badonneli* (Danks)

-- Known from both sexes. Thoracic pleura marked with single broad, dark band above leg bases (Fig. 262) .... *A. maculosa* (Aaron)

4. Known in study area only from females. Forewing markings distinct (Fig. 263) and including radial series of spots in cells R3, R5, M1, M2, and M3 .... *P. picicornis* (Fabricius)

-- Known from both sexes in study area. Forewing markings faint, never with radial series of spots (Fig. 757) .... *P. kwakiutl* Mockford

Genus *Aaroniella* Mockford

Aaroniella Mockford, 1951:102.

Diagnosis. Nymphs and adults found on tree trunks and branches, single or in small groups under dense webbing. Distal flagellomere (Fig. 258) elongate via a slender distal process bearing a terminal seta; the flagellomere partially or completely fused with f10. Veins of distal half of hindwing moderately setose. Phallosome (Fig. 745) with sclerotized strip of sides continuing around to form base, with little or no basal widening; external parameres not over one-third length of entire phallosome; endophallic sclerites usually conspicuous but not closely joined to distal end of aedeagal arch. Sense cushions of male paraproct rounded (Fig. 745). Male epiproct wider than long (Fig. 747). Subgenital plate (Fig. 748): distal piece usually no longer than wide; pigmented arms broad, usually not over 2.5X as long as their basal width. Ovipositor valvulae: v2 with short spiculate process or none; v3 triangular or rounded.

Generotype. *A. maculosa* (Aaron).

North American species: *A. achrysa* (Banks), *A. badonneli* (Danks), *A. maculosa* (Aaron).

Note: all of the North American species have the forewing marked with a radial series of spots: one spot in each of cells R1, R3, R5, M1, M2, and M3.

Aaroniella achrysa (Banks)

Graphocaecilius achrysus Banks, 1941:391.
Aaroniella achrysa (Banks). Mockford, 1974a:133.

Recognition features. Known only from females. Forewing length ~ 2.25 mm. Forewing markings (Fig. 749). Distal piece of subgenital plate bearing several short setae; pigmented arms ~ 2.2X as long as their greatest width (Fig. 750). V2 with short spiculate process; v3 rounded distally (Fig. 751).

Relationships. This species appears to be part of a complex of small species found on the southern Gulf Coast, the Caribbean islands, and Florida.

Distribution and habitat. Within the study area, the species occurs throughout Florida. It inhabits trunks and branches of oaks, cabbage palms, strangler figs, birches, and probably other trees, and occasionally dead leaves of cabbage palms. Outside the study area, it is known from the Dominican Republic, Cuba, and Puerto Rico.

Aaroniella badonneli (Danks)

Philotarsus badonneli Danks, 1950:1.
Aaroniella badonneli (Danks). Mockford, 1951:104.
Aaroniella eertmoedi Mockford, 1979a:38.

Note. Lienhard (1990b) established the synonymy of *A. eertmoedi* with *A. badonneli*.

Recognition features. Known only from females. Forewing length ~ 2.45 mm. Thoracic pleura marked with slender, pale longitudinal band between two dark bands (Fig. 752). Distal piece of subgenital plate (Fig. 812) without setae; pigmented arms ~ 2.4X as long as their greatest width. Second valvula (Fig. 753) with short spiculate process; v3 triangular, somewhat concave on postero-dorsal surface.

Relationships. This species appears to be very close to the following species.

Distribution and habitat. Within the study area, the species occurs throughout much of eastern United States, from northern Florida north along the Atlantic Coast to southern Maine and in the interior north to southern Ontario, and west to central Illinois and eastern Texas. It has been found on a great variety of trees and shrubs (Mockford, 1979a) and on shaded stone outcrops and bridges. Outside the study area, it has been found on the Black Sea Coast in the Soviet state of Georgia (Danks, 1950) and in the Azores (Baz, 1988).

Aaroniella maculosa (Aaron)

Elipsocus maculosus Aaron, 1883:40.
Philotarsus maculosus (Aaron). Chapman and Nadler, 1928:62.
Aaroniella maculosa (Aaron). Mockford, 1951:103.

Recognition features. Known from both sexes. Female forewing length ~ 2.7 mm, that of male somewhat longer. Males with phallosome readily visible through cuticle postero-ventrally on abdomen. Thoracic pleura marked with single broad dark band above leg bases (Fig. 262). Distal piece of subgenital plate (Fig. 748) without setae; pigmented arms ~ 1.6X as long as greatest width. Second valvula with short spiculate process (Fig. 754); v3 triangular, postero-dorsal edge straight.

Relationships. These are as noted for *A. badonneli*.

Distribution and habitat. The species is found in a band from central Illinois eastward across the northern half of Indiana, most of Ohio and probably on eastward to the Philadelphia region. A seemingly separate population is found along the southern border of Tennessee and North Carolina. The webs of this species are found mostly on the trunks of bottomland forest trees: sycamore, cottonwood, silver maple, hackberry, and honey locust, also to a lesser extent on oaks, elms, and *Carpinus*.

Genus *Philotarsus* Kolbe

Philotarsus Kolbe, 1880:116.

Diagnosis. Nymphs and adults found on tree branches, associated with little or no webbing. Distal flagellomere (f1 1) short, distinctly free from f10, not bearing a terminal seta. Veins of distal half of hindwing without setae (Fig. 259). Hypandrium somewhat keeled towards distal end (Fig. 755). Phallosome (Fig. 120) with heavy, triangular base; external parameres nearly half length of entire phallosome; fused (or separate) pair of endophallic sclerites closely joined to distal end of aedeagal arch. Sense cushions of male paraprocts elongate (Fig. 260). Male epiproct longer than wide (Fig. 756). Subgenital plate: distal piece longer than wide, bearing terminal setae; pigmented arms slender, at least 3X longer than wide. Ovipositor valvulae: v2 with at least a short spiculate process (Fig. 760); v3 rounded.

Generotype. *P. picicornis* (Fabricius).

North American species: *P. kwakiutl* Mockford. *P. picicornis* (Fabricius).

Philotarsus kwakiutl Mockford

Philotarsus kwakiutl Mockford, 1951:104.

Recognition features. Known from both sexes. Forewing markings (Fig. 757) faint, never including radial series of spots in Rs and M cells (except specimens from Baywood Park, San Luis Obispo County, California with the radial spots but not otherwise different). Endophallic sclerites separate or joined by membrane. Male epiproct (Fig. 756) approximately parallel-sided, the base only slightly narrower than widest region. Subgenital plate (Fig. 758): distal piece slightly tapering towards distal end; pigmented arms not widened anteriorly. V2 with prominent spiculate process (Fig. 759).

Relationships. The species is closest to an undescribed species in the southern Rocky Mountains. The latter is very close to *P. picicornis*.

Distribution and habitat. The species is found on the Pacific Coast from San Francisco north to Vancouver, British Columbia, and inland in the north to central British Columbia, northern Idaho, western Montana (Flathead County), and in the south to Yosemite National Park. It inhabits branches of most of the conifers of the region, and also oaks, maples, alders, and probably other broad-leaf trees.

Philotarsus picicornis (Fabricius)

Psocus picicornis Fabricius, 1793:86.
Psocus flaviceps Stephens, 1836:124.
Psocus striatulus Stephens, 1836:124.
Caecilius iroratus Curtis, 1837:648.

Psocus lasiopterus Burmeister, 1839:777.
Psocus pusillus Zetterstedt, 1840:1053.
Caecilius lasiopterus (Burmeister). Hagen, 1866:205.
Caecilius flaviceps (Stephens). Hagen, 1866:205.
Elipsocus flaviceps (Stephens). McLachlan, 1867:10.
Philotarsus picicornis (Fabricius). Kolbe, 1880:117.
Mesopsocus poecilopterus Navás, 1913:87.
Philotarsus flaviceps (Stephens). Badonnel, 1943:71.

Recognition features. Known from the study area only from females. Forewing markings (Fig. 263) generally distinct and including a radial series of spots, one in each Rs and M cell. European males: endophallic sclerites fused distally; epiproct ~ 2.2X broader in middle than at base. Subgenital plate essentially as described for P. kwakiutl. Second valvula with low, rounded spiculate process (Fig. 760).

Relationships. These are as noted for *P. kwakiutl.*

Distribution and habitat. Within the study area, the species has been found at Vancouver, British Columbia; Toronto, Ontario; Tracadie, Prince Edward Island; Digby, Lockeport, and Shelburne, Nova Scotia, and Indian Point, Sagadahoc County, Maine. Collections from Vancouver and Tracadie are extensive enough to assure that the species is established at these localities; they consist only of females. Collections from the other localities are small but consist only of females. The field collections at Vancouver were made on conifers and on Scotch broom. Those at Digby and Indian Point were made on dwarf mistletoe (*Arceuthobium pusillium*). The species has also been taken at New York City as an introduction from Europe. Outside the study area, it is known throughout Europe.

Family Elipsocidae

Diagnosis. Small to medium-sized forms (length from head to tip of closed wings, or tip of abdomen in short-winged or apterous forms 1.0 - 4.5 mm). Wing dimorphism, usually related to sex, frequent but absent in some genera. Macropterous adults with wings membranous. Lacinial tip bicuspid with lateral cusp generally wider than median, both generally denticulate. Distal inner labral sensilla: usually 11 as noted for the family group, but some genera (designated below in generic accounts) with fewer. Setae usually sparse on body and wings, very sparse or absent on hindwings. Cell Cu1a usually present in forewing but absent in some genera; when present, its limiting vein usually free from M, but joined in some genera. Adults of most species with tarsi three segmented, but some genera with tarsi two segmented. Phallosome with external parameres of moderate length, about one-third length of entire phallosome. Ovipositor valvulae usually complete.

North American genera: *Cuneopalpus* Badonnel, *Elipsocus* Hagen, *Nepiomorpha* Pearman, *Pulmicola* Mockford, *Propsocus* McLachlan, *Reuterella* Enderlein.

Key to North American Genera and Species of Family Elipsocidae

1.   Adults with tarsi three-segmented 2
--   Adults with tarsi two-segmented 9

2.   Vein Cu1a in forewing at least touching vein M. Forewing with striking banding pattern (Fig. 264) ....
     .... Genus *Propsocus, P. pulchripennis* (Perkins)
--   Vein Cu1a in forewing free from vein M. Forewing with at most a single transverse band, usually no banding, wings sometimes brachypterous or micropterous 3

3.   Mx4 short, < twice as long as its greatest width. Five distal inner labral sensilla (three placoids altering with two trichoids) .....
     .... Genus *Cuneopalpus, C. cyanops* (Rostock)
--   Mx4 > twice as long as its greatest width. Eleven distal inner labral sensilla .... Genus *Elipsocus* 4

4.   Male forewings at least 3500 μm in length. Females either macropterous or micropterous. Preclunial abdominal terga with broad transverse bands of reddish brown or dark brown usually interrupted on dorsal midline 5
--   Male forewings < 3450 μm in length. Females macropterous. Abdominal terga never with broad, pigmented segmental bands
    6

5.   Male forewing not over 3950 μm. Head, thorax, and appendages medium reddish brown. Most females micropterous ....
     ....*E. guentheri* Mockford
--   Male forewing at least 4300 μm. Head, thorax, and antennae mostly dark chestnut brown. Females macropterous ....
     .... *E. obscurus* Mockford

6.   Known from females only. Head dark brown except for (usually) two paler spots on each side between ocelli and antennal base (Fig. 265) .... *E. hyalinus* (Stephens)
--   Known from both sexes. Head with more extensive pale areas than described above 7

7. Preclunial abdominal segments dull white in dorsal view except, in some females, a slender transverse reddish brown band in each segment. Female epiproct pigmented throughout except a small colorless spot on each side at base (Fig. 266) ....
.... *E. abdominalis* Reuter

-- Preclunial abdominal segments with some extensive pigmentation. Epiproct (both sexes) extensively unpigmented, the pigmentation either concentrated in basal two-thirds excluding edges or in two parallel longitudinal bands 8

8. Preclunial abdominal segments uniformly reddish brown in dorsal view. Epiproct with large pigmented area covering basal two-thirds except edges (Fig. 267) .... *E. moebiusi* Tetens

-- Abdominal terga 3 - 6 dark purplish brown; remaining preclunial segments creamy yellow. Epiproct with two broad, longitudinal pigmented bands (as in Fig. 268); remainder unpigmented ..
.... *E. pumilis* (Hagen)

9. Males apterous. Females macropterous or apterous. Apterous individuals with stout spinulose setae on body (Fig. 269); winged individuals with spinulose setae on veins in basal half of forewing . Genus *Nepiomorpha, N. peripsocoides* Mockford

-- Males macropterous, females micropterous or apterous. No stout spinulose setae on body or wings 10

10. Setae present on veins and margin of forewing. Vein Cu1a present in forewing (Fig. 270). Subgenital plate with two distal processes (Fig. 271) ....
.... Genus *Reuterella, R. helvimacula* Enderlein

-- Setae absent on forewing. Vein Cu1a absent in forewing (Fig. 272). Subgenital plate with single distal process (Fig. 273) ..
.... Genus *Palmicola* 11

11. Known from both sexes. Females ~ 1.2 mm in length. Subgenital plate with moderately deep anterior indentation in pigmented area (Fig. 273) .... *P. aphrodite* Mockford

-- Known only from females. Body length ~ 0.9 mm. Subgenital plate with only slight indentation in pigmented area (Fig. 274)
.... *P. solitaria* Mockford

Genus *Cuneopalpus* Badonnel

Cuneopalpus Badonnel, 1943:76.

Diagnosis. Adults macropterous. Lacinial tip slender (Fig. 761). Mx4 short, less than twice its greatest width. Five distal inner labral sensilla; other inner labral sensilla usually found in the family on outer surface. Adults with tarsi three-segmented. Pulvilli much widened at distal ends (Fig. 762). Cell Cu1a present in forewing, its limiting vein free from M. Vein Cu2 of forewing without setae. V2 terminating in a strong, pointed process (Fig. 763).

Generotype. *C. cyanops* (Rostock).
North American species: *C. cyanops* (Rostock).

Cuneopalpus cyanops (Rostock)

Elipsocus cyanops Rostock, 1876:192.
Cuneopalpus cyanops (Rostock). Badonnel, 1943:77.

Recognition features. The only member of its genus. Entirely pale orange yellow. Wings clear.
Relationships. These are not currently understood.
Distribution and habitat. Within the study area, the species is apparently established locally in Contra Costa and San Luis Obispo Counties, California, and in Ocean Park, Brooklyn, New York. At the California localities it was taken on foliage of Monterrey Pine and Bishop Pine. No specific habitat was noted for the New York collection. It has also been taken at Seattle and Boston as introductions from Switzerland and England respectively. Outside the study area, the species occurs throughout Europe.

Genus *Elipsocus* Hagen

Elipsocus Hagen, 1866:203.
Cabarer Navás, 1908:410.

Diagnosis. Medium sized forms (length from head to tip of closed wings 3.9 - 4.5 mm). Both sexes usually macropterous; some females brachypterous. Lacinial tip moderately wide (Fig. 764). Eleven distal inner labral sensilla. Cell Cu1a present in forewing, its limiting vein free from M. Vein Cu2 of forewing usually with setae. Adults with tarsi three-segmented; pulvilli only slightly widened at tip (Fig. 765). V2 rounded distally, with minute pointed process, or rounded process or no process.

Generotype. *E. pumilis* Hagen - *E. westwoodi* McLachlan, new synonym.

North American species: *E. abdominalis* Reuter, *E. guentheri* Mockford, *E. hyalinus* (Stephens), *E. moebiusi* Tetens, *E. obscurus* Mockford, *E. pumilis* (Hagen).

Elipsocus abdominalis Reuter

Elipsocus hyalinus var. *abdominalis* Reuter, 1904:6.
Elipsocus occidentalis Banks, 1907a:166.
Elipsocus mclachlani Kimmins, 1941:528.
Elipsocus abdominalis Reuter. Lienhard, 1985:123.

Recognition features. Known from both sexes. Without wing polymorphism. Vertex of head with distinct median and lateral dotted areas (as in Fig. 773). Female forewing (Fig. 766) marked with transverse band of brown spots: around Rs-M junction, around base of vein Cu1a, and around nodulus; spot in cell Cu1a extending into cell M3; spot over distal half of pterostigma; male forewing with pterostigma dull brown, otherwise unmarked. FW/P index ~ 3.6 (length of forewing/ length of pterostigma). Basal row of forewing ~ 4 - 5 (row of small setae on anterior margin of forewing at base; Mockford, 1980). Preclunial abdominal segments largely dull white in dorsal view, some females with transverse reddish brown band in each segment. Female epiproct (Fig. 266) colored nearly throughout, with intense pigment spot on base to each side of midline and colorless spot to each side of this spot.

Relationships. The species is probably close to *E. pumilis* (Hagen). (Pacific Northwest, New York, Europe).

Distribution and habitat. Within the study area the species is found in the Pacific Northwest from Lane County, Oregon north to Vancouver, British Columbia, and on Vancouver Island. It inhabits branches of western red cedar, Douglas fir, maple, and alder. Outside the study area, the species is found throughout Europe.

Elipsocus guentheri Mockford

Elipsocus guentheri Mockford, 1980:250.

Recognition features. Known from both sexes. Vertex of head (Fig. 767) with median and lateral dotted areas with narrow lightly pigmented area between. Wing polymorphism present: males macropterous, most females micropterous, a few macropterous. Forewings unmarked except pterostigma lightly pigmented. FW/P index ~ 3.2 - 3.4. Cu2 in forewing with few or no setae. Basal row of forewing 0 - 1.

Preclunial abdominal segments dorsally each with broad transverse band of reddish brown interrupted along dorsal midline. Female epiproct (Fig. 768) reddish brown laterally, broadly creamy yellow in middle and distally.

Relationships. The species is rather close to *E. obscurus* Mockford (California) and an un-named species in Wisconsin.

Distribution and habitat. The species occurs throughout the Rocky Mountains from Cariboo County, British Columbia south to Grand Canyon National Park, Arizona. It inhabits branches of pines, spruces, firs, Douglas fir, and aspens.

Elipsocus hyalinus (Stephens)

Psocus hyalinus Stephens, 1836:123.
Psocus bipunctatus Stephens, 1836:123.
Elipsocus abietis Kolbe, 1880:114.
Elipsocus hyalinus (Stephens). Kimmins, 1941:522.

Recognition features. Known only from females in the study area. Without wing polymorphism. Head (Fig. 265) dark brown except for two paler spots on each side between ocelli and antennal base. Forewing (Fig. 769) with large brown spot mostly in cell R basal to Rs - M junction; pterostigma brown. FW/P index 3.4 - 3.8. Basal row ~ 5. Preclunial abdominal segments: terga 4-6 dark brown contrasting with pale terga before and beyond. Epiproct (Fig. 770) with large basal dark spot; edges and distal end pale.

Relationships. The species is probably close to *E. pumilis*. It forms a species-pair with *E. nuptialis* Roesler (Europe).

Distribution and habitat. Within the study area, the species is found on the Pacific Coast from San Luis Obispo County, California north to Vancouver, British Columbia. It inhabits foliage and branches of coniferous trees. Outside the study area, it is found throughout Europe.

Elipsocus moebiusi Tetens

Elipsocus moebiusi Tetens, 1891:372, 379.
Elipsocus brevistylus Reuter, 1894: 15, 33, 44.
Elipsocus pallidus Jentsch, 1938:27.

Recognition features. Known from both sexes. Without wing polymorphism. Vertex of head with median and lateral dotted areas. Male forewing unmarked; that of female (Fig. 771) with obscure spot in distal half of pterostigma, another in cell R before Rs - M junction, and another at nodulus. FW/P index 3.3 - 3.4. Basal row 8-9. Preclunial abdominal segments uniformly reddish brown in dorsal view. Epiproct (Fig. 267) with large pigmented area covering basal two-thirds except edges; remainder pale.

Relationships. The species is probably close to *E. pumilis*.

Distribution and habitat. Within the study area, the species has been found only at Vancouver, British Columbia, where it has been taken on branches of pine, Douglas fir, and Scotch broom. Outside the study area it has been found over much of Europe.

Elipsocus obscurus Mockford

Elipsocus obscurus Mockford, 1980:255.

Recognition features. Known from both sexes. Without wing polymorphism. Vertex of head with median and lateral dotted areas; region between somewhat pigmented. Male forewing unmarked except pterostigma slightly pigmented; female (Fig. 772) with pterostigma dark in distal two-thirds, R and M+Cu and Cu1 veins bordered with dark pigment, likewise most of cell 1A and region around nodulus. FW/P index 3.1 - 3.2 (male), ~ 3.4 (female). Basal row 1-3. Preclunial abdominal segments creamy yellow ringed with broad bands of reddish brown. Female epiproct with two longitudinal dark bands separated by pale middle region (Fig. 268).

Relationships. These are as noted for *E. guentheri*.

Distribution and habitat. The species is found in California, from Riverside County north to Mendocino County. It has been collected on live oak (*Quercus agrifolia* Nee), sumac, *Artemisia, Salvia*, and grasses.

Elipsocus pumilis (Hagen)

Psocus quadrimaculatus Westwood, 1840:19 (preoccupied).
Psocus pumilis Hagen, 1861:9.
Elipsocus pumilis (Hagen). Hagen, 1866:207.
Elipsocus quadrimaculatus (Westwood). Hagen, 1866:207.
Elipsocus westwoodi McLachlan, 1867:274, new synonym.

Note. The types, consisting of two pinned female syntypes, MCZ No. 261, were examined. A slide was made of wings and abdomen of one, here designated lectotype. The abdominal color pattern (visible on the dried specimen before mounting), forewing markings, pigmentation pattern of the epiproct, and details of the spermatheca and ovipositor valvulae all agree with British, French, and Pacific Coast specimens determined as *E. westwoodi*. The median dotted area of the vertex of the paralectotype (the lectotype's head is missing) is broader than usually encountered in *E. westwoodi*, but this character, alone, does not seem to justify specific distinction. The type locality, "New York", suggests that the species was introduced on the Atlantic Coast as well as the Pacific (see Distribution and habitat, below).

Recognition features. Known from both sexes. Without wing polymorphism. Vertex of head with median and lateral dotted areas (Fig. 773). Male forewing unmarked except for dusky pterostigma; female (Fig. 774) with pterostigma dark in distal half, a large dusky spot before and around Rs - M junction; cell 1A dusky with pigment extending around nodulus; cell Cu1a dusky anteriorly around vein. FW/P ~ 3.5 (male), 3.8 (female). Basal row 3-5. Preclunial abdominal segments: terga 3-6 dark purplish brown; remainder creamy yellow. Epiproct (as in Fig. 768) with two broad pigment bands extending from base to distal two-thirds; remainder pale.

Relationships. The species is apparently close to *E. moebiusi*, *E. abdominalis* and the *E. hyalinus- E. nuptialis* species pair.

Distribution and habitat. Within the study area, the species has been found at Vancouver, British Columbia, where it has been collected on western red cedar, maples, pines, and Scotch broom and in "New York" (type locality). Outside the study area, it occurs throughout Europe.

Genus *Neplomorpha* Pearman

Neplomorpha Pearman, 1936b:4.

Diagnosis. Small forms (1.0 - 1.5 mm from head to tip closed wings or tip of abdomen in apterous forms). Males apterous, females macropterous and apterous. Usually in groups on tree trunks with no webbing. Lacinial tip slender. Labral sensilla as described for the family. Cell Cu1a absent in forewing. Veins in basal half of forewing (Fig. 775) with spinulose setae except Cu2 without setae. Body of apterous forms with stout spinulose setae (Fig. 269). Abdomen with dark and light color pattern (Fig. 269). Adults with tarsi two-segmented; pulvilli only slightly widened at tip. V2 with stout denticulate process arising near distal end (Fig. 776). Subgenital plate (Fig. 777) with short, wide distal process.

Generotype. *N. crucifera* Pearman.

North American species: *N. peripsocoides* Mockford.

Neptomorpha peripsocoides Mockford

Neptomorpha peripsocoides Mockford, 1955a:98.

Recognition features. Distal process of subgenital plate short and broad (Fig. 777). Preclunial abdominal segments marked with broad white cross delimited by fuscous quadrangles (Fig. 269).

Relationships. The species is probably close to *N. crucifera* Pearman (Sri Lanka).

Distribution and habitat. The species has been found in Sarasota and Alachua Counties, Florida, and near Brownsville, Cameron County, Texas.

Genus *Palmicola* Mockford

Palmicola Mockford, 1955a:102.

Diagnosis. Small forms (length from head to tip of abdomen or tip of closed wing 0.9 - 1.5 mm). Males macropterous, females apterous. Usually solitary under small dense webs on tree trunks. Lacinial tip slender. Eleven distal inner labral sensilla. Cell Cu1a absent in forewing. Wings without setae. Body with scattered setae. Abdomen brown, more or less uniform in color. Adults with tarsi two segmented; pulvilli only slightly widened at tip. V2 with slender, acuminate process arising near distal end (Fig. 778). Subgenital plate (Fig. 273) with distal process longer than wide. Female epiproct (Fig. 779) at least twice as wide as long.

Generotype. *P. aphrodite* Mockford.

North American species. *P. aphrodite* Mockford, *P. solitaria* Mockford.

Palmicola aphrodite Mockford

Palmicola aphrodite Mockford, 1955a:102.

Recognition features. Adults of both sexes known. Females ~ 1.2 mm in length. Male with wings clear (Fig. 272) except slightly pigmented pterostigma and cell IA. Female subgenital plate with moderately deep indentation in pigmented area (Fig. 273).

Relationships. These are not currently understood.

Distribution and habitat. The species has been found in Alachua and Marion Counties, north-peninsular Florida. Its webs have been found on trunks of pines, *Liquidambar*, *Sabal palmetto*, and the shaded north side of a brick building.

Palmicola solitaria Mockford

Palmicola solitaria Mockford, 1955a:105.

Recognition features. Known only from females. Body length ~ 0.9 mm. Subgenital plate with only slight indentation of pigmented area (Fig. 274).

Relationships. These are not currently understood.

Distribution and habitat. The species occurs in south-peninsular Florida in the following counties: Glades, Hendry, Highlands, Indian River, and Sarasota. It has been collected entirely on trunks of *Sabal palmetto*.

Genus *Propsocus* McLachlan

Propsocus McLachlan, 1866:352.
Tricladus Enderlein, 1906b:410.
Tricladellus Enderlein, 1909c:273.

Diagnosis. Medium-sized forms (length from head to tip of closed wings, or tip of abdomen in short winged forms 2.0 - 2.7 mm). Adults of both sexes macropterous and brachypterous. Lacinial tip (Fig. 780) relatively broad with lateral cusp protruding beyond median. Distal inner labral sensilla (Fig. 781): three placoids alternating with two trichoids; an outer series present, consisting of six trichoids. Cell Cu1a present in forewing, its limiting vein at least touching M. Forewing (Fig. 264) with striking pattern of brown bands along veins and a transverse band partially across wing at one-third distance from base to tip. Adults with tarsi three-segmented; pulvilli broad throughout. V2 with short, slender, acuminate process arising near distal end (Fig. 782). Subgenital plate (Fig. 783) with pair of distal processes.

Generotype. *P. pallipes* (McLachlan).

North American species: *P. pulchripennis* (Perkins).

Propsocus pulchripennis (Perkins)

Stenopsocus pulchripennis Perkins, 1899:83.
Myopsocus nitens Hickman, 1934:81.
Tricladellus nitens (Hickman). Edwards, 1950:113.
Tricladellus nitens var. *brachypterus* Edwards, 1950:115.
Propsocus pulchripennis (Perkins). Smithers, 1963:891.

Recognition features. The only species of its genus in the study area.

Relationships. The species is close to *P. pallipes* (McLachlan) of Australia.

Distribution and habitat. Within the study area, the species has been found in the California counties of Marin, Sacramento, and Humboldt. It inhabits dry grass and leaf litter. Outside the study area, it has been found in Hawaii, Australia, New Zealand, Zambia, South Africa, and Chile.

Genus *Reuterella* Enderlein

Leptella Enderlein, 1901:539 (preoccupied)
Reuterella Enderlein, 1903e:132.

Diagnosis. Small forms (Length from head to tip of closed wings or tip of abdomen 1.5 - 2.0 mm). Wing dimorphism present: males macropterous, females micropterous. Usually solitary under small dense webs on tree trunks and stone outcrops. Lacinial tip moderately wide; low, flat cusps denticulate (Fig. 784). Labral sensilla in two series as described for *Propsocus*. Cell Cu1a present in forewing (Fig. 270), its limiting vein free from M. Ciliation moderate on forewing margin, sparse on veins, in hindwing ciliation limited to margin in cell R3 and adjacent part of cell R5. Adults with tarsi two segmented; pulvilli slender, slightly broadened at tip. V2 with long, slender, pointed process arising near tip (Fig. 785). Subgenital plate (Fig. 271) with pair of distal processes.

Generotype. *R. helvimacula* (Enderlein).
North American species: *R. helvimacula* Enderlein.

Reuterella helvimacula (Enderlein)

Leptella helvimacula Enderlein, 1901:539.
Reuterella helvimacula (Enderlein). Enderlein, 1903e:132.
Caecilius corticis Pearman, 1924:58.

Recognition features. The only species of its genus. Adults in two color forms: one with preclunial region of abdomen yellow; other with this region dark reddish brown.

Relationships. The genus is monotypic. It is probably close to *Pseudopsocus* Kolbe.

Distribution and habitat. Within the study area, the species has been found in New Brunswick, in the states of Maryland, West Virginia, North Carolina, Indiana, Wisconsin, Illinois, California, Oregon, Montana, and the Province of British Columbia. At sites in Maryland, Indiana, Wisconsin, and Illinois it was found on shaded rock outcrops.

Thus it is very restricted in its distribution in these states. In Indiana it has been found only at Turkey Run State Park, Parke County. In Illinois it has been collected only in a few counties in the far southern part of the state where rock outcrops are common. At the West Virginia, California, Oregon, and British Columbia localities it has been found on trees. Outside the study area the species is known throughout Europe and east to Mongolia.

Family Mesopsocidae

Diagnosis. Medium-sized to moderately large forms (length from head to tip of closed wings, or tip of abdomen in short-winged forms 3.0 - 4.5 mm). Wing dimorphism present or absent; when present, males macropterous, females micropterous. Lacinial tip (Fig. 786) bicuspid, of moderate width, outer cusp longer and wider than inner cusp; both rough-surfaced. Inner labral sensilla: nine, with median placoid followed by three tricholds and one placoid on each side. Setae short and sparse on body and appendages, absent on wings. Cell Cu 1a present in forewing (Fig. 787), its limiting vein free from M. Adults with tarsi three segmented. Phallosome (Fig. 278) with external parameres of moderate length, about one-third length of phallosome; endophallus lacking sclerotizations. Subgenital plate with distal process in two parts: a short basal neck containing two heavily sclerotized bars, and a rounded or pointed, somewhat fleshy distal lobe (Fig. 275). Ovipositor valvulae complete; v1 and v2 each bearing distally a pointed, upward-directed spinulose process (Fig. 788).

North American genus: *Mesopsocus* Kolbe.

Key to the North American Species of Family Mesopsocidae, Genus *Mesopsocus*

1.   Wing dimorphism present: males macropterous, females micropterous. Pigmented arms of subgenital plate curved backward at tips (Fig. 277) 2

--   Adults of both sexes macropterous. Pigmented arms of subgenital plate relatively broad, curved to sides at tips (Fig. 275) ...
....*M. laticeps* (Kolbe)

2.   External parameres with inner lobe extending distad of outer lobe most of its length (Fig. 276). Distal piece of subgenital plate tapering to blunt point at tip (Fig. 277) ....
....*M. immunis* (Stephens)

--   External parameres with inner lobe extending only slightly distad of outer lobe (Fig. 278). Distal piece of subgenital plate broad, rounded posteriorly (Fig. 279) ....
.... *M. unipunctatus* (Müller)

Genus *Mesopsocus* Kolbe

Mesopsocus Kolbe, 1880:112.
Trocticus Bertkau, 1883:99.
Holoneura Tetens, 1891:372.
Labocoria Enderlein, 1910:71.

Diagnosis. Same as for the family.
Generotype. *M. unipunctatus* (Müller).
North American species. *M. immunis* (Stephens), *M. laticeps* (Kolbe), *M. unipunctatus* (Müller).

Mesopsocus immunis (Stephens)

Psocus immunis Stephens, 1836:121.
Elipsocus aphidioides (Schrank). Hagen, 1866:214.
Mesopsocus unipunctatus (Müller). Kolbe, 1880:112.
Mesopsocus immunis (Stephens). Badonnel, 1936:26.

Recognition features. Wing dimorphism present. External parameres (Fig. 276) with inner lobe extending distad of outer lobe most of its length; proportions of inner lobe: length about equal to basal width. Pigmented arms of subgenital plate slender, curved backward at tips; distal piece of subgenital plate tapering to blunt point at tip (Fig. 277).
Relationships. This species belongs to a group including another European species (*M. lucitanus* Lienhard) and several North African species (Lienhard 1981).
Distribution and habitat. Within the study area the species has been found only at Victoria and Vancouver, British Columbia, and at Lockeport, Nova Scotia, where it was presumably introduced. In British Columbia it has been collected on branches of Garry Oak, pine, maple, and western red cedar. Outside the study area it is known throughout most of Europe.

Mesopsocus laticeps (Kolbe)

Elipsocus laticeps Kolbe, 1880:114.

Holoneura laticeps (Kolbe). Tetens, 1891:372, 378.
Mesopsocus laticeps (Kolbe). Gurney, 1949:62.

Recognition features. Wing dimorphism absent: adults of both sexes macropterous. Frons with dark brown mark (Fig. 789) covering most of its surface, narrowly divided in middle and not quite reaching anterior margin, but extending posteriorly onto vertex mesad of eyes. External parameres (Fig. 790) with inner lobe distad of outer lobe its entire length; proportions of inner lobe: length about twice basal width. Pigmented arms of subgenital plate relatively broad, curving to sides at tips; distal piece of subgenital plate rounded.

Relationships. These are not clear at present.

Distribution and habitat. Within the study area, the species is known from Malakoff, Ontario; New Haven County, Connecticut; Lake County, Illinois, and Livingston and Midland Counties, Michigan. At the Connecticut locality it was taken on pine; at the Illinois localities it was collected on tamarack and other trees and shrubs. Outside the study area, the species is widely distributed throughout Europe and eastward to Mongolia.

Mesopsocus unipunctatus (Müller)

Hemerobius unipunctatus Müller, 1764:66.
Hemerobius aphidioides Schrank, 1781:314.
Psocus longicornis Stephens, 1836:121.
Caecilius vitripennis Curtis, 1837:648.
Psocus obliteratus Zetterstedt, 1840:1053.
Psocus oculatus Zetterstedt, 1840:1053.
Psocus naso Rambur, 1842:320.
Psocus signatus Hagen, 1861:9.
Elipsocus aphidioides (Schrank). Hagen, 1866:207.
Elipsocus signatus (Hagen). Hagen, 1866:207.
Mesopsocus unipunctatus (Müller). Kolbe, 1880:112.
Trocticus gibbulus Bertkau, 1883:99.
Holoneura unipunctatus (Müller). Tetens, 1891:372.
Elipsocus unipunctatus (Müller). Tetens, 1891:378.

Recognition features. Wing dimorphism present. External parameres (Fig. 278) with inner lobe extending only slightly beyond outer lobe; proportions of inner lobe: basal width slightly greater than length. Pigmented arms of subgenital plate slender, curving posteriorly at distal ends; distal piece of subgenital plate broad, rounded posteriorly (Fig. 279).

Relationships. This is a member of a small group of species, the others entirely European (Badonnel, 1980).

Distribution and habitat. Within the study area, this species occurs across the continent in the Canadian provinces, and the northern tier of U.S. states, south in the Appalachian Mountains to Yancey County, North Carolina, in the lowlands of the Midwest spottily to southern Illinois, in the Rocky Mountains south to Gunnison County, Colorado, and on the Pacific Coast to Riverside County, California and some of the off-shore islands at about the same latitude. The species may be found on a great variety of trees, both broad-leaf and needle-leaf. Outside the study area, it occurs throughout Europe.

Family Group Psocetae

Diagnosis. Pair of sclerotized knobs on labrum, one to each side of distal sensillar field; knobs often connected by partial or complete sclerotized band over sensillar field (Fig. 791). Distal inner labral sensillar field with five, seven, nine, or eleven sensilla; five - median placoid (P) followed by one trichoid (T) followed by one P on each side; seven - either median P followed by two T followed by one P on each side or median P followed by one T followed by one P followed by one T on each side; nine - median P followed by three T, followed by one P on each side; eleven - as nine, but lateral P followed by one T on each side. Mandibles short, not excavated posteriorly. Meso-precoxal bridges narrow; mesotrochantins broad basally, tapering to tip (Fig. 109). Tarsi two or three segmented in adults. Pretarsal claws with preapical denticle. Coxal organ well developed. Adults usually macropterous but occasionally brachypterous or micropterous. Forewing with cell Cu1a present. Male paraproct with a process on free margin. Ovipositor valvulae complete; v3 independent except at base, usually bearing numerous setae. Female paraprocts extended distally as conical projections (Fig. 792).

North American taxa: Families Hemipsocidae, Myopsocidae, Psocidae.

Family Hemipsocidae

Diagnosis. Medium-sized forms (length from head to tip of closed wings 2.5 - 3.5 mm). Five distal inner labral sensilla. Tarsi two segmented in adults. Pretarsal claws with minute preapical denticle, relatively wide pulvillus (Fig. 793). Vein Cu1a in forewing (Fig. 797) joined to M by a crossvein; M in forewing two branched. Hypandrium lacking heavy sclerotizations. Phallosome (Fig. 794) weakly sclerotized, open distally. Female subgenital plate (Fig. 795) simple, rounded on hind margin. Ovipositor valvulae (Fig. 796): v2 slender; v3 broad, triangular, with few setae.

North American genus: *Hemipsocus* Selys-Longchamps.

Genus *Hemipsocus* Selys-Longchamps

Hemipsocus Selys-Longchamps, 1872:146.

Diagnosis. As for the family.
Generotype. *H. chloroticus* (Hagen).
North American species: *H. chloroticus* (Hagen), *H. pretiosus* Banks.

Hemipsocus chloroticus (Hagen)

Psocus chloroticus Hagen, 1858:474.
Hemipsocus chloroticus (Hagen). Selys-Longchamps, 1872:146.
Hemipsocus hyalinus Enderlein, 1906a:311.

Recognition features. Head, thorax, antennae, and legs pale tawny brown. Wings clear. Preclunial abdominal segments creamy yellow, each with a broad transverse pale purplish brown ring, incomplete ventrally. Body length ~ 3.0 mm. (head to tip of abdomen).
Relationships. These are not currently understood.
Distribution and habitat. The species is established locally in Florida and on the coastal plain in Texas and South Carolina. At these localities it inhabits forest ground litter. It has been taken repeatedly at Miami as an introduction on orchid plants from Guatemala; it has also been taken once at Brownsville, Texas, as an introduction from Veracruz State, Mexico and once at Portland, Oregon as an introduction from Japan. Outside the study area, it is known from Sri Lanka, Japan, Vietnam, Philippines, Taiwan, Sarawak, Java, Sumatra, and numerous islands in the central Pacific.

Hemipsocus pretiosus Banks

Hemipsocus pretiosus Banks, 1930a:184.

Recognition features. Head, thorax, antennae, and legs medium brown. Abdomen dark brown. Forewings (Fig. 797) heavily blotched with brown, with dark brown spots at setal bases on veins. Body length ~ 2.5 mm.
Relationships. There appears to be a small complex of blotch-winged forms in the Antilles, of which this species is a member.
Distribution and habitat. Within the study area, the species is restricted to southern Florida north to Highlands County. It is found in woodland ground litter and lower, dead leaves of small palms. Outside the study area, the species is known from Cuba.

Family Myopsocidae

Diagnosis. Medium-sized to moderately large forms (length from head to tip of closed wings 3.0-5.0 mm). Seven inner sensilla of distal labral field (median P followed by two T, one P on each side). Tarsi three-segmented in adults. Pretarsal claws with moderately large preapical denticle; pulvilli relatively slender (Fig. 798). Vein Cu1a in forewing joined to M at a point or by a crossvein, or fused for a short distance; M in forewing three branched. Forewings (Fig. 283) heavily blotched with brown, the margins with alternating brown and colorless banding. Hypandrium moderately well sclerotized but symmetrical and simple in structure. Phallosome (Figs. 804, 816) a pair of rods or a closed frame. Female subgenital plate (Figs. 280, 281) with broad pigmented arms and a distal piece. Ovipositor valvulae (Fig. 800): v2 broad-based, gradually tapering to long, slender tip; v3 rounded or quadrate, bearing numerous setae.

North American genera: *Lichenomima* Enderlein, *Myopsocus* Hagen.

Key to the North America Genera and Species of Family Myopsocidae

1. Veins Rs and M in hindwing joined by a crossvein. Distal piece of subgenital plate transverse, flat or slightly depressed on hind margin (Fig. 285) .... Genus *Lichenomima* 2
-- Veins Rs and M in hindwing fused for a distance. Distal piece of subgenital plate triangular, tapering posteriorly to form slender process bearing setae on tip (Fig. 281) .... Genus *Myopsocus* 4

2. Lateral pigmented arms of subgenital plate not connected by pigment (Fig. 282). Forewing markings: oblique pigmented band of basal half of wing distinctly set off by basal and distal clear areas; a large clear spot in middle of cell R5 (Fig. 283) .... *L. coloradensis* (Banks)
-- Lateral pigmented arms of subgenital plate connected by pigment. Forewing markings not as above 3

3. Lateral pigmented arms of subgenital plate broadly connected by pigment, thus forming a broad u-shaped pigmented area (Fig. 280). Forewing markings: dense brown blotching interrupted only by small colorless spots (Fig. 284) ...... *L. lugens* (Hagen)

--   Lateral pigmented arms of subgenital plate connected by moderately broad pigment band, and each extended laterally (Fig. 285). Forewing markings: oblique band of basal half of wing diffuse, including numerous pale and colorless spots (Fig. 286) .... *L. sparsa* (Hagen)

4.   In forewing, two conspicuous bulges on vein 1A; wing margin slightly protruding at marginal ends of veins M2 and M3 (Fig. 295) .... *M. antillanus* (Mockford)
--   In forewing, bulges on vein 1A only slight; wing margin not at all protruding at marginal ends of veins M2 and M3 (Fig. 296) 5

5.   Forewing 3.10 mm or less in length; v1 over half length of v2; phallosome widest distad of middle ... *M. minutus* (Mockford)
--   Forewing over 3.50 mm in length; v1 ~ one-third length of v2; phallosome widest in middle .... *M. eatoni* McLachlan

Genus *Lichenomima* Enderlein

Lichenomima Enderlein, 1910:66.

Diagnosis. Rs and M in hindwing joined by a crossvein. Distal piece of subgenital plate (Fig. 280) transverse, flat or slightly depressed on hind margin.

Generotype. *L. conspersa* Enderlein.

North American species. *L. coloradensis* (Banks), *L. lugens* (Hagen), *L. sparsa* (Hagen).

Lichenomima coloradensis (Banks)

Myopsocus coloradensis Banks, 1907a:164.
Lichenomima coloradensis (Banks). Mockford, 1982:217.

Recognition features. Known only from female. Fore- and middle femora entirely dark brown. Oblique brown band of forewing (Fig. 283) distinctly set out from surrounding mottling by narrow band of clear spots basally and clear band distally; the latter continuing as clear area through cell R1, interrupted by pale mottling around Rs branching point; a large clear spot in middle of cell R5; distal one-third of wing otherwise largely brown with small clear spots. Subgenital plate (Fig. 282) lacking pigmented connection between lateral pigment arms; distal piece pigmented on edges only, deeply depressed on hind margin in middle, lacking setae on hind margin. Sclerotizations of ninth sternum (Fig. 799): pair of parenthesis-shaped sclerites and short median rod. Ovipositor valvulae: v1 slender throughout; v3 ~ 3X as long as its greatest width (Fig. 800).

Relationships. The species is close to an undescribed species from the mountains of North Carolina and northern Nuevo León State, Mexico.

Distribution and habitat. The species is known from the Boulder and Fort Collins regions of northern Colorado. At Boulder it was collected under stones.

Lichenomima lugens (Hagen)

Psocus lugens Hagen, 1861:9.
Myopsocus lugens (Hagen). Hagen, 1866:210.
Lichenomima lugens (Hagen). Enderlein, 1910:66.
Psocus virginianus Banks, 1900b:239. New synonym.
Myopsocus virginianus (Banks). Enderlein, 1906d:320, new synonym.

Recognition features. Relatively small (length from head to tip of closed wing ~ 4.32 mm [female], ~ 3.55 mm [male]). Forewing (Fig. 284) with dense brown blotching, small colorless areas scattered almost at random except for Rs and M marginal cells. All femora brown with creamy yellow subapical band. Lateral pigmented arms of subgenital plate (Fig. 280) broadly connected by pigment, thus forming a broad u-shaped pigmented area; distal piece pigmented throughout, straight on hind margin, with single marginal seta. Sclerotizations of ninth sternum (Fig. 801): two small diffuse spots and long irregular rod, the latter longer than width of adjacent v2 near base. Ovipositor valvulae (Fig. 802): v1 slightly widened near base; v3 ~ 2X as long as its greatest width. Hypandrium (Fig. 803) not much longer than its basal width, tapering gradually to round distal end. Phallosomal sclerotizations (Fig. 804): pair of rods held together by membrane; each distally with short, thumb-like lateral process and slenderer, rounded distal process.

Relationships. This is a member of a group of species, most of them undescribed, found throughout eastern United States.

Distribution and habitat. The species occurs from southern Quebec south to Great Smoky Mountains National Park, west to Columbia, Missouri and Jackson and Pope Counties in southern Illinois. It is found on shaded rock outcrops, stone and cement structures, and shaded tree trunks.

Lichenomima sparsa (Hagen)

Psocus sparsus Hagen, 1861:8.
Lichenomima sparsa (Hagen). Enderlein, 1910:66.
Lichenomima sparsa (Hagen). Enderlein, 1915:46 (female only).

Recognition features. Relatively large (length from head to tip of closed wing ~ 5.64 mm [female], ~ 4.80 mm [male]). Oblique brown band of forewing (Fig. 286) diffuse; pterostigma and stigmasaum covered with purplish brown pigment; otherwise little pattern to distribution of brown and colorless blotches in forewing except a dark brown spot on vein M just before and around M - Cu1a junction and transverse colorless bands across cells R5, M1 and M2. Femora as described for *L. lugens*. Lateral pigmented arms of subgenital plate (Fig. 285) connected by moderately broad pigment band and extending laterally (but not curving backward) to form broad plate base; distal piece pigmented throughout, straight on hind margin, with two sub-marginal setae. Sclerotizations of ninth sternum (Fig. 805): two diffuse triangular areas and a short, curved rod. Ovipositor valvulae (Fig. 806): v1 relatively wide near base; v3 ~ 1.75X as long as its greatest width. Hypandrium (Fig. 807) slightly longer than its greatest width, tapering to slender apical region. Phallosome (Fig. 808) of two rods joined by membrane, each widened before apex, with lateral ear-like lobe and blunt, rounded distal process.

Relationships. These are as noted for *L. lugens*.

Distribution and habitat. The species is found from the Washington, D.C. area south to Alachua County, Florida, west to Parke County, Indiana, and Hot Springs National Park, Arkansas. It inhabits shaded stone outcrops and tree trunks.

Note. The male illustrated by Enderlein (1915, Fig. 28) is not the male of this species but of an undescribed species common in eastern United States. The only type (a syntype) which Enderlein saw was a female. My determination is based on examination of another syntype female in the Museum of Comparative Zoology.

Genus *Myopsocus* Hagen

Myopsocus Hagen, 1866:203, 210.
Philotodes Enderlein, 1910:67.
Rhaptoneura Enderlein, 1910:68.

Diagnosis. Rs and M in hindwing fused for a distance. Distal piece of subgenital plate triangular, tapering posteriorly to form slender process bearing setae on tip (Fig. 281).

Generotype. *M. unduosus* (Hagen).

North American species: *M. antillanus* (Mockford), *M. eatoni* McLachlan, *M. minutus* (Mockford).

Myopsocus antillanus (Mockford)

Phlotodes antillanus Mockford, 1974a:157.
Myopsocus antillanus (Mockford). Mockford, 1982:215.

Recognition features. Known from female only. Size moderate (forewing 3.30 - 3.70 mm). Forewing (Fig. 287) extensively brown-blotched; oblique brown band moderately well set off by narrow basal and distal colorless bands; distal radial series of clear spots moderately distinct, each spot crossing a vein (R2+3 through M3). Two conspicuous bulges on vein IA in forewing; wing margin protruding at tips of veins M2 and M3. Distal piece of subgenital plate (Fig. 281) tapering to base of slender terminal process; the process bearing two to three lateral setae in addition to two long distal ones. Ovipositor valvulae (Fig. 809); v1 ~ one-third length of v2; v3 relatively slender, slightly over 3X as long as its greatest width. Sclerotizations of ninth sternum (Fig. 810): an elongate, twisted rod subtended by a broad, triangular semimembranous area.
Relationships. The species appears to be close to an undescribed Hispaniolan species which lacks bulges on the postero-distal forewing margin.
Distribution and habitat. Within the study area, the species occurs in peninsular Florida north to Pinellas County. It has been collected by beating live oak, slash pine, and a large bromeliad on live oak. Outside the study area, the species is known from Cuba and the Dominican Republic. All adults collected to date, about 20, have been females, and the possibility exists that the species is parthenogenetic.

Myopsocus eatoni McLachlan

Myopsocus eatoni McLachlan, 1880:103.
Rhaptoneura eatoni (McLachlan). Badonnel, 1935b:201.

Recognition features. Known from both sexes. Size moderate (female forewing ~ 3.75 mm, male slightly less). Forewing (Fig. 288) extensively blotched with brown, but large colorless area in base of cell R5, all cells of distal wing margin only lightly blotched, broad region distal to oblique brown band lightly blotched. Pair of bulges on vein IA in forewing only slightly developed, basal bulge barely perceptible; postero-distal wing margin lacking protrusions at vein ends. Distal piece of subgenital plate (Fig. 811) tapering to long, slender terminal process bearing two long, distal setae. Ovipositor valvulae as described for M. antillanus except v3 relatively wider. Hypandrium short, wide; hind margin with rounded process in middle (Fig. 812). Phallosome an elongate-circular structure with median sclerotized band running through its length (Fig. 813).

Relationships. These are not known at present.

Distribution and habitat. A single series of specimens was intercepted from Portugal at Baltimore, Maryland on virgin cork. Outside the study area, the species occurs throughout southern Europe (Spain, Portugal, southern France, Italy, Greece), North Africa (Algeria, Morocco), and the Canary Islands.

Myopsocus minutus (Mockford)

Phlotodes minutus Mockford, 1974a:155.
Myopsocus minutus (Mockford). Mockford, 1982:215.

Recognition features. known from both sexes. Size small (forewing 2.58-3.10 mm). Forewing (Fig. 814) well blotched with brown but also with extensive colorless areas; oblique brown band well set off by broad basal and distal colorless areas; basal colorless area passing nearly through oblique brown band in cell Cu1b; distal colorless area nearly continuous with some of distal radial clear spots. Bulges on vein 1A in forewing scarcely perceptible; distal wing margin evenly rounded. Subgenital plate (Fig. 815) as described for M. antillanus except terminal process of distal piece wider, its terminal pair of setae shorter. Ovipositor valvulae as described for M. antillanus except v1 slightly over half length of v2. Hypandrium longer than wide, evenly rounded distally. Phallosome (Fig. 816) an elongate frame, broadest distad of middle, with a median sclerotized band running through its length.

Relationships. These are not known at present.

Distribution and habitat. A single specimen was intercepted at Miami, Florida from the Dominican Republic. Outside the study area, the species is known from Cuba and the southern Mexican states of Chiapas, Tabasco, and Veracruz. It has been collected on branches of oaks, *Casuarina* sp., and on the trunk of an orange tree.

Family Psocidae

Diagnosis. Medium-sized to large forms (length from head to tip of closed wings 3.5 - 8.0 mm). Five to eleven distal inner labral sensilla. Tarsi two-segmented in adults. Pretarsal claws with moderately large preapical denticle; pulvilli slender. Vein Cu1a in forewing generally fused to M for a distance, occasionally joined at a point, rarely joined by a crossvein; M in forewing three branched. Forewing markings variable, from none to extensive banding or blotching. Hypandrium usually well sclerotized, often complex, frequently asymmetrical.

Phallosome variable: a pair of rods joined or not at their bases, or a closed frame of various shapes. Ovipositor valvulae: v1 elongate, slender; v2 broad, usually with slender apical process; v3 rounded to long-oval, setose, often with posteriorly directed lobe , the median lobe, arising on inner surface in middle.

This family has been divided by most recent authors into three subfamilies: Psocinae, Amphigerontiinae, and Cerastipsocinae. An analysis of characters shows 1) that Psocinae and Cerastipsocinae stand much closer together than either does to Amphigerontiinae and 2) elements within Psocinae and Cerastipsocinae may be closer together than is indicated by this subfamilial arrangement. Accordingly, I have combined these two subfamilies as subfamily Psocinae, divided into the following four tribes: Psocini, Ptyctini, Metylophorini, and Cerastipsocini. The tribes are diagnosed and the genera from the study area assigned to each are listed below.

Key to North American Subfamilies
and Tribes of Family Psocidae

1.   Vein M forming distal closure of discoidal cell of forewing usually concave or straight (Fig. 289). Hypandrium a single segment with at most two small sclerites or small pigmented arms on preceding segment. Phallosome usually a closed frame. Male paraproctal process tapering to acuminate point (Fig. 290) .. .. Subfamily Psocinae 2

--   Vein M forming distal closure of discoidal cell of forewing usually convex (Fig. 291). Hypandrium of two well sclerotized sternal plates. Phallosome open distally. Male paraproctal process rounded distally. (Fig. 292) ...... Subfamily Amphigerontiinae

2.   Hypandrium (Fig. 293) usually with small pair of sclerites or pigmented arms on segment before principal sclerotization (absent in *Steleops*, where eyes slightly stalked). Interface of male clunium and epiproct either a straight line or epiproct protruding over clunium. V2 terminating in a long, slender process abruptly narrower at its base than valvula producing it (Fig. 294) .... Tribe Ptyctini

--   Hypandrium without pair of sclerites or pigmented arms on segment before principal sclerotization. Interface of male clunium and epiproct either a straight line or clunium extended over epiproct. V2 terminating either in a gradually tapering process or no process 3

3.   Mx4 length/width near base < 4.0. Hypandrium usually symmet-
        rical. Pigmented and sclerotized median region of subgenital
        plate meeting pigmented arms perpendicularly, forming a t-
        shaped pattern (Fig. 295) .... Tribe Cerastipsocini
        (Genus *Cerastipsocus*, two species, see recognition features)
--   Mx4 length/ width near base > 4.0. Hypandrium usually
        asymmetrical. Pigmented arms of subgenital plate usually
        each forming an angle > 90° with median pigmented and
        sclerotized region 4

4.   Ovipositor valvulae (Fig. 829): v2 with at least a short distal
        process; v3 a large plate extending nearly to half length of v2.
        Male paraproct with a shoulder or lobe lateral to distal process
        and distal to sense cushion (Fig. 290) .... Tribe Psocini
--   Ovipositor valvulae (Fig. 296): v2 rounded distally with no trace
        of a process; v3 short, not extending beyond one- fourth length
        of v2. Either no lobe lateral to distal process of male paraproct,
        or lobe bearing sense cushion .... Tribe Metylophorini

Subfamily Psocinae

Diagnosis. Distal inner labral sensilla five or nine. Vein M forming
distal closure of discoidal cell of forewing usually concave, sometimes
straight (Fig. 289). Hypandrium usually a single sclerotized sternal
plate, sometimes with two small basal sclerites or pigmented arms on
preceding segment; usually an asymmetrical central strap or tongue in
hypandrial sclerotization. Phallosomal skeleton usually a closed frame.
Male paraproctal process cylindrical, tapering to an acuminate point.
    North American tribes as noted above.

Tribe Psocini

Diagnosis. Five or nine distal inner labral sensilla. Hypandrium a
single broad, asymmetrical, sclerotized plate lacking paired sclerites or
pigmented arms on segment before the plate. V2 termination a gradually tapering process of moderate to short length. Female ninth
sternum with either a large sclerotized plate or a semimembranous
region with small rounded or keyhole-shaped plate around spermapore.
Male paraproct with a shoulder or lobe lateral to distal process. Male
clunial-epiproctal interface either straight or clunium overlapping
epiproct.
    North American genera: *Atropsocus* new genus, *Hyalopsocus*
Roesler, *Psocus* Latreille.

Key to the North American Genera
and Species of Tribe Psocini

1.   Forewings extensively banded (Fig. 361). Male clunium expand-
       ed as broad shelf over epiproct (Fig. 362). Five sensilla in distal
       inner labral field .... Genus *Atropsocus, A. atratus* (Aaron)
--   Forewings not extensively banded, at most a diagonal band plus
       a few spots. Male clunium only slightly extended over base of
       epiproct if at all. Nine sensilla in distal inner labral field 2

2.   Male clunium not extending over base of epiproct. Phallosome
       distally with single lobate process on right and pair of such
       processes on left (Fig. 363). V3 (in North American species)
       lacking median lobe .... Genus *Hyalopsocus* 3
--   Male clunium extending slightly over base of epiproct. Phallo-
       some variable distally but not as described above. V3 with
       median lobe .... Genus *Psocus* 4

3.   In forewing, cell IA mostly clear in distal half except at nodulus;
       no brown spots at marginal ends of branch veins of Rs and M
       or at end of Cu1a (Fig. 364) .... *H. striatus* (Walker)
--   In forewing cell IA pigmented all or nearly all of its length. A
       cloudy brown spot at marginal end of each branch vein of Rs
       and M and at end of Cu1A (Fig. 365) .. *H. floridanus* (Banks)

4.   In forewing a distinct rounded dark brown spot completely
       surrounding branching point of veins M and Cu1 (Fig. 289).
       Male clunium expanded as two lateral lobes each covering base
       of paraproct plus a small median lobe covering part of base of
       epiproct (Fig. 366) .... *P. leidyi* Aaron
--   In forewing a diffuse brown spot surrounding branching point of
       veins M and Cu1 and extending posteriorly along vein Cu1 (Fig.
       367). Male clunium expanded as single short median shelf
       covering base of epiproct (Fig. 368) .... *P. crosbyi* Chapman

Genus *Atropsocus* New Genus

Diagnosis. Adults of both sexes macropterous, relatively small,
forewing length ~ 2.7 mm. Forewing (Fig. 361) with extensive banding.
Five distal inner labral sensilla. Hypandrium (Fig. 817) with broad
median strap narrowing and bending upward distally, lacking basal
tooth or lobe. Phallosomal skeleton thin and truncated anteriorly,
expanded as large plate posteriorly (Fig. 818). Male epiproct mostly
covered by large clunial shelf (Fig. 362). Male paraproct (Fig. 819) with

crest lateral to distal process. Subgenital plate (Fig. 820): egg guide short, rounded, with numerous setae distally; pigmented arms wide, curving antero-laterally and expanded distally as broad quadrate areas; a heavily pigmented band through each arm and into base of quadrate area. Ovipositor valvulae (Fig. 821): v2 tapering gradually to form moderately long distal process; v3 a large transversely quadrate body with numerous long setae on hind margin; relatively short, slender median lobe. Female ninth sternum without sclerotization.

Generotype. '*Psocus*' *atratus* Aaron, the only included species at present.

Atropsocus atratus (Aaron)

Psocus atratus Aaron, 1883:39.
Trichadenotecnum atratum (Aaron). Enderlein, 1925:105.

Recognition features. Forewing markings (Fig. 361): pterostigma dark in distal two-thirds, the dark pigment continuing through distal half of broad stigmasaum, across rest of cell R1 to spot in basal half of cell R5; a marginal band covering M cells except clear marginal spot in each cell; band extending as distal spot into cell R5 and covering distal half of cell Cu1a; most of basal two-thirds of wing brown except small colorless spot in middle and larger one at distal end of cell Cu1b, colorless spot before nodulus in cell Cu2. Other details as noted in generic diagnosis.

Relationships. The genus appears to be closest to *Psocus* s. str.

Distribution and habitat. The species is known in the east from southern Ontario (Ottawa region) and southwestern New York (Steuben County) south to Alachua County, Florida. Farther west it is recorded from Ohio, Michigan (Livingston County), Indiana (several localities), and central Illinois (McLean County). Its seeming scarcity is probably due to the fact that it is rarely collected except by careful examination of tree trunks. It is known to inhabit trunks of cherry, elm, hackberry, and oak. Additional searching will probably show it to be much more widely distributed.

Genus *Hyalopsocus* Roesler

Hyalopsocus Roesler, 1954:572.
Tillapsocus Smithers, 1983:79.
Pictopsocus Lienhard, 1983:9.

Diagnosis. Adults of both sexes macropterous. Nine distal inner labral sensilla. Forewing marking (Fig. 364): at least pigmentation distally in pterostigma and stigmasaum; frequently a spot at M-Cu1 branching and one at nodulus; these spots sometimes connected by a partial or complete diagonal crossband; sometimes a spot in base of cell R5. Hypandrium with asymmetrical median strap bending upward and forward at apex. Phallosome (Fig. 363) with short, broad anterior apodeme, or broad-based; distally a single lobe on right and pair of lobes, one sometimes pointed, on left. Male epiproct flat, not extending anteriorly over clunium. Male paraproct with shoulder lateral to distal process (Fig. 822). Subgenital plate (Fig. 824) with distal process variable: relatively short and broad to long and slender; pigmented arms relatively broad, directed laterally. Ovipositor valvulae: apical process of v2 short or absent; v3 large, quadrate, lacking median lobe in North American species.

Generotype: *H. contrarius* (Reuter). Roesler, 1954:572.

North American species. *H. floridanus* (Banks), *H. striatus* (Walker).

Hyalopsocus floridanus (Banks)

Psocus floridanus Banks, 1905:2.
Hyalopsocus floridanus (Banks). Roesler, 1954:572.

Recognition features. Forewing markings (Fig. 365): pterostigma and stigmasaum brown in distal half, except a colorless spot on postero-distal margin; diffuse spotting along veins M and Cu1a in region of their junction; cell 1A largely brown, the pigment covering parts of adjacent cell Cu2, including its distal one-fifth; an obscure spot in cell R5 near base; distal ends of Rs, M veins and vein Cu1a each with a small brown spot ; some females with spot around Rs - M junction, brown clouding in distal Rs and M cells. Phallosome (Fig. 823) with anterior apodeme about twice as long as its width before its subapical expansion; right distal lobe cylindrical, rounded at tip, directed posteriorly; left distal lobes: more median one short, blunt pointed; more lateral one longer, expanded distally. Distal process of subgenital plate slender, spatulate (Fig. 824). Second valvula with apical process minute (Fig. 825).

Relationships. The species is very close to *H. striatus* and several undescribed species from southwestern United States.

Distribution and habitat. The species occurs over most of eastern United States and in southeastern Canada (North County, New Brunswick; southern Quebec), south to Dade County, Florida, west to central Illinois, southeastern Missouri, and southwestern Arkansas (Clark County). It inhabits trunks and branches of broad-leaf trees and pines.

Hyalopsocus striatus (Walker)

Psocus striatus Walker, 1853:486.
Psocus frontalis Harris, 1869:330.
Clematostigma striatum (Walker). Enderlein, 1925:102.
Hyalopsocus striatus (Walker). Roesler, 1954:572.

Recognition features. Forewing markings variable (Fig. 364): pterostigma and stigmasaum brown except basal one- to two-thirds and narrow distal margin or posterior part of it; a diffuse to concentrated spot bordering M-Cu1 branching point, mostly posterior to branching point; cell 1A brown in basal half and before nodulus; cell Cu2 brown before nodulus; a diffuse brown spot around Rs - M junction. Some females with diffuse brown band connecting 1A - Cu2 markings with M - Cu1 and Rs - M markings. Phallosome (Fig. 363) with anterior apodeme as described for H. floridanus; right distal lobe broad in middle, tapering somewhat to rounded tip; left distal lobes: more median one short, blunt pointed; more lateral one longer, cylindrical, rounded distally. Female genitalic characters as described for H. floridanus except distal process of subgenital plate more heavily sclerotized and darkly pigmented (paler in H. floridanus).

Relationships. These are as noted for *H. floridanus*.

Distribution and habitat. The species occurs throughout northeastern United States and in southeastern Canada (Ottawa region, Ontario; Kouchibouguac National Park, New Brunswick). It appears to reach its southern limit in southern Illinois, central Kentucky (Fayette County), and the Washington, D.C. region. It occurs spottily in western United States. Specimens are on hand from Utah County, Utah, and Tuolumne County, California. It inhabits trunks and branches of broad-leaf trees and pines, as well as shaded wooden pilings and shaded sides of wooden buildings where the wood is well weathered and supports an algal growth.

Genus *Psocus* Latreille

Psocus Latreille, 1794:85.
Psochus Latreille, 1796:99.

Diagnosis. Adults of both sexes macropterous. Nine distal inner labral sensilla. Forewing marking (Fig. 289): pterostigma with brown spot, very variable in size, likewise stigmasaum; a brown mark associated with M-Cu1 branching point; a brown spot at nodulus. Hypandrium lacking basal sclerites; with broad median strap bending upward distally and bearing basal tooth or lobe. Phallosome somewhat tapering and rounded anteriorly or with short anterior apodeme; distally variable. Male epiproct flat, not extending anteriorly over clunium; clunium extending at least slightly over epiproct in middle (Fig. 366). Male paraproct with shoulder lateral to distal process (Fig. 290). Subgenital plate (Fig. 832) with distal process relatively long and slender; pigmented arms directed postero-laterally, bearing a shallow to moderate trench and row of setae along their length; arms expanded distally. Ovipositor valvulae (Fig. 833): apical process of v2 short; v3 large, rounded or quadrate, its median lobe short to moderate.

Generotype. *P. bipunctatus* (Linnaeus).

North American species: *P. crosbyi* Chapman, *P. leidyi* Aaron.

Psocus crosbyi Chapman

Psocus crosbyi Chapman, 1930:235.

Recognition features. Forewing markings (Fig. 367): pterostigma brown except basal one-third and small distal margin; likewise stigmasaum; cloudy brown spot around Rs before junction with M; females with cloudy brown connecting region between spot at M - Cu1b branching and spot at nodulus, also brown wash in all cells in distal one-third of wing. Hypandrium (Fig. 826) with short basal lobe on median strap; the strap relatively slender, its basal width about one-third of its length. Phallosome (Fig. 827) bluntly tapered anteriorly; aedeagal region with two broad, rounded lobes, denticulate on their outer surfaces and separated in middle by a short, broad tooth. Male clunium extending as rounded shelf over base of epiproct (Fig. 368). Subgenital plate (Fig. 828): distal piece with broad base narrowing abruptly to tongue-like spatula slightly more than 2X as long as its basal width; pigmented arms very broad with shallow groove subtended by dense row of long setae. Ovipositor valvulae (Fig. 829): v1 broadened before distal process; v3 quadrate, broader on median than lateral margin; median lobe protruding slightly beyond margin of valvula.

Relationships. These are not understood at present. The species does not appear to be very close to *P. leidyi*.

Distribution and habitat. The species is found on the Pacific Coast from Langford, British Columbia south to San Luis Obispo County, California. All collections have been within 80 Km of the seashore, on several kinds of broad-leaf trees, shrubs, and vines including *Cornus*, *Sambucus*, and *Vitis*.

Psocus leidyi Aaron

Psocus leidyi Aaron, 1886:15.
Psocus bilobatus Banks, 1918:4.

Recognition features. Forewing markings (Fig. 289): brown spot in pterostigma occupying distal half or less but separated from margin anteriorly and distally; distal brown spot in stigmasaum; brown mark around M- Cu1 branching completely separate (male) or connected by brown wash (female) to spot at nodulus. Hypandrium (Fig. 830): median strap relatively broad with blunt basal process, a clear but well sclerotized rounded lobe to each side. Phallosome (Fig. 831) rounded basally; aedeagus with thick neck region heavily denticulate on right side, expanding into rounded distal lobe bearing forward-directed, denticulate process on left side. Subgenital plate (Fig. 832): distal piece gradually tapering to spatula of proportions described for *P. crosbyi*; pigmented arms lightly pigmented, each with moderately deep groove subtended by row of scattered setae; arms greatly expanded distally. Ovipositor valvulae (Fig. 833): v1 as described for *P. crosbyi*; v3 rounded medially, squared disto-laterally; median lobe not extending to margin of valvula.

Relationships. The species appears to be close to *P. bipunctatus* (L.) of Europe.

Distribution and habitat. The species occurs from Maine and New Brunswick west to northern Minnesota, south to Highlands County, Florida, and southwest to central Arkansas (Saline County). An outlier population occurs in northern California (Humboldt, Trinity, and Mendocino Counties). The species inhabits trunks and branches of broad-leaf trees and pines, and occurs occasionally on stone outcrops.

Tribe Ptyctini, new tribe

Diagnosis. Five or nine distal inner labral sensilla. Hypandrium an asymmetrical sclerotized plate usually with curved median strap or movable median tongue, usually with small pair of sclerites or pigmented arms on preceding segment. V2 terminating in an abruptly tapered slender process. Sclerotization of female ninth sternum usually a two-part structure with basal piece relatively large. Male paraproct lacking shoulder or lobe lateral to distal process. Male clunial-epiproctal interface either straight or epiproct overlapping clunium.

North American genera: *Camelopsocus* Mockford, *Indiopsocus* Mockford, *Loensia* Enderlein, *Ptycta* Enderlein, *Steleops* Enderlein, *Trichadenotecnum* Enderlein.

Key to the North American Genera and Species of Tribe Ptyctini

1. Adults strongly sexually dimorphic: males macropterous with slightly raised tubercle dorsally on abdomen; females micropterous with middle segments of abdomen bearing a raised hump (Fig. 297) .... Genus *Camelopsocus* 2

-- Adults not as above, rarely sexually dimorphic and never with a dorsal abdominal hump or tubercle 6

2. Hypandrium with at least a few tubercle-based setae before base of median strap (Fig. 298). Female abdomen with a major dorsal turret on segment 5 and a minor one on segment 4, occasionally with lesser ones on segments 3 and 2 (Fig. 297) 3

-- Hypandrium with no tubercle-based setae before base of median strap. Female abdomen with a major dorsal turret on segment 5 and no other turrets 5

3. Median strap of hypandrium parallel-sided most of its length (Fig. 298). Distal process of phallosome short, not longer than its basal width, truncated apically (Fig. 299). Pigmented arms of subgenital plate parallel at bases then diverging abruptly; region between arm bases containing a slender pigment band (Fig. 300) .... *C. bactrianus* Mockford

-- Median strap of hypandrium widest in middle, tapering to each end. Distal process of phallosome decidedly longer than its basal width and bluntly pointed apically. Pigmented arms of subgenital plate diverging from their bases; no pigmented band between arms 4

4. Median strap of hypandrium ~ 1.3X as wide at its greatest width as at base (Fig. 301). Pigmented arms of subgenital plate (Fig. 302) relatively narrow with relatively broad bases; greatest width of an arm ~ 1.2X basal width .... *C. tucsonensis* Mockford

-- Median strap of hypandrium ~ 2X as wide at its greatest width as at base (Fig. 303). Pigmented arms of subgenital plate (Fig. 304) relatively broad with relatively narrow bases; greatest width of an arm ~ 2.1X basal width ..... *C. hiemalis* Mockford

5. Edges of phallic frame broad in distal half (Fig. 305). Pigmented arms of subgenital plate with anterior ends truncated and directed laterally (Fig. 306) .... *C. similis* Mockford

-- Edges of phallic frame slender in distal half (Fig. 307). Pigmented arms of subgenital plate with anterior ends tapering and curved backward (Fig. 308) .... *C. monticolus* Mockford

6. Phallosome broad at base. Male epiproct flat, meeting clunium in straight line. Distal half of male (and usually female) forewing unmarked except for a spot in pterostigma, one in stigmasaum, and sometimes a spot in base of cell R5. Five distal inner labral sensilla .... Genus *Indiopsocus* 7

--   Phallosome slender or tapering towards base. Male epiproct
        extending over clunium in middle. Distal half of forewing either
        extensively spotted or banded, or if marked as above, then nine
        distal inner labral sensilla present 12

7.   Forewings (Fig. 846) dusky gray, darker on pterostigma and
        stigmasaum, but not otherwise marked ....
        .... *I. infumatus* (Banks)
--   Forewings not as above, generally clear with several pigment
        spots other than those in pterostigma and stigmasaum 8

8.   Frame of phallosome parallel-sided bearing distally two long
        processes, each as long as the frame (Fig. 309). Hypandrium
        with denticulate ridge to right of midline dividing it into more
        sclerotized right-hand region and more membranous left-
        hand region (Fig. 310). Pigmented arms of subgenital plate
        directed antero-laterally at base, then bent abruptly forward,
        then expanded to sides as large quadrate areas (Fig. 311) ...
        .... *I. campestris* (Aaron)
--   Frame of phallosome with sides converging distally, usually with
        a single distal process much shorter than frame (Fig. 312).
        Hypandrium with an asymmetrical central strap. Pigmented
        arms of subgenital plate not as described above 9

9.   Sexually dimorphic: most females brachypterous, males macrop-
        terous; rare macropterous females with banding in distal half
        of forewing (Fig. 313). Phallosome essentially symmetrical with
        relatively long median distal process (Fig. 312) ....
        .... *I. coquilletti* (Banks)
--   Sexual dimorphism absent. Phallosome either clearly asymmetri-
        cal, or, if nearly symmetrical, then distal process short, subap-
        ical on an apical plate 10

10.   Forewing with complete diagonal pigment band from base of
        pterostigma to nodulus (Fig. 314). Phallosome with distal
        process relatively long, arising left of midline 11
--   Forewing (Fig. 315) with partial transverse pigment band con-
        sisting of spot in cell R near distal end, spot behind branching
        point of M and Cu 1, and spot at nodulus. Phallosome with
        distal process short, pointed, approximately on midline of
        expanded distal plate (Fig. 316) .... *I. bisignatus* (Banks)

11. Phallosome with distal process slender, rounded at tip, separated by deep excavation from right distal lobe (Fig. 317). Pair of sclerotized lines running lengthwise through egg guide separated at bases by about one-third width of egg guide (Fig. 318) .... *I. ceterus* Mockford
-- Distal process of phallosome broader, truncated apically, separated by shallow excavation from right distal lobe (Fig. 319). Pair of sclerotized lines running through egg guide separated at bases by only about width of a line (Fig. 320) ....
.... *I. texanus* (Aaron)

12. A pair of pigmented arms extending forward from principal sclerotized region of hypandrium (Fig. 874). Forewings extensively marked, always with radial series of spots, one in each of cells R1 through M3 .... Genus *Trichadenotecnum* 13
-- No pigmented arms extending forward from principal sclerotized region of hypandrium. Forewings variously marked but usually lacking series of radial spots as designated above 22

13. V3 lacking median lobe. Pigmented arms of subgenital plate directed anteriorly or antero-laterally 14
-- V3 with median lobe well developed. Pigmented arms of subgenital plate directed laterally, or if anteriorly at base, curving abruptly laterally 15

14. Spots of radial series large and distinct; those of M cells rendered more distinct by each being juxtaposed with a distal clear spot (Fig. 322). Egg guide bilobed at apex (Fig. 323 ....
.... *T. circularoides* Badonnel
-- Spots of radial series not large, those of cells M2 and M3 no larger than surrounding spots (Fig. 324). Egg guide slightly flattened at apex (Fig. 325) .... *T. pardus* Badonnel

15. All spots of radial series rendered obscure by surrounding spots (Fig. 326). Pigmented arms of subgenital plate abruptly broadened near bases to form large quadrate areas lying close together on midline (Fig. 327) .... *T. desolatum* (Chapman)
-- Spots of radial series distinct although irregular in outline in some cases. Pigmented arms of subgenital plate not as above, usually terminating in quadrate or triangular areas not close together on midline 16

16. Pigment of distal forewing margin forming a continuous band from cell R3 to cell M3 or at least some spots bordering veins of this region connected by cross band (Fig. 328) 17

-- Pigment of distal forewing margin not a continuous band, concentrated as elongate spots along veins (Fig. 329) .... *T. alexanderae* complex 19

17. Diagonal band of forewing incomplete, interrupted or greatly constricted on vein M immediately beyond M-Cu1 branching (Fig. 328). Female forewing length > 4.2 mm, male somewhat shorter (~ 4.0 mm). Male epiproct approximately quadrate (Fig. 330) .... *T. majus* Kolbe

-- Diagonal band of forewing not interrupted. Female forewing length not over 3.8 mm, that of male not exceeding 3.7 mm. Male epiproct greatly expanded anteriorly (Fig. 331) 18

18. Spots of radial series of forewing relatively smooth in outline (Fig. 332). Pigmented arms of subgenital plate each terminating laterally in large triangular area (Fig. 333) .... *T. slossonae* (Banks)

-- Spots of radial series of forewing irregular in outline (Fig. 334). Pigmented arms of subgenital plate each terminating laterally in quadrate area longer than broad (Fig. 335) .... *T. quaesitum* (Chapman)

19. Sclerites of ninth sternum: a distinct clear crescent in anterior region; posterior region pigmented throughout except extreme edges (Fig. 336). In forewing, vein M limiting discoidal cell decidedly concave (Fig. 337) .... *T. innuptum* Betz

-- Sclerites of ninth sternum: central clear structure in anterior region not crescent-shaped; posterior region with unpigmented area around small spermapore pigmentation (Fig. 338). In forewing, vein M limiting discoidal cell approximately straight (Fig. 339) 20

20. In forewing, pigmentation bordering vein limiting cell M3 anteriorly narrowly restricted to edges of vein (Fig. 339). Clear region in middle of egg guide frequently indistinct at one or both ends (Fig. 877) .... *T. castum* Betz

-- In forewing, pigmentation bordering vein limiting cell M3 anteriorly broader (Fig. 340). Clear region in middle of egg guide usually distinct at both ends (Fig. 341) 21

21. Sclerites of ninth sternum: pigmented horns of posterior region extending beyond spermapore (Fig. 342). Known from females only .... *T. merum* Betz

-- Sclerites of ninth sternum: pigmented horns of posterior region not reaching level of spermapore (Fig. 343). Known from both sexes *T. alexanderae* Sommerman

22. Eyes at least slightly stalked (Fig. 344). No pair of small sclerites anterior to principal sclerotized region of hypandrium. Outer tine of lacinial tip slender, spatulate, with no apical denticles (Fig. 345) Genus *Sieleops* 23
-- Eyes not stalked. Pair of small sclerites anterior to principal sclerotized region of hypandrium present or absent. Outer tine of lacinial tip broad, denticulate at apex (Fig. 346) 24

23. Body pale tawny brown. Antennae banded brown and white. Forewings extensively brown marked over most of basal half, with brown band extending to apex covering all of cell R5; all brown areas sprinkled with colorless lacunae (Fig. 347) ....
.... *S. lichenatus* (Walsh)
-- Body extensively white. Antennae not banded. Forewing markings restricted to diagonal band, distal half of pterostigma and band through bases of cells M1 through Cu1a (Fig. 348) ....
.... *S. elegans* (Banks)

24. Forewing extensively spotted with brown. Five distal inner labral sensilla .... Genus *Loensia* 25
-- Forewing either mostly unmarked or with distal submarginal band and a few spots in and around base of cell R5. Nine distal inner labral sensilla .... Genus *Ptycta* 28

25. Forewing with sparse scattered small spots; only larger spots in pterostigma, middle of cell R5, nodulus, and basal half of cell 1A (Fig. 349). Hypandrium lacking pair of sclerites on segment before principal sclerotization. Phallosome terminating apically in spearhead shaped process (Fig. 350) ....
.... *L. conspersa* (Banks)
-- Forewing more extensively marked. Hypandrium with pair of rounded sclerites on segment before principal sclerotization. Apex of phallosome single broader structure or paired structures 26

26. Forewing with diagonal band and distal submarginal band plus scattered spots (Fig. 351). Left strap of hypandrium longer than right one (Fig. 352) .... *L. fasciata* (Fabricius)
-- Forewing extensively spotted; some spots running together to form larger blotches, but without bands (Fig. 353). Left process of hypandrium not over half as long as right one (Fig. 354) ....
.... 27

27. Spotting dense over most of forewing (Fig. 353). Left strap of hypandrium about half length of right strap (Fig. 354). Phallosome terminating in simple process slightly bilobed at tip (Fig. 355) .... *L. moesta* (Hagen)

-- Spotting relatively sparse over much of forewing (Fig. 356) leaving large clear areas, but spots coalescing in cells M1, M3, and middle of Cu1b, forming large blotches. Left process of hypandrium a short lobe much less than half length of right strap (Fig. 357). Phallosome terminating in pair of processes (Fig. 358) .... *L. maculosa* (Banks)

28. Forewing largely unmarked except for brown spot in pterostigma and stigmasaum and light spotting in and around distal end of cell R (Fig. 359) ....*P. lineata* Mockford

-- Forewing much more extensively marked: a nearly complete diagonal band; spot in pterostigma continuing through stigmasaum into cell R1, two spots before middle in cell R5, and a distal marginal band through radial and medial cells into cell Cu1a (Fig. 360) .... *P. polluta* (Walsh)

Genus *Camelopsocus* Mockford

Camelopsocus Mockford, 1965c:3.

Diagnosis. Strongly sexually dimorphic, males macropterous with cylindrical abdomen bearing slightly raised tubercle dorsally on middle segments (Fig. 834); female micropterous with abdomen rounded and greatly raised as a hump in middle segments (Fig. 297). Nine distal inner labral sensilla. Pterostigma shallow. Forewing (Fig. 835) unmarked except for pigmentation of pterostigma and stigmasaum. Hypandrium (Fig. 298) lacking basal sclerites, with asymmetrical median strap bending upward and forward at apex. Phallosome (Fig. 299) with a long, slender anterior apodeme. Male epiproct chair-shaped (Fig. 836), capable of flexing forward over clunium. Subgenital plate (Fig. 300) with broad, rounded distal process; pigmented arms slender, usually curved. Ovipositor valvulae (Fig. 294): apical process of v2 extremely slender; v3 with broad, rounded median lobe.

Generotype. *C. monticolus* Mockford.

North American species: *C. bactrianus* Mockford, *C. hiemalis* Mockford, *C. monticolus* Mockford, *C. similis* Mockford, *C. tucsonensis* Mockford.

Camelopsocus bactrianus Mockford

Camelopsocus bactrianus Mockford, 1984b:197.

Recognition features. Hypandrium (Fig. 298) with small field of tubercle-based setae anterior to median strap; median strap approximately parallel-sided most of its length. Distal process of phallosome short, truncated apically (Fig. 299). Female abdomen with a major turret on segment five and a minor one on segment four (Fig. 297). Pigmented arms of subgenital plate parallel at bases then diverging abruptly; region between parallel bases of arms containing a pigment band (Fig. 300).

Relationships. All of the species of this genus are very close to each-other, and more precise relationships among them cannot be dealt with at present.

Distribution and habitat. The species is known only from Riverside County, California, where nymphs and adults are found on chaparral shrubs during winter and spring (November to May).

Camelopsocus hiemalis Mockford

Camelopsocus hiemalis Mockford, 1984b:199.

Recognition features. Hypandrium (Fig. 303) with well developed field of tubercle-based setae anterior to median strap; median strap broadest before middle, tapering towards both ends. Distal process of phallosome (Fig. 837) moderately long, tapering to blunt point. Female abdomen with a major turret on segment five and a minor one on segment four. Pigmented arms of subgenital plate (Fig. 304) diverging from their bases; region between arms without a pigment band.

Relationships. These are as noted for *C. bactrianus*.

Distribution and habitat. The species has been found at several localities in Riverside County, California. It inhabits chaparral shrubs, including jojoba, *Artemisia californica*, and *Salvia apiana*.

Camelopsocus monticolus Mockford

Camelopsocus monticolus Mockford, 1965c:4.

Recognition features. Hypandrium (Fig. 838) lacking tubercle-based setae anterior to base of median strap; median strap broadest before middle, tapering towards both ends. Distal process of phallosome (Fig. 307) moderately long, tapering to point, with scaly sculpturing basally, without flanking lobes. Female abdomen with major turret on segment five and no other turrets. Pigmented arms of subgenital plate (Fig. 308) with anterior ends tapering and curved backwards.

Relationships. These are as noted for *C. bactrianus.*

Distribution and habitat. Within the study area, the species is known from Larimer County, Colorado south through western New Mexico and eastern Arizona, where it has been collected on branches of oaks, pines, and junipers. Outside the study area, it is known from the Mexican states of Durango, Nuevo León, Oaxaca, and Zacatecas. Generally, it is found at elevations around 2000 m.

Camelopsocus similis Mockford

Camelopsocus similis Mockford, 1965c:6.

Recognition features. Hypandrium (Fig. 839) lacking tubercle-based setae anterior to base of median strap; median strap widest basally, tapering towards tip. Distal process of phallosome (Fig. 305) broad basally, tapering to pointed tip. Pigmented arms of subgenital plate (Fig. 306) with anterior ends truncated and directed laterally.

Relationships. These are as noted for *C. bactrianus.*

Distribution and habitat. Within the study area, the species is known from Teton and Sweetwater Counties, Wyoming, and Chaffee and Gunnison Counties, Colorado, where it has been collected on sagebrush, oaks, small heath plants, and Douglas fir. Outside the study area, it is known from the Mexican states of Coahuila, Durango, Nuevo León, Oaxaca, and San Luis Potosí. It is found at elevations of 1800 - 2700 m.

Camelopsocus tucsonensis Mockford

Camelopsocus tucsonensis Mockford, 1984b:204.

Recognition features. As described for *C. hiemalis* except greatest width of median strap of hypandrium (Fig. 301) only ~ 1.4X its basal width (vs. 2.0X in C. hiemalis); frame of phallosome (Fig. 840) tapering towards its base (vs. parallel-sided in *C. hiemalis*); pigmented arms of subgenital plate (Fig. 302) only ~ 1.2X as broad at their greatest width as at their base (vs. ~ 2.1X in *C. hiemalis*).

Relationships. The species is very close to *C. hiemalis.*

Distribution and habitat. The only known material was taken in the vicinity of the Desert Museum, Tucson Mountains, Pima County, Arizona, on jojoba shrubs.

Genus *Indiopsocus* Mockford

Indiopsocus Mockford, 1974a:165.

Diagnosis. Adults of both sexes usually macropterous; females rarely brachypterous. Five distal inner labral sensilla. Forewing markings variable, but usually a dark mark in pterostigma and stigmasaum and a partial or complete brown band across wing from distal end of cell R to nodulus. Hypandrium usually with pair of rounded basal sclerites; asymmetrical median strap bending upward and forward at apex. Phallosome broad based; apex usually asymmetrical; North American species with a pointed process usually to one side of midline, rarely symmetrical with pointed process on midline. Male epiproct flat, not extending anteriorly over clunium. Subgenital plate with distal process tapering from base to tip or parallel-sided, usually 1-2X as long as its basal width; pigmented arms usually slender basally, becoming broader distally. Ovipositor valvulae: apical process of v2 relatively long and slender; v3 transversely elongate, with well developed median lobe.

Generotype. *I. texanus* (Aaron).

North American species: *I. bisignatus* (Banks), *I. campestris* (Aaron), *I. ceterus* Mockford, *I. coquilletti* (Banks), *I. infumatus* (Banks), *I. texanus* (Aaron).

Indiopsocus bisignatus (Banks)

Psocus bisignatus Banks, 1904b:203.
Psocidus bisignatus (Banks). Smithers, 1967:107.
Indiopsocus bisignatus (Banks). Mockford, 1985b:21.

Recognition features. Both sexes macropterous. Forewing markings (Fig. 315): pterostigma mostly brown except basal one-third and distal margin; stigmasaum brown except basal one-third; a nearly continuous brown band from distal end of cell R to nodulus, narrowly interrupted before vein Cu2; most of cell 1A brown except clear spot at distal two-thirds; a pale brown spot in cell R5 near base. Hypandrium (Fig. 841) with pair of rounded basal sclerites. Phallosome (Fig. 316): distally a broad, nearly symmetrical sclerotized area with two broad, short distal lobes and a pointed process nearly on midline. Distal process of subgenital plate (Fig. 842) with base broad, slightly broader than length of process; sides of process tapering from base to relatively broad, slightly curved distal margin.

Relationships. The species is closest to *I. infumatus* (Banks).

Distribution and habitat. Within the study area, the species occurs extensively in eastern United States from northern Minnesota (Voyageurs National Park) and western New York (Ithaca region) and New Jersey, south to West Florida (Franklin County). West of the Mississippi River it occurs spottily to Wyoming (Converse County) and Colorado (Gunnison County) and in the Texas counties of Jim Wells and Edwards. It inhabits trunks and branches of broad-leaf trees and conifers. Outside the study area, it is known from the Mexican states of Oaxaca, San Luis Potosí, Sinaloa, Sonora, Tamaulipas, and Veracruz, also from Honduras.

Indiopsocus campestris (Aaron), new combination

Psocus campestris Aaron, 1886:14.
Psocus insulanus Chapman, 1930:244, new synonym.
Psocidus insulanus (Chapman). Smithers, 1967:108, new synonym.
Indiopsocus insulanus (Chapman). Mockford, 1974a:175, new synonym.

Recognition features. Adults of both sexes macropterous. Forewing markings (Fig. 850): a diffuse pale brown spot over distal half of pterostigma except extreme distal margin clear; stigmasaum pale brown; a pale brown spot in distal end of cell Cu2; a diffuse pale brown spot occupying basal one-third of cell IA. Hypandrium (Fig. 310) with basal sclerites present but weakly sclerotized; median strap broad, triangular, bent slightly leftward at apex; a row of large denticles running to apex of strap in middle, dividing it into left, weakly sclerotized region and right well sclerotized region; latter region with field of smaller denticles distally. Phallosome (Fig. 309) with two long distal lobes separated by deep invagination; right lobe bilobed at apex; left lobe with short pointed process on median side. Subgenital plate (Fig. 311): distal process broad basally, tapering abruptly to short spatula, the latter not quite as long as its basal width; pigmented arms arising far apart, directed anteriorly, abruptly broadened laterally to form each a quadrate area.

Relationships. The species appears to be most closely related to *I. affinis* Mockford (Cuba).

Distribution and habitat. The species occurs throughout Florida, north on the Atlantic Coast to Long Island, New York, and around the Gulf Coast to southern Texas. It has been collected on branches of live oak, turkey oak, several other kinds of broad-leaf trees and shrubs in scrub and flatwoods vegetation, and on pine branches.

Indiopsocus ceterus Mockford

Indiopsocus ceterus Mockford, 1974a:189.

Recognition features. Both sexes macropterous. Forewing markings (Fig. 314): pterostigma with brown spot in distal one-third, not reaching distal margin; stigmasaum brown in distal one-third; a nearly complete slender brown band across wing from base of pterostigma to nodulus; a brown spot in cell IA near its base. Hypandrium (Fig. 843) with pair of rounded basal sclerites; median strap of hypandrium broad basally, dividing into two straps at about half distance from base to tip: major right-hand strap and minor left-hand strap. Right-hand strap with two rows of denticles running its length; left-hand strap with single row of denticles. Phallosome (Fig. 317) with right distal lobe relatively broad and short, separated by broad, rounded invagination from left distal lobe; latter lobe bearing a long, somewhat left-curved finger-like process arising from broad base. Distal process of subgenital plate (Fig. 318) broad-based, narrowing abruptly to parallel-sided spatula ~ 2.5X as long as its basal width; median longitudinal sclerotized strips of the spatula farther apart than distance between one of them and edge of spatula. Pigmented arms of subgenital plate projecting antero-laterally from broad median pigmented area; arms slender at their bases, greatly widened and rounded distally.

Relationships. The species is close to *I. texanus* (Aaron) of southeastern United States and Cuba, also to several Cuban species -- *I. cubanus* (Banks), *I. ubiquitus* Mockford, and *I. camagueyensis* Mockford.

Distribution and habitat. Within the study area, the species occurs in southeastern Georgia (Camden and Candler Counties), and in the south-central Florida counties of Hernando, Lee, Sarasota, and Highlands. It has been collected on sand pine, saw palmetto, and several kinds of scrub oaks. Outside the study area it is known from Mayari, Oriente Province, Cuba (type locality).

Indiopsocus coquilletti (Banks)

Psocus coquilletti Banks, 1920:305.
Psocidus coquilletti (Banks). Smithers, 1967:107.
Indiopsocus coquilletti (Banks). Mockford, 1985b:14.

Recognition features. Males macropterous; females dimorphic: macropterous and brachypterous. Male forewing markings (Fig. 844): pterostigma brown in distal two-fifths; stigmasaum brown throughout; a spot along vein M + Cu1 before branching; a spot along vein Cu1 after branching; most of cell IA and distal end of cell Cu2 brown. Macropterous female forewing markings (Fig. 313): a small spot in distal end of pterostigma, another in distal end of stigmasaum; a spot covering most of basal half of cell R3 and most of distal half of cell R5, another in basal half of cell R5; another along vein M before junction with Cu1a, continuing along Cu1a to wing margin and covering most of cell M3 except for clear marginal spot; veins M1 posteriorly, M2 on both sides, and M3 anteriorly edged in brown; a spot along vein M + Cu1 before branching, another along vein Cu1 after branching; a spot covering most of distal two-thirds of cell Cu2 and extending into cell lA in basal half and at distal end. Brachypterous female forewing largely brown except for irregular colorless blotches in basal half of wing and a colorless marginal spot in each distal cell. Hypandrium (Fig. 845) lacking pair of basal sclerites; median strap of only moderate width, bending leftward, upward, and forward distally. Phallosome (Fig. 312) symmetrical distally with median blunt-tipped process and a short lobe on each side. Distal process of subgenital plate broad-based, not longer than its basal width; pigmented arms of plate only slightly widened beyond their bases.

Relationships. The species appears to be closest to *I. bisignatus* and *I. infumatus* (eastern United States).

Distribution and habitat. Within the study area the species is known only from Los Angeles and Riverside Counties, California, where it inhabits chaparral shrubs. Outside the study area, it is recorded from the Mexican states of Baja California and Baja California Sur.

Indiopsocus infumatus (Banks)

Psocus infumatus Banks, 1907a:165.
Psocidus infumatus (Banks). Smithers, 1967:108.
Indiopsocus infumatus (Banks). Mockford, 1985b:22.

Recognition features. Adults of both sexes macropterous. Forewing markings (Fig. 846): pterostigma and stigmasaum medium brown; remainder of wing pale smoky gray-brown, slightly darker around nodulus. Hypandrium (Fig. 847) with pair of small basal sclerites; median strap tapering abruptly from somewhat swollen base, curved left, upward, and forward distally. Phallosome (Fig. 848): right distal lobe short, broad, rounded, with slightly lobed margin; separated from left lobe by narrow invagination; left lobe with long, pointed process near midline, otherwise nearly mirror image of right lobe. Subgenital plate (Fig. 849): distal process broad-based, abruptly tapering to broad, short spatula, the latter ~ 1.2X as long as its basal width; pigmented arms arising on midline, each abruptly broadening to form triangular distal region.

Relationships. The species appears to be very close to an undescribed species from Durango State, Mexico, and another undescribed species from Oaxaca State, Mexico.

Distribution and habitat. The species is known from the eastern and central states of New York, Maryland, Virginia, North Carolina, (western mountains), Tennessee, Kentucky, Indiana, Illinois, and Wisconsin also from Kouchibouguac National Park, New Brunswick. It inhabits dead branches of broad-leaf trees.

Indiopsocus texanus (Aaron)

Psocus texanus Aaron, 1886:16.
Psocidus texanus (Aaron). Smithers, 1967:111.
Indiopsocus texanus (Aaron). Mockford, 1974a:169.

Recognition features. Both sexes macropterous. Forewing markings (Fig. 851): nearly identical to those of *I. ceterus*, but distal spot of pterostigma tending to be larger, occupying up to entire distal half of pterostigma (except for clear distal and anterior margin); as much as distal half of stigmasaum brown; basal spot in cell 1A occupying as much as basal half of cell. Hypandrium (Fig. 293) with distinct, rounded basal sclerites; median strap relatively broad basally, gradually tapering to slender apex, where twisted leftward and curved upward and forward; left side of strap with field of denticles near base and two ridges, more basal one bearing row of denticles joined to the basal field. Phallosome (Fig. 319) with right distal lobe relatively broad, tapering towards apex, separated from left distal lobe by broad shallow invagination; left distal lobe with broad, left-curved, tapering process near midline and broader, shorter tapering process laterally. Distal process of subgenital plate (Fig. 320) broad based, narrowing abruptly to form spatula, the latter slightly constricted before middle then somewhat broadened towards apex; pair of sclerotized strips of distal process joined on midline, or nearly so, at base. Pigmented arms of subgenital plate arising somewhat farther apart than width of an arm at base, directed antero-laterally, each forming quadrate region at its anterior end.

Relationships. These are as discussed for *I. ceterus*.

Distribution and habitat. Within the study area the species occurs throughout Florida, up the Atlantic Coast to southern Virginia, and on the Gulf Coast of Texas. It is probably continuous around the Gulf Coast from Florida to Texas, but there are no collection records. The record from Long Island, New York (Mockford, 1974a) was an error. It is found on sand pine branches, and branches of various broad-leaf trees and shrubs, especially in coastal scrub vegetation, and in Florida also on inland dunes. Outside the study area. It is recorded from the Cuban provinces of Oriente and Habana.

Genus *Loensia* Enderlein

Loensia Enderlein, 1924:35.

Diagnosis. Adults of both sexes macropterous. Five distal inner labral sensilla. Forewing marking: scattered spotting and blotching over most of wing surface with no particular pattern. Hypandrium with or without basal sclerites; with asymmetrical median strap based on right side, trending leftward, and bent upward and forward at apex; in some species strap not bending leftward but a lobe or process on left side sometimes as long as or longer than strap. Phallosome with short anterior apodeme varying greatly in width among species; distally with a single process, usually symmetrical. Male epiproct chair-shaped, extending anteriorly over clunium. Subgenital plate with distal process short and wide, about as long as, or slightly shorter than, its basal width; pigmented arms slender, directed forward, abruptly widening to form broad, quadrate lateral areas. Ovipositor valvulae: apical process of v2 long and slender; v3 relatively slender, transverse, with moderate to long median lobe.

Generotype: *L. fasciata* (Fabricius).

North American species: *L. conspersa* (Banks), *L. fasciata* (Fabricius), *L. maculosa* (Banks), *L. moesta* (Hagen).

Loensia conspersa (Banks), new combination

Psocus conspersus Banks, 1903a:237.
Psocidus conspersus (Banks). Smithers, 1967:107.

Recognition features. Forewing markings (Fig. 349): sparse, scattered small spots over most of wing, densest in basal cells R, Cu1b, and Cu2; sparsest in distal M cells; a large spot covering about middle two-thirds of pterostigma, extending to distal end along front margin and extending posteriorly through stigmasaum; a large spot in cell R5 before middle; a large spot around nodulus; a spot covering basal two-thirds of cell 1A. Hypandrium (Fig. 852) without basal sclerites; median strap tending leftward at base then curved posteriad; a row of denticles along each edge of strap; left half of hypandrium opposite base of strap with a short, rounded lobe. Phallosome (Fig. 350) with short, slender basal apodeme; slender, spear-head shaped distal process. Subgenital plate with distal process very short and wide, ~ 0.6X as long as its basal width. V3 with long, slender median process.

Relationships. The species does not appear to be very close to the other North American or the European species. Its relationships are not understood at present.

Distribution and habitat. The species is known from Inyo, Mariposa, and Modoc Counties, California; Deschutes County, Oregon; Coconino and Pima Counties, Arizona; Catron and Torrance Counties, New Mexico; Cache County, Utah, and Dawes County, Nebraska. These records suggest a more nearly continuous range in western United States, probably excluding the northern tier of states. It inhabits primarily branches and foliage of ponderosa pine and junipers.

Loensia fasciata (Fabricius)

Hemerobius fasciatus Fabricius, 1787:247.
Psocus fasciatus (Fabricius). Fabricius, 1798:203.
Psocus pilicornis Latreille, 1799:13.
Amphigerontia fasciata (Fabricius). Enderlein, 1906a:75.
Loensia fasciata (Fabricius). Enderlein, 1924:35.
Trichadenotecnum (*Loensia*) *fasciatum* (Fabricius). Roesler, 1943:5.

Recognition features (male characters from literature). Forewing markings (Fig. 351): some scattered brown spotting, but dominant pattern in bands: a submarginal band from cell R1 to cell Cu1a, reaching margin on veins; pterostigma brown in distal half, the pigment continuing across wing as a loose band to cell Cu1a; diagonal band broad from before pterostigmal base to distal two-fifths of cell 1A; a spot in basal half of cell 1A extending into cell Cu2. Hypandrium (Fig. 352): median strap nearly straight before apex, curving upward and leftward at apex; on left side a second strap longer than that of right, directed postero-laterally. Phallosome symmetrical with relatively long basal apodeme, pair of short lobes disto-laterally, single broad distal process. Subgenital plate as described for *L. maculosa*.

Relationships. These are as noted for *L. maculosa*.

Distribution and habitat. Within the study area, the species has been taken at Dover, Delaware, and an unspecified New Jersey locality in cargo areas of United States Air Force planes from Germany. Outside the study area, it occurs throughout Europe, on trunks and branches of broad-leaf trees and conifers.

Loensia maculosa (Banks)

Myopsocus maculosus Banks, 1908:258.
Psocus maculosus (Banks). Chapman, 1930:252.
Trichadenotecnum (*Loensia*) *maculosum* (Banks). Thornton, 1961:2.

Recognition features. Forewing markings (Fig. 356): scattered discrete brown spots over most of wing, the spots coalescing to form large blotches covering most of distal one-third of cell R5, basal half and anterior margin of cell M1, most of cell M3; a large spot covering distal half of pterostigma and distal half of stigmasaum, a large pale brown spot before middle in cell R5; a dark brown spot around M - Cu1 branching point. Hypandrium (Fig. 357) with pair of rounded basal sclerites; median strap elongate, curving slightly leftward at base, then running straight most of its length,then curving slightly leftward and bent upward and forward at apex; sides of strap regularly denticulate; left side of hypandrium opposite base of strap with a short, rounded lobe separated from strap by a narrow declivity. Phallosome (Fig. 358) nearly symmetrical, with a short, broad basal apodeme; pair of short subapical lobes, the right lobe arising lower than the left; pair of apical processes. Male epiproct (Fig. 853) with short, rounded lobe in back part of "seat." Subgenital plate (Fig. 854) with distal process moderately wide at base, ~ 1.5X as long as basal width; pigmented arms with short, slender bases abruptly widening to sides to form pair of very broad quadrate areas.

Relationships. The species is part of a group including *L. moesta* and the three European species of the genus.

Distribution and habitat. The species occurs in the Pacific Coastal states from Riverside County, California north to King County, Washington. From there, it extends eastward to western Idaho (Valley County). It inhabits primarily foliage and branches of coniferous trees but is occasionally found on broad-leaf trees.

Loensia moesta (Hagen)

Psocus moestus Hagen, 1861:11.
Amphigerontia moestus (Hagen). Banks, 1892:344.
Loensia moesta (Hagen). Mockford, 1950:201.
Trichadenotecnum moestum (Hagen). Smithers. 1967:116.

Recognition features. Forewing markings (Fig. 353): dense brown spotting over entire wing surface; spots coalescing to form larger blotches behind base of pterostigma, in distal one-third of pterostigma, before middle and at distal end of cell R5, over most of cells M3 and Cu1a and in distal half of cell Cu2; a large dark spot around branching point of veins M and Cu1. Hypandrium (Fig. 354) with pair of rounded basal sclerites; median strap as described for *L. maculosa*; left side of hypandrium opposite base of strap with a shorter strap, slightly more than half length of right-hand strap and separated from it by a narrow, deep declivity. Phallosome (Fig. 355) with basal apodeme broad at base, tapering anteriorly; pair of short subapical lobes, the right lobe projecting farther than left; apical process simple but with slight indication of bilobing at apex. Male epiproct (Fig. 855) with back part of seat short, somewhat narrower than rest of epiproct. Subgenital plate (Fig. 856) much as described for *L. maculosa*, but basal pigmented region of distal process somewhat wider, distal quadrate regions of pigmented arms not so wide laterally: lateral width/antero-posterior length ~ 1.35 vs. ~ 1.53 in *L. maculosa*.

Relationships. These are as noted for *L. maculosa*.

Distribution and habitat. The species occurs from Kouchibouguac National Park, New Brunswick west to Voyageurs National Park, Minnesota, south in the east to Sarasota County, Florida and in the west to Bowie County, Texas. Populations assumed to be isolated are found in the Rocky Mountains in Franklin County, Idaho, and Cache and Utah Counties, Utah. It inhabits trunks and branches of broad-leaf trees and conifers.

Genus *Ptycta* Enderlein

Ptycta Enderlein, 1925:102.

Diagnosis. Adults of both sexes macropterous. Nine distal inner labral sensilla. Forewing markings extremely variable: usually a dark mark in pterostigma and dark marks on diagonal from pterostigma base to nodulus, but some species with extensive wing banding. Hypandrium with or without basal sclerites; with asymmetrical median strap based in middle or right, some species with processes at base of strap. Phallosome slightly tapering and rounded anteriorly; distally with pair of semi-membranous (presumed) external parameres and median aedeagal process of varying width, usually bifid at apex. Male epiproct chair-shaped, extending anteriorly over clunium. Male paraproct with distal process long, slender, acuminate; no bulge on shoulder lateral to process; sense cushion near base of paraproct; basal articular region short and broad. Subgenital plate with distal piece short and wide, about as long as its basal width; pigmented arms directed antero-laterally, somewhat widened distally. Ovipositor valvulae: v2 with long, slender distal process; v3 ovoid; median lobe variable -- prominent to absent.

Generotype. *P. haleakalae* (Perkins).

North American species: *P. lineata* Mockford, *P. polluta* (Walsh).

Ptycta lineata Mockford

Ptycta lineata Mockford, 1974a:192.

Recognition features. Forewing markings (Fig. 359): pterostigma brown except colorless in basal one-fourth and at distal extreme; stigmasaum brown in distal half; light brown spotting before and around Rs - M junction, behind M - Cu1 branching, beyond middle in cell Cu2, in middle and distal end of cell 1A. Hypandrium (Fig. 857) lacking basal sclerites; median strap slender, curving slightly leftward at base, then rightward and upward distally; without basal processes. Phallosome (Fig. 858) tapering to rounded base anteriorly; posteriorly with elongate, membranous external parameres and slender, bifid aedeagal process, the two terminal points tending leftward. Male epiproct rounded anteriorly. Subgenital plate as for the genus. Ovipositor valvulae (Fig. 859): v3 lacking median lobe.

Relationships. These are not understood at present.

Distribution and habitat. Within the study area, the species has been found in the southern Florida counties of Monroe and Indian River. It inhabits branches of trees and shrubs, including mangroves, live oaks, and *Rapanea*. Outside the study area, it has been found in the Cuban provinces of Oriente, Pinar del Río, and Habana.

Ptycta polluta (Walsh)

Psocus pollutus (Walsh), 1862:361.
Psocidus pollutus (Walsh). Pearman, 1934:123.
Ptycta polluta (Walsh). Eertmoed and Eertmoed. 1983:125.
Psocus texanus var. *submarginatus* Aaron. 1886:16. New synonym.
Psocus submarginatus Aaron. Chapman, 1930:279. New synonym.

Recognition features. Forewing markings (Fig. 360): pterostigma with brown spot over distal half except antero-distal angle; stigmasaum brown in distal one-third, the spot extending back half-way through cell R1; a brown marginal band from cell R3 through cell Cu1a becoming darker from front to back, separated from margin by clear areas in M cells; three brown spots in basal half of cell R5, two continuing into discoidal cell; a brown diagonal band from base of pterostigma to nodulus, narrowly interrupted along vein Cu2; a brown spot on vein M + Cu1 at about half its length; a brown spot in cell Cu2 before nodulus; cell 1A brown in basal half at apex. Hypandrium (Fig. 860) lacking basal sclerites; median strap broad, of two parts -- right part much broader than left, curved leftward at base, then right and up, denticulate on right side, its base with three larger denticles. Phallosome (Fig. 861) tapering anteriorly to rounded base; external parameres moderately long, membranous; aedeagal process stout, bifid apically, the two points directed posteriorly. Male epiproct (Fig. 862) abruptly narrowed in anterior half, anterior margin bilobed. Subgenital plate (Fig. 863): pigmented arms abruptly widened distally to form quadrate areas; plate otherwise as for the genus. Ovipositor valvulae: v3 with well developed median lobe protruding beyond body of valvula.

Relationships. This species is part of a complex found in Mexico and Central America. Other members of the complex remain undescribed.

Distribution and habitat. The species occurs from Nova Scotia south to Key Largo, Florida, west to northern Minnesota and Cameron County, Texas. It inhabits branches of various kinds of broad-leaf and coniferous trees and shrubs. It is thelytokous throughout most of its range. Males are restricted to the southern Appalachians in Tennessee and North Carolina, also Florida and the southern Atlantic Coast north to Hatteras Island, North Carolina, also south-central Missouri and northwestern Arkansas, and southern Texas.

Genus *Steleops* Enderlein

Diagnosis. Adults of both sexes macropterous. Eyes at least slightly stalked (Fig. 344). Five distal inner labral sensilla. Forewing markings extremely variable: from a few spots to a complex pattern of spots and blotches. Hypandrium without basal sclerites: median sclerotization broad, in three lobes (in North American species) -- laterals somewhat to much shorter than median and with denticles or large teeth on their bases, sides, or both. Phallosome tapering to blunt point or with short apodeme anteriorly; distally with short, wide external parameres or these fused into sides of aedeagal arch; aedeagal region otherwise variable. Male epiproct extending anteriorly over clunium. Male paraproct (Fig. 864) with distal process long and slender; no bulge lateral to process; basal articular region short and broad. Subgenital plate (Fig. 865) with distal piece short and wide, about as long as its basal width; pigmented arms vague, directed antero-laterally, slightly expanded distally. Ovipositor valvulae (Fig. 866): v2 with long, slender distal process; v3 transversely ~ 3-4X its antero-posterior dimension with a moderate median lobe or none.

Generotype. *S. punctipennis* Enderlein.

North American species: *S. elegans* (Banks), *S. lichenatus* (Walsh).

Steleops elegans (Banks), new combination

Psocus elegans Banks. 1904b:203.
Psocidus elegans (Banks). Smithers, 1967:107.

Recognition features. Body and legs mostly white; front of head and mesonotum brown. Forewing (Fig. 348) marked with bold pattern: pterostigma brown in distal one-third; stigmasaum brown at distal end; brown spot in basal half of cell M1 extending into cell M2, brown spot in anterior half of cell M3 extending into and covering distal two-thirds of cell Cu2 before nodulus; brown clouding around ends of veins from R2+3 through M3. Hypandrium (Fig. 867) with median sclerotizations broad, complex, consisting of field (basally) and ridge (more distally) of large teeth on right, a median strap curved upward distally and a closely approximated right-hand ridge of teeth; a tapering, blunt-tipped process on left with a field of slender teeth and spines on its left side. Phallosome (Fig. 868) with blunt-pointed base lacking apodeme; external parameres fused to sides of aedeagal arch and to broad distal plate. Male epiproct (Fig. 869) bilobed basally, with small raised area in middle. Subgenital plate (Fig. 865): distal piece ending in short square-tipped spatula; a row of long setae on each side marking junction of distal piece and pigmented arms; arms pigmented only lightly at their bases. Third valvulae triangular, lacking median lobe (Fig. 866).

Relationships. These are poorly understood at present. Several species from various parts of tropical America show similar color patterns and genitalic characters. The species appears not to be close to its North American congener, *S. lichenatus*.

Distribution and habitat. The species occurs throughout eastern United States from New York (Tompkins County) south to Collier County, Florida, and west to central and southern Illinois. Collection records farther southwest are currently lacking. It inhabits tree trunks and branches and rock outcrops.

Steleops lichenatus (Walsh), new combination

Psocus lichenatus Walsh. 1863:183.
Amphigerontia lichenatus (Walsh). Banks, 1892:344.
Psocidus lichenatus (Walsh). Smithers, 1967:109.

Recognition features. Body and legs pale tawny brown, slightly darker on abdominal sterna and small dark brown spots on legs. Antennae banded medium brown and white. Forewing markings (Fig. 347): pterostigma and stigmasaum brown with scattered colorless lacunae; basal three-fourths of cell R1, basal half of cell R3, all of cell R5, and all of cell Cu 1a brown with scattered colorless lacunae; anterior margins of M cells and basal margin of cell M3 brown; a broad diagonal band of brown with scattered colorless lacunae from base of pterostigma to nodulus, basal one-fourth of wing brown-clouded with scattered colorless lacunae. Hypandrium (Fig. 870) with median sclerotizations broad, nearly symmetrical, consisting of two short lateral lobes bearing heavy denticles on their outer surfaces and a slender, straight median strap curved upward distally. Phallosome (Fig. 871) with short, blunt anterior apodeme; short, broad, partially sclerotized external parameres, and a broad, apically bifid aedeagal process. Subgenital plate (Fig. 872): distal piece as described for S. elegans; pigmented arms distinct, relatively wide, broadening to form quadrate areas distally. Ovipositor valvulae: v3 transversely ~ 4X its antero-posterior length; with a well developed median lobe (Fig. 873).

Relationships. These are not understood at present.

Distribution and habitat. The species is known from the states of Illinois, Indiana, and Ohio, where it is found on tree trunks and rock outcrops.

Genus *Trichadenotecnum* Enderlein

Trichadenotecnum Enderlein, 1909a:329.
Trichadenopsocus Roesler (subgenus), 1943:5.

Diagnosis. Adults of both sexes macropterous. Five distal inner labral sensilla. Forewing markings usually including a radial series of brown spots in cells R1 through M3; other markings variable. Angle formed by veins Cu 1a and M + Cu 1a segment acute (obtuse or right in other genera of the subfamily). Hypandrium of two segments; basal segment consisting of sclerotized and pigmented arms seemingly unique in the subfamily; distally, hypandrium usually with a movable sclerotized tongue in middle surrounded by other sclerotizations. Phallosome tapering anteriorly; without external parameres; aedeagal region variable. Male clunium on each side with a process running towards base of paraproct. Male epiproct extending anteriorly over clunium. Male paraproct with slender, acuminate distal process, lacking bulge lateral to process; articular region short, weakly sclerotized. Subgenital plate: distal piece short and broad, usually not as long as its basal width; pigmented arms directed laterally or antero-laterally, usually broadened into quadrate or triangular regions distally. Ovipositor valvulae: v2 with long, slender distal process; v3 with well developed median lobe or none.

Generotype. *T. sexpunctatum* (Linnaeus).

North American species: *T. alexanderae* complex: *T. alexanderae*
Sommerman, *T. castum* Betz, *T. irruptum* Betz, *T. merum* Betz; other
species: *T. circularoides* Badonnel, *T. desolatum* (Chapman), *T. majus*
(Kolbe), *T. pardus* Badonnel, *T. quaesitum* (Chapman), *T. slossonae*
Banks.

Trichadenotecnum alexanderae Complex

Betz (1983b) recognized as a complex the sexual species *T.
alexanderae* Sommerman and three closely similar parthenogenetic
forms possibly derived from the sexual species. The species share
many characters and differ in few. Therefore, the complex is first
characterized followed by brief characterizations of the species. Relationships of the species will not be discussed, as they are all very close
to each other.

Recognition features of the complex. Forewing markings (Fig.
329): pterostigma dark in distal one-third; radial spots distinct but
irregular in shape; each distal vein R2+3 through M3 with a cloudy
brown border; basal one-third of cell R3 and most of cell Cu1a cloudy
brown; a brown diagonal band from base of pterostigma to cell Cu2
before nodulus, interrupted along vein M immediately beyond M - Cu1
branching; two brown spots separated by a clear spot in basal half of
cell 1A; nodulus brown. Subgenital plate with wide pigmented arms
directed laterally, broadening distally into roughly triangular areas.
Ovipositor valvulae: v3 with large median lobe, basal width of lobe
about one-third width of the valvula.

Trichadenotecnum alexanderae Sommerman

Trichadenotecnum alexanderae Sommerman, 1948:169.

Recognition features. Known from both sexes. Forewing markings (Fig. 329): numerous discrete spots both before and beyond
diagonal ('nodular') band; a distinct spot around vein in "roof" of cell M3.
Vein M closing discoidal cell generally concave. Hypandrium (Fig. 874):
basal arms directed laterally, expanded into irregular quadrate areas
distally; a short forward-directed lobe of pigmentation between arms;
distal sclerotizations: a central tongue articulated at base, expanded
from base to tip, arising on a broad, symmetrical, sclerotized area
terminating on each side in a laterally directed horn bearing a tooth
near its base. Phallosome (Fig. 875) with aedeagal region slightly
thickened, bilobed. Male clunial process broad, semi-membranous,
terminating in an upward-directed point or pair of points (Fig. 876).
Male epiproct broad and trilobed posteriorly. Subgenital plate: setae in

proximal field of egg guide arranged roughly in triangle; clear region in middle of egg guide distinct at both ends. Sclerotizations of ninth sternum (Fig. 343): posterior region with large clear area between spermapore and basal pigmentation; anterior region with cowling only slightly pigmented on sides, the pigment concentrated near articulation with posterior region.

Distribution and habitat. The species is known throughout central United States (Illinois, southern Wisconsin, Ohio, West Virginia, Maryland, New Jersey, and southern New Hampshire). It inhabits bark of trees and shaded stone outcrops.

Trichadenotecnum castum Betz

Trichadenotecnum castum Betz, 1983b:1341.

Recognition features. Known from females only. Forewing markings (Fig. 339): spot on "roof" of cell M3 extremely narrow along the vein (not usually as illustrated by Betz, 1983, Fig. 2); numerous discrete spots before and beyond diagonal band. Vein M closing discoidal cell generally straight. Subgenital plate: setae in proximal field of egg guide arranged roughly in triangle; clear region in middle of egg guide distinct at both ends (Fig. 877) or indistinct at one or both ends. Sclerotizations of ninth sternum: posterior region as described for *T. alexanderae* but the clear area tending to be not quite so large (pigmented 'horns' tending to be longer); anterior region with cowling well pigmented on sides to its anterior margin (Fig. 338).

Distribution and habitat. Within the study area, the species occurs from central Illinois, central Indiana, and southern Ohio south to northern Florida (Alachua County), central Mississippi, and southwestern Arkansas. It inhabits trunks and branches of broad-leaf and coniferous trees. Outside the study area, it was recorded from the Azores Islands (Baz, 1988).

Trichadenotecnum innuptum Betz

Trichadenotecnum innuptus Betz, 1983b:1349.

Recognition features. Known from females only. Forewing markings: spot on "roof" of cell M3 well developed; relatively few discrete spots before and beyond diagonal band. Vein M closing discoidal cell decidedly concave (Fig. 337). Subgenital plate: setae in proximal field of egg guide arranged roughly in triangle; clear region in middle of egg guide distinct its entire length. Sclerotizations of ninth sternum: posterior region pigmented from base to and including spermapore;

anterior region (Fig. 336) with distinct clear crescent in middle (median hyaline sheath of Betz, 1983); pigmentation of cowling variable.

Distribution and habitat. Within the study area, the species is known only from southern Michigan (Livingston County) and from northern Indiana (Kosciusco County), where it has been taken on trunks of broad-leaf trees. Outside the study area, it has been found in the Geneva region of Switzerland and in Hungary (Lienhard, 1986a, 1986b).

Trichadenotecnum merum Betz

Trichadenotecnum merum Betz, 1983b:1348.

Recognition features. Known from females only. Forewing markings (Fig. 340): a distinct large spot on "roof" of cell M3; relatively few distinct spots before and beyond diagonal band. Vein M closing discoidal cell straight. Subgenital plate: setae in proximal field of egg guide (Fig. 341) arranged in ellipsoid (fewer setae) or triangle (more setae); clear area of middle of egg guide distinct throughout. Sclerotizations of ninth sternum: posterior region approximately as described for *T. alexanderae* except pigment extending to level of spermapore on sides or beyond (Fig. 342); anterior region with large, distinct lateral pigmented lobes (transverse ridges of Betz, 1983), little pigmentation of cowling.

Distribution and habitat. The species is found in the central part of eastern United States from Michigan (Mason County) south to southern Tennessee (Shelby and Marion Counties) and southwestern North Carolina (Cherokee County). It inhabits trunks and branches of broad-leaf and coniferous trees.

Other Species of *Trichadenotecnum*

Of the remaining six species, none shows close relationships to any other or to the *T. alexanderae* complex. Three (*T. circularoides*, *T. quaesitum*, and *T. slossonae*) have close relatives in South America and Mexico. *T. pardus* is a widespread tropical waif, probably introduced into southern United States. *T. majus* is widespread in Europe, has European close relatives, and may have been introduced into North America. *T. desolatum* appears to stand apart and is probably an old endemic of the Mexican mountains and the southern Rockies, but its position is not as isolated as suggested by Roesler's (1943) creation of a separate subgenus for it, which I have synonymized.

Trichadenotecnum circularoides Badonnel

Trichadenotecnum circularoides Badonnel, 1955:229.

Recognition features. Known from female only. Forewing markings (Fig. 322): pterostigma brown in most of distal half; likewise stigmasaum; radial spots large and relatively smooth-edged, those in M cells each followed distally by a clear spot set off by more distal brown clouding; diagonal band consisting of a major spot in each of cells R1, R, Cu1b, and Cu2 plus a few minor spots; a spot in middle of cell R1, two in basal half of cell R5; discoidal cell with irregular spotting; spot in cell Cu1a; most of cell 1A brown. Subgenital plate (Fig. 323): egg guide broader than long, bilobed distally; pigmented arms directed anterolaterally, expanding into rounded lateral areas distally. Ovipositor valvulae: v2 gradually tapering to pointed tip; v3 transversely about twice its antero-posterior length, lacking median lobe.

Relationships. The species is close to *T. gonzalezi* (Williner) of South America.

Distribution and habitat. Within the study area, the species is widely distributed in Florida (Hendry, Highlands, Alachua, and Leon Counties), has been collected in Camden and Clarke Counties, Georgia, and up the Atlantic Coast to Cape Hatteras, North Carolina. Outside the study area, the species is known from Angola (type locality) and northern Brazil (Mockford, in press).

Trichadenotecnum desolatum (Chapman)

Psocus desolatus Chapman, 1930:236.
Trichadenotecnum (Trichadenopsocus) desolatum (Chapman). Roesler, 1943:4.

Recognition features. Known from both sexes. Forewing markings (Fig. 326): numerous small spots scattered over entire wing surface; pterostigma bordered in dark brown (male), or distal half bordered in medium brown (female); radial spots irregular in shape and obscured by surrounding smaller spots; a large spot (female) or smaller one (male) including "roof" of cell M3; diagonal band distinct but slender in female, less distinct due to more breaking up in male. Hypandrium (Fig. 878): basal segment a transverse bar; distal sclerotizations broad, complex, semi-symmetrical: on each side a rounded, well pigmented lobe; from base of left lobe a slender arm passing diagonally across towards lobe of other side; in middle a triangular region terminating distally in two points, the left longer than right. Male clunial process (Fig. 879) broad basally, tapering to upturned blunt point distally. Phallosome (Fig. 321) with short anterior apodeme; aedeagal region greatly widened, bilobed, with deep declivity between lobes; a short tongue directed anteriorly arising between lobes. Male epiproct relatively slender and rounded at anterior end, gradually broadening to articular region posteriorly. Subgenital plate (Fig. 327): egg guide broad and short, about as long as its basal width, straight on distal margin; pigmented arms broad throughout, directed laterally. Ovipositor valvulae (Fig. 880): v2 with distal process gradually tapering to slender tip; v3 with transverse measurement ~ 2.5X longitudinal measurement; median lobe well developed, broad basally. Ninth sternum with large, sclerotized posterior region.

Relationships. These are not currently understood.

Distribution and habitat. Within the study area, the species occurs from central Colorado south in the mountains to southern New Mexico. It inhabits primarily branches of pines. Outside the study area, it is found in the Mexican mountains south to the Distrito Federal.

Trichadenotecnum majus (Kolbe)

Psocus sexpunctatus var. *major* Kolbe, 1880:109.
Psocus major Kolbe. Loens, 1890:6.
Trichadenotecnum majus (Kolbe). Enderlein, 1909a:330.
Trichadenotecnum (Trichadenotecnum) majus (Kolbe). Thornton, 1961:1.

Recognition features. Known from both sexes. Forewing markings (Fig. 328): somewhat subdued; pterostigma mostly brown in distal one-third; stigmasaum brown in distal one-fourth; radial spots irregular in outline; a cloudy marginal band of pale brown, slightly darker along veins, from cell R3 through cell M3; a spot in base of cell R3; two spots before middle in cell R5; discoidal cell with irregular spotting; diagonal band relatively narrow, interrupted along vein M just beyond M-Cu1 forking; two brown spots in basal half of cell 1A. Hypandrium (Fig. 881): basal arms in form of broad arc with ends directed antero-laterally; distal sclerotizations broad, semi-symmetrical, with immovable median tongue of irregular shape on apex and denticulate surface; above and lateral to the tongue on each side a broad projection divided before apex into heavily sclerotized median prong and more lightly sclerotized lateral lobe; in middle immediately above tongue a sclerotized band curved upward at apex bearing numerous teeth. Male clunial process (Fig. 882) semi-membranous in base, more sclerotized distally, bearing a long strongly sclerotized apical point directed forward. Phallosome (Fig. 883) with short basal apodeme; somewhat broadened and bilobed apically. Male epiproct (Fig. 330) broader than long, protruding only slightly over clunium, with three sclerotized longitudinal, denticulate ridges in basal half, the middle one nearly twice as long as the laterals. Subgenital plate (Fig. 884): egg guide short and broad, its pigmentation extending antero-laterally as pair of arms separated from usual pair of pigmented arms by a narrow clear area; the latter arms slender and forward-directed at base, curving laterally and broadening to form each a lateral quadrate area. Sclerotizations of ninth sternum inconspicuous except a field of tubercles along anterior margin of anterior region. Ovipositor valvulae as described for T. desolatum.

Relationships. The species appears to be close to *T. germanicum* Roesler (central Europe).

Distribution and habitat. Within the study area, the species occurs in several of the northern states of eastern United States (Vermont, Pennsylvania, Michigan, Wisconsin, Minnesota) in the mountains of West Virginia (Randolph County), in southern and central Ontario (Belleville and St. Joseph's Island) and in New Brunswick, (Frederickton). It has been intercepted at Dover, Delaware in an aircraft from Germany. It inhabits branches of both coniferous and broad-leaf trees. Outside the study area, it is widely distributed in Europe.

Trichadenotecnum pardus Badonnel

Trichadenotecnum pardus Badonnel, 1955:231.

Recognition features. Known from females only. Forewing markings (Fig. 324): small spots scattered over entire wing; pterostigma heavily spotted in all but basal one-fourth; spots of radial series distinct except those of cells M2 and M3 sometimes no larger than surrounding spots; a large spot on roof of cell M3, also covering roof of cell Cu1a and part of cell M2; diagonal band a narrow border along veins Rs and M before becoming broader across cells Cu1b and Cu2. Subgenital plate (Fig. 325): egg guide short, about a third as long as its width; pigmented arms directed forward from base of egg guide, then curving antero-laterally and expanded to form quadrate areas; cuticle thickened in series of concentric rings in area of expansion of pigmented arms. Ovipositor valvulae (Fig. 885): v2 abruptly narrowed distally to form long, slender distal process: v3 ovoid, lacking median lobe.

Relationships. *T. pardidum* Thornton (Hong Kong) is extremely similar and may prove to be the same species.

Distribution and habitat. Within the study area, the species has been found in Collier, Sarasota, Highlands, and Alachua Counties, Florida. It has been collected most frequently on living and dead palm leaves, but also on vine *Ficus* on the side of a building. Outside the study area, the species has been found in Angola (type locality),

Madagascar, the Mexican states of Veracruz and Tabasco, also Cuba, Puerto Rico, Trinidad, Brazil (Belem) and Suriname.

Trichadenotecnum quaesitum (Chapman)

Psocus quaesitus Chapman, 1930:270 (male).
Not *P. quaesitus* Chapman, 1930:270 (female).
Trichadenotecnum slossonae (Banks). Sommerman, 1948:273, new synonym.
Trichadenotecnum (*Trichadenotecnum*) *slossonae* (Banks). Thornton, 1961:1, new synonym.

Recognition features. Known from both sexes. Forewing markings (Fig. 334): pterostigma mostly dark in distal half; radial series of spots irregular in outline; a pale brown marginal band from cell R5 through cell M2, entering cells R3 and M3; numerous small spots in basal half of cell R1, basal one-fifth of cell R5, and discoidal cell; diagonal band continuous from base of pterostigma to cell 1A. Hypandrium (Fig. 886): basal arms a transverse band broadly attached to distal sclerotized area; latter area broad, bearing near apex a heavily sclerotized movable tongue in middle with broad, rounded base and narrower, squared-off distal margin; on each side of base of tongue a pointed prong directed towards opposite side, the two prongs crossing above tongue, that of right side longer than left; more laterally on each side a blunt process directed postero-laterally. Male clunial process (Fig. 887) short, curving forward, the tip blunt, bearing numerous long spines; a hyaline lobe near base. Phallosome (Fig. 888) pointed anteriorly; on each postero-lateral corner a broad rounded lobe with a small field of denticles on median margin. Male epiproct (Fig. 889) slightly bilobed anteriorly with pointed process between lobes. Subgenital plate (Fig. 335): egg guide wide at base, tapering towards end, about as long as basal width; pigmented arms directed laterally from median pigmented area, abruptly widened distally to form quadrate areas. Ovipositor valvula: v2 abruptly tapered distally to form long, slender process; v3 triangular with well developed median lobe. Ninth sternum (Fig. 890): posterior region large, darkly pigmented over most of its surface.

Relationships. A closely related undescribed species is found in northern Mexico.

Distribution and habitat. The species occurs from Kouchibouguac National Park, New Brunswick and southern Quebec (Montreal area) south to Alachua County, Florida and west to central Ontario (Echo Bay) and southern Missouri (Carter County). It inhabits primarily branches of broad-leaf trees, but also hemlocks and bald cypresses.

Trichadenotecnum slossonae (Banks)

Psocus slossonae Banks, 1903b:236.
Psocus slossonae Banks. Chapman, 1930:273 (male).
Psocus quaesitus Chapman, 1930:270 (female), new synonym.
Trichadenotecnum unum Sommerman, 1948:167, new synonym.
Trichadenotecnum (Trichadenotecnum) unum Sommerman. Thornton,
        1961:1, new synonym.

Note. Although Chapman apparently looked at the types of this species, it is evident that he did not study them critically. The types, two females in the MCZ, represent the species described by Sommerman (1948) as *T. unum* Sommerman. Sommerman (1948) correctly noted that Chapman (1930) had the female of this species mis-associated with the male of his species '*Psocus*' *quaesitus*.

Recognition features. Known from both sexes. Forewing markings (Fig. 332): pterostigma brown in distal half; stigmasaum brown in distal one-third; radial spots distinct, relatively smooth-edged; each distal vein from R2+3 through Cu1a surrounded by pale brown spot before wing margin and point of junction with margin emphasized by darker spot; roof of cell M3 and Cu1a with a dark brown spot; scattered spotting in basal half of cells R1 and R5, and throughout discoidal cell; diagonal band relatively broad throughout; cell IA with two brown spots in basal half. Hypandrium (Fig. 891): arms of basal segment forming a broad u fused into distal segment; distal sclerotized region broad, lightly pigmented; in middle with a slender, clear, movable tongue; disto-laterally each side with a hook with pointed tip directed mesad. Male clunial process (Fig. 892) broad, curved upward, clear on outer margin, apex truncate and irregularly serrate. Distal process of male paraproct long, slender, curved, about half length of paraproct. Male epiproct (Fig. 331) bilobed, one lobe at each antero-lateral edge. Phallosome (Fig. 893) slightly bowed anteriorly, posteriorly bearing a pair of pointed projections. Subgenital plate (Fig. 333): egg guide slightly longer than its basal width, very shallowly bilobed on distal margin; pigmented arms directed laterally, widening to form triangular areas distally. Ovipositor valvulae: v2 gradually tapering distally to slender distal process; v3 about twice as wide as its longitudinal measurement, with conspicuous median lobe. Posterior region of ninth sternum roughly triangular.

Relationships. *T. decui* Badonnel (Venezuela, northern Brazil) is close to this species.

Distribution and habitat. Within the study area, the species is found from Kouchibouguac National Park, New Brunswick south to Hendry County, Florida, west to northern Minnesota (Voyageurs

National Park) and south-central Texas (Bexar County). Outlier populations have been found in Mineral County, Montana and Alexandria, British Columbia. Outside the study area, it has been found in the Mexican states of Nuevo León, Hidalgo, and Veracruz, also in Guatemala. It inhabits branches of broad-leaf trees and conifers.

Tribe Metylophorini

Diagnosis. Relatively large psocids; forewing > 4 mm in length. Nine distal inner labral sensilla. Hypandrium a broad, asymmetrical sclerotized region of one segment. V2 rounded distally. Sclerotizations of female ninth sternum a rounded or semicircular plate of moderate size around spermapore (Fig. 894). Sense cushion of male paraproct (Fig. 895) on a lobe lateral to distal process or nearly so. Male clunial-epiproctal interface usually with clunium protruding over base of epiproct as bilobed shelf.

North American genus: *Metylophorus* Pearman.

Genus *Metylophorus* Pearman

Metylophorus Pearman, 1932:202.

Diagnosis. Mx4 length divided by basal width > 4.0. Male clunial shelf bilobed. Pigmented arms of subgenital plate directed antero-laterally, not forming a t-shaped structure with egg guide. Spermapore sclerite long-semicircular. Nymphs and adults usually found singly or in loose aggregations of a few individuals.

Generotype. *M. nebulosus* (Stephens).

North American species: *M. barretti* (Banks), *M. novaescotiae* (Walker), *M. purus* (Walsh).

Key to the North American Species of Tribe Metylophorini, Genus *Metylophorus*

1.   Head with no prominent dark marks on vertex or frons. Forewing with band of spots from pterostigma to cell Cu1a, continuous with marginal/submarginal band running forward from cell Cu1a to cell R5 (Fig. 369) .... *M. barretti* (Banks)

--   Head with four to six prominent dark marks on vertex and frons. Forewing with or without banding pattern described above 2

2.   Head markings: pair of dark brown spots on vertex, another pair
      including and to sides of lateral ocelli, and third pair on frons
      bordering postclypeus. Forewing banded as described above,
      or with some reduction of it (Fig. 370), or with no banding but
      pterostigma dark throughout. Hypandrium with two length-
      wise denticulate ridges (Fig. 371) ... *M. novaescotiae* (Walker)
--   Head white with four large dark brown spots: one in each parietal
      region, one covering ocellar field, and one on postclypeus
      bordering its dorsal margin. Hypandrium (Fig. 372) with two
      lengthwise ridges: right ridge with no teeth but a distal horn-
      like projection; left ridge with large basal tooth and two smaller
      distal teeth .... *M. purus* (Walsh)

Metylophorus barretti (Banks)

Psocus barretti Banks, 1900b:239.
Psocus hoodi Chapman, 1930:239, new synonym.
Psocidus barretti (Banks). Smithers, 1967:106.
Psocidus hoodi (Chapman). Smithers, 1967:108, new synonym.
Metylophorus hoodi (Chapman). Badonnel, 1986:719, new synonym.

   Recognition features. No prominent dark spots on vertex or frons.
Forewing markings variable (Fig. 369): pterostigma pale to medium
brown in distal half; a more or less complete brown band formed of
spots, from behind pterostigma, across cell R5 to cell Cu1a; a more or
less complete marginal/submarginal band from cell Cu1a to cell R5,
slightly separated from margin in cells M1 and M2; band continuous
from cell Cu1a back across discoidal cell to cell Cu1b, on to nodulus
and across cell Cu2 distad of middle. Hypandrium (Fig. 896) with broad
sclerotized band running lengthwise through middle bearing denticulate ridge on right side, the ridge curved mesad from base and laterad
towards tip; left of middle another broad sclerotized band with ridge of
denticles along its left side. Phallosome (Fig. 897) with long asymmetrical distal process in form of scoop with edges denticulate. Male clunial
shelf (Fig. 898) with each lateral edge lobed, the lobes only slightly
bulging. Subgenital plate (Fig. 899) with egg guide arising from broader
basal piece; egg guide about twice as long as its basal width; pigmented
arms arising from basal piece before egg guide, expanded but only
weakly pigmented at their distal ends.
   Relationships. The species is probably close to *M. novaescotiae*.
   Distribution and habitat. Within the study area the species has
been found only in extreme southern Arizona (Santa Cruz and Cochise
Counties). Outside the study area, the species is known from the

Mexican states of Michoacán, Puebla, Morelos, Guerrero, Sinaloa, Nayarit, Veracruz, Oaxaca, and Chiapas. It inhabits trunks and branches of broad-leaf trees and conifers, and also is found occasionally on rock outcrops.

Metylophorus novaescotiae (Walker)

Psocus novaescotiae Walker, 1853:72.
Psocus contaminatus Hagen, 1861:10.
Psocus perplexus Walsh, 1862:361.
Psocus hageni Banks, 1904b:202.
Metylophorus novaescotiae (Walker). Mockford, 1961:137.

Recognition features. Head marked with pair of dark brown spots on vertex, another pair including and to sides of lateral ocelli, and third pair on frons bordering postclypeus. Forewing markings variable: most complex (Fig. 370) as described for M. barretti but lacking connection from pigment in cell Cu 1 a to that in cell Cu 1 b; simplest with pterostigma dark and remainder of wing unmarked. Males never with complete banding pattern, often only with the dark pterostigma. Hypandrium (Fig. 371) with broad sclerotized band running lengthwise through right half, bearing longitudinal, straight denticulate ridge to the right of its middle; another sclerotized band on left side not as broad, with denticulate ridge along its left side. Phallosome (Fig. 900) symmetrical, terminating distally in process of moderate length slightly bifid at tip. Male clunial shelf (Fig. 901) with distal corners lobed, the lobes decidedly protruding. Subgenital plate (Fig. 902): egg guide arising from tapered basal piece; egg guide long and slender, ~ 4X as long as its width beyond short basal neck; pigmented arms extending antero-laterally from median pigmented region; their distal ends expanded but weakly pigmented.

Relationships. These are as noted for *M. barretti*.

Distribution and habitat. Within the study area, the species occurs in the east from northern Nova Scotia (White Point) and coastal Maine (Mount Desert Island) south to Highlands County, Florida, and in the west from Voyageurs National Park, Minnesota south to Liberty County, Texas. It inhabits branches of broad-leaf trees and shrubs, and conifers. Outside the study area, it is known from the Mexican states of Hidalgo, Puebla, and Veracruz.

Metylophorus purus (Walsh)

Psocus purus Walsh, 1862:361.
Psocus lucidus Harris, 1869:328.

Psocus genualis Banks, 1903b:236.
Metylophorus purus (Walsh). Mockford, 1961:137.

Recognition features. Head white with four large dark brown spots: one in each parietal region, one covering ocellar field, one on postclypeus bordering its dorsal margin. Forewing markings (Fig. 903, sexes same): a brown spot at base of pterostigma; vein R1 brown bordered in distal two-thirds of pterostigma; a brown spot at nodulus; remainder of wing unmarked. Hypandrial sclerotization (Fig. 372): a large quadrate plate occupying middle, bearing a ridge on each side with large basal tooth and two smaller distal teeth on left ridge; no teeth on right ridge; a horn-like projection arising at distal end of right ridge; left of quadrate plate an ear-shaped plate bearing rabbit-ear shaped projection distally. Phallosome (Fig. 904) symmetrical, its rounded base directed ventrally, at right angle to rest of structure; distal process broad, elongate, flat at apex. Male clunial shelf of two broad lobes with deep declivity between (Fig. 905). Subgenital plate (Fig. 906): egg guide arising on small, rounded basal piece; egg guide long and slender, ~ 4X as long as its basal width; pigmented arms extending antero-laterally from small median pigmented region; the arms only faintly pigmented.

Relationships. These are not understood at present.

Distribution and habitat. The species has been found in the east from Cambridge, Massachusetts south to Alachua County, Florida and in the west from Rock Island, Illinois south to Austin Texas. It is known from the Belleville area of southern Ontario. It inhabits trunks and branches of broad-leaf trees and conifers, also fence posts, sides of old buildings, and occasionally stone surfaces.

Tribe Cerastipsocini

Diagnosis. Large psocids; forewing usually 4.5 mm in length or more. Nine distal inner labral sensilla. Hypandrium a broad, symmetrical sclerotized plate on a single segment. V2 terminating in a point but not prolonged as a process. Sclerotization of female ninth sternum a small key-hole shaped plate around spermapore. No lobe or shoulder lateral to distal process of male paraproct, the sense cushion well removed from the distal process. Male clunial-epiproctal interface a short, simple clunial shelf over base of epiproct.

North American genus: *Cerastipsocus* Kolbe.

Genus *Cerastipsocus* Kolbe

Cerastis Kolbe, 1883:65 (preoccupied).
Cerastipsocus Kolbe, 1884:38.
Titella Navás, 1912:196.

Diagnosis. Mx4 length divided by basal width < 4.0. Male clunial shelf simple, not bilobed. Pigmented arms of subgenital plate transverse, forming cross-piece of t-shaped structure with egg guide the stem (Fig. 295). Nymphs and teneral adults usually found in herds of up to several hundred individuals.

Generotype. *C. fuscipennis* (Burmeister).

North American species: *C. trifasciatus* (Provancher), *C. venosus* (Burmeister).

Cerastipsocus trifasciatus (Provancher)

Psocus trifasciatus Provancher, 1876:186.
Psocus speciosus Aaron, 1883:40.
Cerastis nigrofasciatus Kolbe, 1883:70.
Psocus tolteca Banks, 1903a:237.
Cerastipsocus trifasciatus (Provancher). Banks, 1907b:10.

Recognition features. Forewing marked with a striking banding pattern (Fig. 907): three transverse bands in distal half of wing converging in cells M3 and Cu1a; one transverse band in basal half of wing from cell R before M - Cu1 branching to nodulus; cell Cu2 brown interrupted by two transverse colorless bands, one before middle, the other after middle.

Relationships. The species seems to stand apart from other known members of its genus.

Distribution and habitat. Within the study area, the species is known from southern Quebec, and from the states of Virginia, North Carolina, Georgia, Alabama, Illinois (Pope County), Missouri, Kentucky, Arkansas, and Arizona. Outside the study area, it is widely distributed in Mexico (states of Nuevo León, San Luis Potosí, Hidalgo, Puebla, Morelos, Guerrero, Sinaloa, Veracruz, Tabasco, Oaxaca, and Chiapas), and is also known from Guatemala, Costa Rica, and northern Colombia. It inhabits tree trunks and branches, and shaded rock outcrops.

Cerastipsocus venosus (Burmeister)

Psocus venosus Burmeister, 1839:778.
Psocus magnus Walker, 1853:484.
Psocus gregarius Walker, 1853:484.
Psocus gregarius Harris, 1869:329.
Cerastis venosa (Burmeister). Kolbe, 1883:69.
Psocus gossypii Ashmead, 1894:29.
Cerastipsocus venosus (Burmeister). Enderlein, 1906e:317.

Recognition features. Forewing membrane uniformly deep fumose brown, nearly black (Fig. 908). Veins in basal half of forewing yellow; pterostigma usually white, occasionally dark brown.

Relationships. Several species of similar size and wing color are found in parts of the American Tropics.

Distribution and habitat. Within the study area, the species occurs from southern Quebec (Montreal area) and southern Maine (Cumberland County) south to southern Florida (Lee and Palm Beach Counties) in the east, and in the west from Fond du Lac County, Wisconsin south to Cameron County, Texas, but is absent from the prairie region of central Illinois. A single record from "Tunjony Lake, B.C." in the Canadian National collection is probably a label error. Outside the study area, it has been identified from the Mexican states of Nuevo León, San Luis Potosi, Veracruz, and Chiapas. It inhabits trunks and branches of broad-leaf trees and conifers.

Subfamily Amphigerontiinae

Diagnosis. Five distal inner labral sensilla. Vein M forming distal closure of discoidal cell in forewing usually convex. Hypandrium of two heavily sclerotized sternal plates, distal one usually symmetrical. Phallosomal skeleton a pair of parameres joined at their bases either directly or by membrane, open distally. Male epiproct overlapping clunial margin or not. Male paraproctal process broad, spatulate, rounded distally. Second valvula terminating in slender process.

North American genera: *Amphigerontia* Kolbe, *Blaste* Kolbe, *Blastopsocus* Roesler.

Key to the North American Genera and Species of Subfamily Amphigerontiinae

1.   Parameres joined at bases only by membrane. Male epiproct slightly overlapping posterior margin of clunium (Fig. 373). Pigmented arms of subgenital plate directed laterally 2

--   Parameres fused at bases. Male epiproct and clunium meeting along a straight line with no overlap (Fig. 374). Pigmented arms of subgenital plate directed antero-laterally . Genus *Blaste* 8

2.   Veins Rs and M in forewing joined by a crossvein longer than preceding Rs segment (Fig. 910). Hypandrium symmetrical distally. A pigmented projection, sometimes forked, between arms of subgenital plate (Fig. 375) ... Genus *Amphigerontia* 3

--   Veins Rs and M in forewing joined at a point, or by a minute
        crossvein, or by a short fusion. Hypandrium asymmetrical
        distally. A pigmented projection present or not between arms
        of subgenital plate; if present, not forked ....
        .... Genus *Blastopsocus* 6

3.   Median lobe of hypandrium slender, terminating in a single point
        (Fig. 376). Male compound eyes large, IO/d ~ 1.20. Median
        pigmented projection of subgenital plate short, not longer than
        its basal width (Fig. 375) .... *A. bifasciata* (Latreille)
--   Median lobe of hypandrium either terminating in two points or a
        broad-based triangle terminating in a single point. Pigmenta-
        tion between terminal quadrate areas of arms of subgenital
        plate not as above: either a longer projection ≥ 1.5X its basal
        width or diffuse pigment usually tending to concentrate in
        middle 4

4.   Median lobe of hypandrium a triangle, terminating in a single
        point (Fig. 377). Male compound eyes very small, IO/d ~ 2.15.
        Pigmentation between terminal quadrate areas of arms of
        subgenital plate diffuse, tending to concentrate in middle (Fig.
        378) .... *A. petiolata* (Banks)
--   Median lobe of hypandrium not triangular, terminating in two
        points (Fig. 379). Male compound eyes not quite so small, IO/
        d ~ 1.90-2.00. Pigmentation between arms of subgenital plate
        a single median projection 5

5.   Median lobe of hypandrium bulbous, widest towards its poste-
        rior end (Fig. 380). Median pigmented projection of subgenital
        plate long and slender, terminating at about level reached by
        lateral quadrate areas of the arms (Fig. 381) ....
        .... *A. contaminata* (Stephens)
--   Median lobe of hypandrium relatively slender (Fig. 379), median
        projection of subgenital plate not so long, ~ 1.5X its basal width
        and usually not reaching half length of lateral quadrate areas
        of the arms (Fig. 382) .... *A. montivaga* (Chapman)

6.   Preclunial abdominal segments 2 through 6 each either with a
        median dorsal reddish- to purplish brown spot (spots also on
        segments 7 and 8) vaguely connected on each side with a band
        extending down below the spiracle, or these segments (not
        beyond six) each with a ring of the same color 7
--   Preclunial abdominal segments marked with five dark purplish
        brown spots on white background: two dorsally on segments
        1 through 3, three dorsally and dorso-laterally on segments 5

through 6 (Fig. 384). Subgenital plate lacking anterior median pigment projection between arms (Fig. 385) ....
....*B. variabilis* (Aaron)

7.   Subgenital plate with anterior median pigment projection be-
     tween pigmented arms (Fig. 383); preclunial abdominal seg-
     ments with ringed pattern described above...........................
     ....*B. lithinus* (Chapman)

--   Subgenital plate lacking median pigment projection between
     pigmented arms; preclunial abdominal segments with pattern
     of dorsal spots and lateral bands described above ....
     ....*B. semistriatus* (Walsh)

8.   Known from female (type) only. Macropterous; ninth sternum
     with heavily sclerotized, dark, rounded sclerite on each side of
     spermapore (Fig. 386) .... *B. cockerelli* (Banks)

--   Known from male or both sexes. Females macropterous or
     micropterous. Ninth sternum not as above; structures flank-
     ing spermapore never darkly pigmented 9

9.   Apex of hypandrium with paired upper prongs, single lower
     prong and laterial processes (Fig. 387). Female forewings
     usually longer than those of male for any local population ...
     10

--   Apex of hypandrium with single median prong plus lateral
     processes (Fig. 388). Female forewings (where known) either
     equal to or shorter than those of male............................. 12

10. Lateral processes of hypandrium terminating in pair of points
     (Fig. 389). Distal hooks of phallosome denticulate at their
     apices (Fig. 390). Female ninth sternum: pigmented zone
     around spermapore decidely longer than wide; sclerite of
     posterior membranous zone bilobed (Fig. 391) ....
     .... *B. garciorum* Mockford

--   Lateral processes of hypandrium terminating in a single point
     (Fig. 392). Distal hooks of phallosome lacking denticles at their
     apices. Female ninth sternum: pigmented zone around
     spermapore varying from slightly longer than wide to decidedly
     wider than long; sclerite of posterior membranous zone either
     not bilobed or absent 11

11. Ventral median process of hypandrium bluntly pointed at tip
     (Fig. 392). V3 triangular, median lobe not set off from rest of
     valvula (Fig. 393) .... *B. osceola* Mockford

--   Ventral median process of hypandrium slightly widened and truncated or bifid at tip (Fig. 387). V3 with median lobe distinctly set off from rest of valvula (Fig. 394) ....
....*B. posticata* (Banks)

12.   Median distal process of hypandrium minute, flanked on each side by longer structure bearing outer pointed (blade-like) process and more mesal truncated process (Fig. 395). Parameres at base joined by membrane to median apodeme (Fig. 396) .
.... *B. persimilis* (Banks)

--   Median distal process of hypandrium larger, flanked on each side by a simple process. Parameres with sclerotic connection at base 13

13.   Cell Cu1a of forewing largely brown in both sexes (Fig. 397). Egg guide about twice as long as its basal width (Fig. 398) ....
....*B. opposita* (Banks)

--   Macropterous individuals with cell Cu1a of forewing colorless (cell not visible in micropterous females). Egg guide less than twice as long as its basal width 14

14.   Adults of both sexes with forewings about same length. Short distal median process of hypandrium flanked on each side by a straight process bent slightly outward (Fig. 399) ....
.... *B. quieta* (Hagen)

--   Male forewings longer than those of female. Distal median process of hypandrium flanked on each side by a process curved mediad (Fig. 404) 15

15.   Females micropterous; males macropterous with wings 4.5 - 5.0 mm. Distal lateral processes of hypandrium wide basally with slender, sharply pointed apices; median process likewise sharply pointed (Fig. 400) .... *B. subapterous* (Chapman)

--   Both sexes macropterous but female forewings shorter than those of male. Distal lateral processes of hypandrium either wide nearly to apex or, if with more slender distal region its apex blunt, apex of median process also blunt 16

16.   Male forewings ~ 3.5 mm in length. Posterior submembranous region of female ninth sternum a clear crescent-shaped region with middle thickened and arms extended (Fig. 401) ....
.... *B. subquieta* (Chapman)

--   Male forewings 4.5 mm or more in length. Posterior submembranous region of female ninth sternum not as above 17

17.   Median distal process of hypandrium slender, elongate, extending beyond lateral processes (Fig. 402). Posterior submembranous region of female ninth sternum with an opaque semicircular sclerite underlying a clear crescent (Fig. 403) ..
.... *B. longipennis* (Banks)

--   Median distal process of hypandrium broad at base, short, abruptly tapered to acuminate point; lateral processes extending beyond it (Fig. 404). Posterior submembranous region of female ninth sternum an arc-shaped opaque sclerite underlying a clear crescent (Fig. 405) .... *B. oregona* (Banks)

Genus *Amphigerontia* Kolbe

Amphigerontia Kolbe, 1880:104.

Diagnosis. Forewing with an Rs - M crossvein usually at least one-third length of preceding Rs segment. Hypandrium symmetrical distally with a median lobe flanked on each side by a curved or bent arm-like process. Parameres (Fig. 909) joined only by membrane at their bases. Male epiproct (Fig. 373) overlapping clunial margin, somewhat bilobed anteriorly. Subgenital plate (Fig. 375): egg guide moderately long; pigmented arms directed laterally, well sclerotized, forming quadrate areas at expanded distal ends; a short median pigment projection from basal region of arms.

Generotype. *A. bifasciata* (Latreille).

North American species: *A. bifasciata* (Latreille), *A. contaminata* (Stephens), *A. infernicola* (Chapman), *A. montivaga* (Chapman), *A. petiolata* (Banks).

Amphigerontia bifasciata (Latreille)

Psocus bifasciatus Latreille, 1799:144.
Psocus subfasciatus Zetterstedt, 1840:1053.
Psocus quadrimaculatus Westwood, 1840:19.
Psocus semistriatus Walsh, 1862:361 (in part).
Amphigerontia subnebulosa Kolbe, 1880:104.
Psocus confraternus Banks, 1905:2.
Psocus moderatus Banks, 1907a:165.
Psocus additus Banks, 1918:3.
Amphigerontia confraterna (Banks). Enderlein, 1924:35.
Amphigerontia bifasciata (Latreille). Ball, 1926:332.

Recognition features. Male compound eyes moderately large, IO/d ~ 1.20. Forewing markings (Fig. 291): entire wing faintly brown-washed; pterostigma brown over distal two-thirds except extreme margin; stigmasaum brown in distal half; a pale brown spot in cell R5 behind Rs fork; a medium brown spot in distal end of cell R extending into cell R1 and along veins M and Cu1; a series of three brown spots, one at distal two-thirds of cell R, one behind it in cell Cu1b, and one behind that in cell Cu2; a brown spot at nodulus and one in cell 1A near base. Median lobe of hypandrium somewhat constricted at its base, terminating in single point (Fig. 376). Male epiproct (Fig. 373) with slightly projecting, rounded antero-lateral corners, two small lobes projecting upward on hind margins of heavily sclerotized outer rim. Distance from sense cushion to base of male paraproct nearly twice diameter of sense cushion. Subgenital plate (Fig. 375): egg guide about same length as a pigmented arm; hind margin of each pigmented arm slightly curved latero-distad; median pigment projection extremely short, no longer than its basal width.

Relationships. The species is very close to *A. montivaga* (United States, Mexico).

Distribution and habitat. Within the study area, the species is found throughout the mountains of western United States north to the area of Dawson, Yukon Territory, also Exmoult Lake, Northwest Territory, and in the Pacific Coastal lowlands from Riverside County, California north to Vancouver, British Columbia. In eastern North America the species is found in Nova Scotia, New Brunswick, Maine, New Hampshire, New York, Pennsylvania, Michigan, Wisconsin, Minnesota, and south in the mountains to Tennessee, North Carolina, and northern Georgia. It inhabits branches of broad-leaf and coniferous trees. Outside the study area, it occurs throughout Europe and in Mongolia.

Amphigerontia contaminata (Stephens)

Psocus contaminatus Stephens, 1836:120.
Psocus megastigmus Stephens, 1836:120.
Amphigerontia bifasciata (Latreille). Kolbe, 1880:104.
Amphigerontia contaminata (Stephens). Badonnel, 1943:54.

Recognition features. Male compound eyes small, IO/d ~ 2.00. Forewing markings (Fig. 910): wing membrane clear; pterostigma brown in distal two-thirds; a pale brown spot below forking of Rs; a brown mark bordering veins in distal end of cell R; another brown spot in cell R bordering vein M + Cu1 before forking of that vein; faint brown mark in cell 1A near base; nodulus brown. Median lobe of hypandrium bulbous (Fig. 380), widest before apex, with two apical points. Male epiproct with anterior corners squared, a pair of very slightly raised areas near posterior margin. Distance from sense cushion to base of male paraproct nearly twice diameter of sense cushion. Subgenital plate (Fig. 381): egg guide as described for *A. bifasciata*; hind margin of each pigmented arm approximately straight; median pigmented projection reaching as far forward as front edges of lateral quadrate areas of pigmented arms.

Relationships. The species appears to be close to *A. intermedia* (Tetens) of Europe.

Distribution and habitat. Within the study area, the species is known only from Vancouver, British Columbia, where it was probably introduced. There it is found primarily on branches of *Larix* and *Thuja*. Outside the study area it is found throughout Europe and east to central Asia.

Amphigerontia infernicola (Chapman), new comb.

Psocus infernicolus Chapman, 1930:240.

Recognition features. I have not seen this species, which was based on a single male collected at West Thumb, Yellowstone National Park, Wyoming. The stated forewing length, 6.0 mm, makes it considerably larger in this character than any other known species of the genus. Forewing markings appear to be typical of the genus. The median apical lobe of the hypandrium terminates in two points. Compound eyes are large. The species is probably closely related to *A. montivaga* and to several un-named Mexican species.

Amphigerontia montivaga (Chapman)

Psocus montivagus Chapman, 1930:255.
Amphigerontia montivaga (Chapman). Mockford, 1950:201.

Recognition features. Male compound eyes relatively small, IO/d ~ 1.90. Forewing markings (Fig. 911): wing membrane faintly brown-washed; pterostigma dark over distal two-thirds and at base; stigmasaum brown in distal half; a pale brown spot behind Rs forking; a brown spot bordering veins in distal end of cell R; some specimens with faint brown spot in cell R on vein M + Cu1 before its forking; another behind it in cell Cu1b and another behind that in cell Cu2; basal half of cell 1A and nodulus pale brown. Median lobe of hypandrium (Fig. 379) somewhat contracted at its base, terminating in two points. Male epiproct projecting at each antero-lateral corner as a blunt point;

more posteriorly bearing a pair of rounded lobes projecting upward. Distance from sense cushion of male paraproct to base of paraproct slightly more than diameter of sense cushion. Subgenital plate as described for A. bifasciata except egg guide slightly longer than a pigmented arm, median pigmented projection ~ 1.5X as long as its basal width (Fig. 382).

Relationships. These are as noted for A. *bifasciata.*

Distribution and habitat. Within the study area, the species occurs from northern New York (Lewis County) south to Dooley County, Georgia, west to northern Wisconsin (Douglas County), and Refugio County, Texas. An outlier population occurs in the mountains in Gunnison County, Colorado. Outside the study area, the species occurs in the Mexican states of Hidalgo, Veracruz, (western highlands), México, Puebla, the Distrito Federal, and Durango. It also is known from the mountains of Guatemala. It inhabits branches of broad-leaf trees and conifers.

Amphigerontia petiolata Banks

Psocus petiolatus Banks, 1918:4.
Amphigerontia petiolata (Banks). Mockford, 1950:201.

Recognition features. Male compound eyes small; IO/d ~ 2.15. Forewing unmarked except for a spot in distal half of pterostigma extending into stigmasaum. Median lobe of hypandrium (Fig. 377) a broad-based triangle terminating in a single point. Male epiproct with front margin nearly straight, slightly concave; a pair of small lobes on outer edge towards hind margin. Distance from sense cushion of male paraproct to base of paraproct ~ 1.5X diameter of sense cushion. Subgenital plate (Fig. 378): egg guide about same length as pigmented arm; hind margin of each lateral arm approximately straight to near end of its distal quadrate area, there curving posteriorly following edge of projection of quadrate area; dusky pigment dispersed between the quadrate areas, slightly concentrated in middle, or in some specimens a distinct, broad pigmented projection in middle.

Relationships. Several probable close relatives, as yet undescribed, occur in the Mexican and Guatemalan highlands.

Distribution and habitat. Within the study area, the species occur from southern Ontario (Barrie, Brentwood) and northern New York (Lewis County) south to Levy County, Florida, west to Ramsey County, Minnesota and Liberty County, Texas. It inhabits branches of broad-leaf trees, especially oaks, also junipers and pines. Outside the study area, it has been found in the Mexican states of Nuevo León, Durango, Guerrero, Oaxaca, and Veracruz.

Genus *Blaste* Kolbe

Blaste Kolbe, 1883:79.

Diagnosis. Adults macropterous or brachypterous, rarely micropterous. In forewing, Rs and M joined at a point, or by a short fusion, or by a short crossvein. Hypandrium symmetrical distally with medially a pointed process (occasionally absent), flanked by two processes of variable shape and length. Male parameres fused at their bases. Male epiproct with straight anterior margin, not overlapping clunial margin. Subgenital plate with egg guide relatively short, usually somewhat arrow-head shaped; pigmented arms directed antero-laterally, only slightly expanded at their distal ends; no anterior median projection from basal junction of arms.

Generotype. *B. quieta* (Hagen).

North American species: *B. cockerelli* (Banks), *B. garciorum* Mockford, *B. longipennis* (Banks), *B. opposita* (Banks), *B. oregona* (Banks), *B. osceola* Mockford, *B. persimilis* (Banks), *B. posticata* (Banks), *B. quieta* (Hagen), *B. subapterous* (Chapman), *B. subquieta* (Chapman).

Blaste cockerelli (Banks), new combination

Psocus cockerelli Banks, 1904a:100.

Recognition features. Known only from type, female. Vertex and frons with dark brown ocellar field; remainder creamy yellow lightly dotted with brown spots to sides of median ecdysial line and parallel to median eye margins. Postclypeus with two upper and two lower dark spots; remainder creamy yellow with faint brown striations. Forewing markings (Fig. 912): compact dark brown spot at hind angle of pterostigma extending onto adjacent part of stigmasaum; two faint brown spots in cell R -- one on vein M + Cu1 at about middle of length of vein; the other on vein M immediately beyond forking of M and Cu1; a dark brown spot at nodulus. Forewing length 2.72 mm. Subgenital plate (Fig. 913): egg guide short with sharp lateral angles. Ovipositor valvulae (Fig. 914): v3 with long median lobe, nearly as long as basal length of valvula. Ninth sternum (Fig. 980): spermapore surrounded by small, darkly pigmented circular area; the latter surrounded by field of vermiculate lines; to each side of latter a darkly pigmented rounded sclerite; posterior membranous region containing a crescent-shaped clear sclerite.

Relationships. The species is part of a complex of southwestern United States including *B. longipennis*, *B. oregona*, *B. subapterous*, and at least one undescribed species. Also included here are the eastern species *B. quieta* and *B. subquieta*. The complex is characterized by moderate to extreme sexual dimorphism in wing length, four dark spots on postclypeus (some species variable in this character), egg guide of short to moderate length with relatively sharp lateral angles, a relatively long median lobe of v3 and a clear crescent-shaped sclerite on the ninth sternum posterior to the spermapore.

Distribution and habitat. The species is known only from the type locality, Whitewater by White Sands, New Mexico, where it was collected on aster.

Blaste garciorum Mockford

Blaste garciorum Mockford, 1984c:558.

Recognition features. Adults macropterous. Male compound eyes relatively small, IO/d ~ 1.59 - 2.15. Forewing markings sexually dimorphic: male with brown spot covering distal two-thirds of pterostigma and stigmasaum, otherwise unmarked; female with pterostigma and stigmasaum as in male, and a brown band bordering both sides of vein Cu1a beyond its junction with M (Fig. 915). Hypandrium terminating (Fig. 389) distally in slender medio-ventral prong, deeply divided medio-dorsal prong and a broad process on each side with cleft apex. Parameres (Fig. 390) with short distal process, no lateral lobe. Subgenital plate (Fig. 916): pigment not quite reaching free edges of egg guide. Female ninth sternum (Fig. 491): posterior submembranous region containing a bilobed sclerotized zone.

Relationships. The species forms a close group with *B. osceola*, *B. posticata*, and *B. pusilla* Mockford (Mexico).

Distribution and habitat. Within the study area, the species is confined to the lower Rio Grande Valley of Texas (Hidalgo and Cameron Counties). Outside the study area, it continues south through the Mexican states of Tamaulipas, Nuevo León, Veracruz, Oaxaca, and Chiapas, to Honduras. It is found on branches of small trees and shrubs in arid woodlands.

Blaste longipennis (Banks)

Psocus longipennis Banks, 1918:3.
Blaste longipennis (Banks). Smithers, 1967:93.

Recognition features. Adults of both sexes macropterous, but sexually dimorphic in size and marking of forewings. Male forewing ~ 5.2 mm long, unmarked except all of pterostigma but basal one-fourth and stigmasaum but basal one-third pale brown; female forewing (Fig. 917) ~ 3.8 mm long; pterostigma and stigmasaum marked as in male; a broad pale brown transverse band from pterostigma to cell Cu1a and another, broader pale brown band covering distal halves of cells R, Cu1b (except extreme distal end), and Cu2. Distal end of hypandrium (Fig. 402) with long, slender, blunt-tipped median process, lateral processes broad basally, slender and somewhat curved distally. Parameres (Fig. 918) with long, curved distal process; shorter, broader lateral lobe bearing small pendant sclerite near tip. Subgenital plate (Fig. 919): egg guide of moderate length with sharp lateral angles. Ovipositor valvulae (Fig. 920): v3 with long median lobe, tapering somewhat distally. Ninth sternum (Fig. 403): small, round pigmented area around spermapore, flanked on each side by clear circular area; slender clear crescent behind spermapore partially underlain by a wider semicircular sclerite.

Relationships. These are as noted for *B. cockerelli*.

Distribution and habitat. The species is found in the Rocky Mountains from Yellowstone National Park, Wyoming, south to the Chiricahua Mountains in Cochise County, Arizona. It has been collected on dwarf mistletoe, Arceuthobium vaginatum, presumably growing on conifers, and probably inhabits the conifer branches as well.

Blaste opposita (Banks)

Psocus oppositus Banks, 1907a:165.
Psocus interruptus Banks, 1920:306.
Blaste opposita (Banks). Smithers, 1967:94.

Recognition features. Both sexes macropterous; wings about equal in length in both sexes. Forewing markings (Fig. 397): pterostigma and stigmasaum dark brown except at base and along anterior margin; cell Cu1a brown, usually completely so in female, distal end pale in male; a pale brown spot in cell R5 before middle; obscure brown marks in middle and distal end of cell R and in distal end of cell Cu1b; nodulus brown. Distal end of hypandrium (Fig. 921) with tapering, blunt-tipped median process; long, slender, slightly curved, blunt-tipped process on each side. Parameres (Fig. 922) with long, curved distal process, short lateral lobe. Male epiproct transverse, slightly thickened on sides. Subgenital plate (Fig. 398): egg guide relatively long and slender, about twice as long as its basal width. Female ninth sternum (Fig. 923):

posterior semimembranous region continuous around most of
spermapore except for narrow sclerotized anterior area.

Relationships. These are not understood at present.

Distribution and habitat. The species is found from southern
Quebéc (Ste. Agathe des Montes), south to Palm Beach and Highlands
Counties, Florida, west to southern Missouri (Ozark County) and
eastern Texas (Liberty County). An outlier population is found in
Sycamore Canyon, Santa Cruz County, Arizona. It inhabits branches
of broad-leaf trees and conifers.

Blaste oregona (Banks)

Psocus oregonus Banks, 1900b:239.
Psocus californicus Banks, 1905:2.
Blaste oregona (Banks). Smithers, 1967:94.

Recognition features. Adults of both sexes macropterous, but
forewings strongly sexually dimorphic in length and markings. Male
forewing ~ 4.7 mm; female forewing ~ 2.9 mm. Male forewing with
pterostigma pale brown to medium brown throughout except clear in
basal one-third and along postero-distal margin; stigmasaum brown in
distal half to two-thirds; wing otherwise unmarked. Female forewing
(Fig. 924): pterostigma and stigmasaum brown in distal half; veins R
and M + Cu 1 and their branches up to a line across wing from middle
of pterostigma to branching point of vein Cu 1 bordered with brown, the
border widening to form a spot along M + Cu 1 distad of its middle; a faint
brown border along vein Cu 1a beyond its separation from M; a brown
spot in middle and another in distal end of cell Cu2. Distal end of
hypandrium (Fig. 404) with median process short, tapering abruptly
from broad base to slender, pointed tip; lateral processes longer,
curved inward. Parameres (Fig. 925) with long, curved distal process,
short lateral lobe bearing pendant sclerite from tip. Subgenital plate
(Fig. 926): egg guide relatively short and wide with sharp lateral angles.
Ovipositor valvulae (Fig. 927): v3 with long median lobe, ~ two-thirds
basal length of v3. Ninth sternum (Fig. 405): small pigmented area to
sides and behind spermapore; posterior submembranous region containing arc-shaped clear to opaque sclerotized zone behind and to sides
of spermapore.

Relationships. These are as noted for *B. cockerelli*.

Distribution and habitat. The species is found from southwestern
Washington (Thurston County) south to Los Angeles, California. The
farthest inland record is near Fort Rock (Lake County), Oregon, about
180 Km from the sea. It inhabits junipers and small shrubs of the semi-
arid vegetation (*Artemisia, Chrysothamnus*).

Note. The synonymy of *Psocus californicus* Banks is accepted, despite the existence of one or more closely related species in California, on the basis of good agreement between this species and Banks' description of the forewing of *P. californicus*.

Blaste osceola Mockford

Blaste osceola Mockford, 1984c:561.

Recognition features. Adults macropterous; wings of both sexes about same length. Forewing markings and their sexual dimorphism as described for *B. garciorum* except brown band bordering vein Cu1a beyond its junction with M at least slightly developed in male. Distal end of hypandrium (Fig. 392) with medio-ventral process broad at base, abruptly tapering in middle to form slender distal prong; two dorsal processes widely separated and partially flanking ventral process; a process on each side broad-based, tapering to blunt tip. Parameres as described for B. garciorum; likewise egg guide. Female ninth sternum: posterior submembranous region without distinct sclerotized zone.

Relationships. These are as noted for *B. garciorum*.

Distribution and habitat. The species occurs throughout northern and central Florida south to Manatee County, and in the southern one-third of Georgia. It inhabits branches of broad-leaf trees and shrubs, pines, and junipers in most of the vegetation types of the area.

Blaste persimilis (Banks), new combination.

Psocus persimilis Banks, 1908:257.

Recognition features. Known from male only. Forewing markings (Fig. 928): pterostigma and stigmasaum pale gray-brown in distal two-thirds; pale brown clouding distally in cells R and Cu1b; wing otherwise unmarked. Distal end of hypandrium (Fig. 395) with short, pointed median process seated in a declivity and longer, complex structure on each side bearing outer pointed process and more mesal truncated process. Parameres (Fig. 396) with distal curved, pointed process, short lateral lobe bearing pendant sclerite on end; basally the two parameres joined by membrane to median apodeme.

Relationships. The species may be close to several undescribed Mexican species with similar hypandrium.

Distribution and habitat. The species is found along the Gulf Coast of Texas from Brownsville north to Refugio County. I have collected a single specimen, one of four known, on oaks in a small grove.

Blaste posticata (Banks)

Psocus posticatus Banks, 1905:3
Psocidus posticatus (Banks). Smithers, 1967:110.
Blaste posticata (Banks). Mockford, 1984c:552.

Recognition features. Adults macropterous. Male compound eyes relatively large, IO/d = 1.23-1.50. Forewing markings and their sexual dimorphism as described for *B. garciorum*. Distal end of hypandrium (Fig. 387) with medio-ventral prong truncated or slightly bifid at tip, pair of dorsal prongs set apart by about width of ventral prong at their bases; on each side a broad-based process tapering to a point at tip. Parameres as described for B. garciorum but with shorter basal apodeme (about half length of an arm versus about equal to length of an arm). Subgenital plate with pigment reaching free edges of egg guide or nearly so (Fig. 929). Female ninth sternum: posterior submembranous region with a transverse, slightly curved sclerotized zone (Fig. 930).

Relationships. These are as noted for *B. garciorum*.

Distribution and habitat. Within the study area, the species is represented in field collections only from three specimens, two from Brazoria County, Texas, and one from adjacent Matagorda County. It is also represented by one specimen taken at Hidalgo, Texas on pineapple fruit from Veracruz State, Mexico. Outside the study area, it is known from the Mexican states of Tamaulipas, Nuevo León, San Luis Potosí, Hidalgo, Puebla, Veracruz, Tabasco, and Chiapas, also in Belize, Honduras, and Guatemala. It inhabits branches of broad-leaf trees and shrubs and pines.

Blaste quieta (Hagen)

Psocus quietus Hagen, 1861:12.
Psocus bifasciatus Walsh, 1863:183
Blaste juvenilis Kolbe, 1883:80.
Psocus stigmosalis Banks, 1914:611.
Blaste (Blaste) quieta (Hagen). Roesler, 1943:3.
Psocus inornatus Aaron, 1883:39. New synonym.

Recognition features. Adults of both sexes macropterous with no sexual dimorphism in wing length. Male forewing markings: pterostigma with pale to dark brown spot occupying about middle one-third of pterostigma and extending into stigmasaum; wing otherwise unmarked. Female forewing markings (Fig. 931): pterostigma and stigmasaum as in male but brown marks darker; a pale brown spot in middle of cell R extending back into cell Cu1b, another in distal end of cell R extending back through discoidal cell into cell Cu1b and joining the more basal spot in that cell. Distal end of hypandrium (Fig. 399) with moderately long, blunt-tipped median prong flanked on each side by a short, pointed process directed disto-laterally. Parameres (Fig. 932) with heavy, pointed distal prong, short, rounded lateral lobe. Subgenital plate (Fig. 933): egg guide of moderate length with sharp lateral angles. Ovipositor valvulae (Fig. 934): v3 with median lobe relatively long ~ two-thirds basal length of v3. Ninth sternum (Fig. 935): small pigmented area around spermapore, domed anteriorly, straight posteriorly, flanked on each side by semi-clear rounded area; posterior semimembranous zone broad, crescent-shaped, extending to sides of spermapore plate proper, bearing small, rounded pigmented sclerite in middle.

Relationships. The species appears to belong to the same complex as *B. cockerelli* (q. v.).

Distribution and habitat. The species is found from the Montréal region of Quebéc and Kouchibouguac National Park, New Brunswick, south to Highlands County, Florida, and west to Boundary Waters National Park, Minnesota; McCook County, South Dakota; Garland County, Arkansas, and Liberty County, Texas. Outlier populations have been found in Boundary County, Idaho, and Williams Lake, British Columbia. These insects inhabit branches of broad-leaf trees and conifers.

Blaste subapterous (Chapman)

Psocus subapterous Chapman, 1930:278.
Blaste subaptera (Chapman). Smithers, 1967:94.

Recognition features. Sexually dimorphic: males macropterous, females micropterous. Male forewing with pterostigma pale brown in distal half; wing otherwise unmarked. Distal end of hypandrium (Fig. 400): in middle a broad process, mostly clear with a heavily sclerotized slender central rib, pointed at apex; on each side a broad-based structure tapering abruptly in middle to form a slender, incurved distal process with pointed apex. Parameres (Fig. 936) with curved distal prong; short, blunt-pointed lateral lobe bearing small pendant sclerite. Subgenital plate (Fig. 937): egg guide short with lateral points directed forward; a short neck between egg guide and pigmented arms; the arms relatively short. Ovipositor valvulae (Fig. 938): v3 with relatively long median lobe, nearly as long as basal length of valvula. Ninth sternum as described for *B. oregona*.

Relationships. These are as noted for *B. cockerelli*.

Distribution and habitat. This is a species of the Sierra Nevada Mountains of California. The only records to my knowledge are from Eldorado, Mariposa, and Tulare Counties. It inhabits branches of junipers, pines, and firs.

Blaste subquieta (Chapman)

Psocus subquietus Chapman, 1930:279.
Blaste subquieta (Chapman). Smithers, 1967:95.

Recognition features. Adults of both sexes macropterous, but sexually dimorphic in wing length: male forewings reaching beyond tip of abdomen by about one-third of a wing length; female forewings not or scarcely exceeding tip of abdomen, at most by about one-fifth of a wing length. Head marked (Fig. 939) with large dark brown spot postero-mesad to each compound eye (females, some males) and another on postclypeus before ocellar field (both sexes). Forewing markings: male pterostigma pale brown with diffuse darker spot in middle bordering hind margin and covering distal two-thirds of stigmasaum; pale brown clouding in cell R distally and in middle and near distal end of cell Cu1b; nodulus medium brown; female wing markings (Fig. 940) essentially the same but markings darker in general and pale brown clouding in middle of cell Cu2. Distal end of hypandrium (Fig. 401) with slender median prong tapering to pointed tip; short, broad, lateral processes with short, incurved, pointed tips. Parameres (Fig. 941) with curved pointed distal prong; short, rounded lateral lobe; basal apodeme broad at area of junction of parameres, tapering to blunt point anteriorly. Subgenital plate (Fig. 942): egg guide short, broad, with short lateral angles pointing somewhat forward. Ovipositor valvulae (Fig. 943): v3 with median lobe of moderate length, ~ two-thirds length of (relatively short) valvula. Ninth sternum (Fig. 401): slender antero-posteriorly oriented pigmented area around sperma-pore, flanked on each side by rounded semi-clear area; posterior membranous area containing crescent-shaped clear sclerite, relatively slender with tapering arms.

Relationships. The species may be somewhat close to *B. quieta*.

Distribution and habitat. The species has been recorded from the states of Illinois, Indiana, Michigan, Ohio, New York, and Tennessee, and from Ottawa and Bancroft, Ontario. A single teneral male from Cedar Key, Levy County, Florida may represent this species. It inhabits trunks and branches of broad-leaf trees.

Genus *Blastopsocus* Roesler

Blastopsocus [subgenus of *Blaste*] Roesler, 1943:3.
Generic status. Mockford, 1961:138.

Diagnosis. Adults macropterous. In forewing, Rs-M junction at a point or by a short fusion; pterostigma rounded or smoothly curved on hind margin, opaque but without pigment spot. Hypandrium asymmetrical distally with large, rounded lobe in or slightly off middle flanked by two asymmetrical processes. Male parameres joined by membrane at their bases. Male epiproct with anterior margin overlapping hind margin of clunium at least slightly (Fig. 944). Subgenital plate with egg guide short, little if any longer than its basal width; pigmented arms directed laterally, somewhat expanded at tips; median pigment projection between arms present or not.
 Generotype. *B. variabilis* (Aaron).
 North American species: *B. lithinus* (Chapman), *B. semistriatus* (Walsh), *B. variabilis* (Aaron).

Blastopsocus lithinus (Chapman)

Psocus lithinus Chapman, 1930:249.
Blaste (Blastopsocus) lithinus (Chapman). Roesler, 1943:2.
Blastopsocus lithinus (Chapman). Mockford, 1961:138.

 Recognition features. Forewing (Fig. 945) faintly brown washed, unmarked or with diffuse pale brown pigmentation in pterostigma, or distinct brown spot in distal half of pterostigma. Preclunial abdominal segments 2-6 each with a purplish brown ring of subcuticular pigment on creamy yellow background. Distal end of hypandrium (Fig. 946) with large, rounded median lobe directed slightly leftward; the two flanking processes folded above median lobe; that of right side outermost, directed medio-distad, bearing row of denticles on outer carina; that of left side bent near base, directed mesad beyond bend, bearing row of denticles on distal half. Parameres (Fig. 947) with beak-like distal process bearing denticles near base anteriorly, wattle-like lateral lobe. Male epiproct (Fig. 944) slightly bilobed at base, the lobes extending over clunial margin. Subgenital plate (Fig. 383) with median pigment projection.
 Relationships. Several closely related, undescribed species are found in Mexico.
 Distribution and habitat. The species is found in the east from the Montréal area of Québec south to Collier County, Florida, in the west from Ramsey County, Minnesota south to Bowie County, Texas. It

inhabits trunks and branches of broad-leaf trees and conifers, as well
as shaded rock outcrops. Individuals which may represent this species
have been collected in the Mexican states of Veracruz and Oaxaca, also
in the Guatemala City region. Further investigation is needed to decide
the specific status of these forms.

Blastopsocus semistriatus (Walsh), new combination

Psocus semistriatus Walsh, 1862:361.

Note. Chapman (1930) synonimized this name with '*Psocus*'
quietus Hagen (= *Blaste quieta* (Hagen) based on examination of
paratypes. Critical study of the holotype shows this synonymy to be
incorrect.

Recognition features. Known from females only. Forewing (Fig.
952) clear, unmarked except for slight brown clouding in cells R and
Cu1b and in some individuals in distal two-thirds of pterostigma and
stigmasaum, also brown spot at nodulus. Preclunial abdominal segments 2-8 marked with reddish brown on white background in
following pattern: each segment with spot on dorsal midline vaguely
connected with lateral band extending down slightly beyond spiracle
except on segment 8, where much abbreviated. Subgenital plate (Fig.
953) lacking median pigment projection; lateral arms terminating as
quadrate areas.

Relationships. An apparently undescribed sexual sister species
occurs spottily in eastern and southern Illinois and eastern Indiana.
The latter may be the species which Chapman (1930) called "*Psocus
inornatus* Aaron" (female only).

Distribution and habitat. The species is known from several
localities in the northern half of Illinois (type locality, Rock Island), in
Indiana from Bloomington in the south to Winona Lake and Indiana
Dunes State Park in the north, and from southern Michigan (Kalamazoo
and Livingston Counties). It inhabits branches and trunks of broad-leaf
trees and occurs occasionally on conifers.

Blastopsocus variabilis (Aaron)

Psocus variabilis Aaron, 1883:38.
Psocus medialis Banks, 1907a:165.
Euclismia variabilis (Aaron). Enderlein, 1925:100.
Blaste (Blastopsocus) variabilis (Aaron). Roesler, 1943:2.
Blastopsocus variabilis (Aaron). Sommerman, 1956b:151.

Recognition features. Forewing (Fig. 948) clear, unmarked in both sexes. Preclunial abdominal segments marked (Fig. 384) with five dark purplish brown spots on white background -- two dorsally on segments 1-3, three dorsally and dorso-laterally on segments 5-6; three slender, purplish-brown vertical bands on sides occupying segments 3-5. Distal end of hypandrium (Fig. 949) with large median lobe bent upward and leftward distally, darkly pigmented except pale in its left basal one quarter. Flanking structures decidedly asymmetrical: right-hand structure divided near base; outer shaft producing broad, thumb-like process with denticulate surface; inner shaft cylindrical with tricuspid tip, one cusp pointed and two rounded. Left flanking structure smaller than right, also divided near base; outer shaft producing short, upward-directed hook; inner shaft producing short, round-tipped lobe. Parameres (Fig. 950) with distal process short, abruptly tapering to acute point; lateral lobe about same length as distal process, rounded. Male epiproct (Fig. 951) tapering to blunt point at base, the point extending over clunial margin. Subgenital plate lacking median pigment projection (Fig. 385).

Relationships. A close relative, *B. mockfordi* Badonnel, occurs in Colombia.

Distribution and habitat. The species is found in the east from the Philadelphia region of Pennsylvania (type locality) south to Big Pine Key, Monroe County, Florida, and in the west from Vermillion County, Illinois south to Monroe County, Arkansas. It inhabits trunks and branches of broad-leaf trees and pines.

Chapter 5.
Distribution Patterns

Numerous works have recorded distributions of species in the study area. In addition to those cited in the species accounts, these include Chapman and Nadler (1928) Eertmoed and Eertmoed (1983), Mockford (1952, 1961, 1979b) and New and Loan (1972). Here, a first attempt is made to understand the broad distribution patterns.

An examination of distribution patterns requires that a distinction be made between native species, i.e., species that have been in the area for thousands to millions of years, and introduced species, i.e., species that have arrived through human agency, presumably since the European colonization of North America. It is not easy to make this distinction in every case. Here, the following criteria have been used: (1) species commonly associated with human commerce are regarded as introduced unless they have an extensive out-door distribution in the study area; (2) species for which introduction into the study area has been documented and which are not otherwise present are regarded as introduced; (3) species widely distributed elsewhere and with a very limited, coastal distribution in the study area are regarded as introduced unless the distribution in the study area appears to be part of a natural distribution largely outside the study area. All other species are regarded as native. Table 1 lists the introduced species and indicates by which of the above criteria each is regarded as introduced. By these criteria, 55 species, i.e., 19.0% of the total fauna, are thought to be introduced. Of the remaining species, there are still some uncertain cases. *Soa flaviterminata, Proentomum personatum, Echmepteryx madagascariensis, Embidopsocus femoralis, Epipsocus* sp., *Trichadenotecnum circularoides* and *T. pardus* may have been introduced on the southeastern coastal plain but the Florida records of each of these species may represent the northern end of a natural distribution. *Pteroxanium kelloggi, Trichopsocus acuminatus, Elipsocus abdominalis,* and *E. hyalinus* may have been introduced on the Pacific Coast but each of them has a wide distribution on the coast. These species, then, do not meet the criteria for introduction discussed above.

Five distribution patterns are readily discernible, all of which probably have to do with the geologic and climatic history of the study area. A sixth category is recognized which consists of patterns not fitting into the other five and which apply to very few species.

The five readily discernible patterns are defined as follows:

1) The Eastern Deciduous Forest Pattern. The northern limit is poorly understood due to scanty collection records, but appears to be from central coastal New Brunswick, southern Quebec, and southern to central Ontario, west to northern Minnesota. The southern limit is somewhat better understood. In the east it lies in northern to central peninsular Florida, and in the west, where collection records are again scanty, it appears to lie along the northern Gulf Coast of Texas.

This area corresponds closely with the Eastern Deciduous Forest, as delimited by Braun (1950). Braun notes that deciduous forest has been extensive in eastern North America since the upper Cretaceous, and early included numerous genera of trees still present in the area plus others now confined to warmer regions or to the Southern Hemisphere.

Sixty-three species of psocids show this distribution pattern (26.9% of the native species of the study area). It is notable that 27 of these belong to Family Psocidae. It does not follow, however, that the Eastern Deciduous Forest has been a major center of evolution of the Psocidae. Most of the genera are small and have Neotropical affinities. The principal role of the region seems to have been as a receiving area for species spreading out from the Tropics. Many of these seem to have found a suitable niche, become established, and in some cases speciated apart from their tropical predecessor.

(2) The Southeastern Subtropical Pattern. This pattern is based on climate rather than a vegetational formation. The pattern is coextensive with the coastal plain of the Gulf States and the southern Atlantic States, including all of Florida, extending up the Mississippi Embayment, for some species, as far as southern Illinois, and up the Atlantic Coast probably to southern Virginia. A total of 85 species (Table II) shows this pattern, i.e., 36.3% of the native species. Numerically, then, this is the most important pattern in the study area. Fifty of these species (Table II N) are in the study area at the northern end of a range which extends well into the Tropics, either on the mainland or in the Antilles, or both. The remaining thirty-three species (Table II) are in tropical genera but are not, themselves, known to range south of the study area, or do so only to a slight extent. However, 21 of the latter 33 species (Table II, species designated by C) have close relatives in the American Tropics and are probably products of a speciation separating them from their close tropical relatives.

Table I. List of introduced species and the criteria used to decide their introduced status (see text).

Species	Criterion 1	Criterion 2	Criterion 3
Psoquilla marginepunctata	X		
Balliella ealensis		X	
Rhyopsocus disparilis	X	X	
Cerobasis annulata	X		
Lepinotus inquilinus	X		
Lepinotus patruelis	X		
Trogium pulsatorium	X		
Dorypteryx domestica	X		
Dorypteryx pallida	X		
Pseudodorypteryx mexicana	X		
Psocatropos microps	X		
Nanopsocus oceanicus	X		
Pachytroctes aegyptius		X	
Sphaeropsocopsis argentinus		X	
Embidopsocus thorntoni	X		
Liposcelis entomophila	X		
Liposcelis fusciceps		X	
Liposcelis bicolor	X		
Liposcelis decolor	X		
Liposcelis pearmani	X		
Liposcelis rufa	X		
Liposcelis mendax	X		
Liposcelis corrodens	X		
Liposcelis paeta	X		
Stimulopalpus japonicus		X	
Graphopsocus cruciatus		X	
Caecilius gonostigma		X	
Caecilius disjunctus		X	
Caecilius atricornis		X	
Lachesilla cintalapa		X	
Lachesilla alpajia		X	
Lachesilla denticulata		X	
Lachesilla fuscipalpis		X	
Lachesilla greeni		X	
Lachesilla quercus	Single record from a warehouse and one introduction.		
Ectopsocus peterst		X	
Ectopsocus pumilis	X	X	
Ectopsocus richardsi	X		
Ectopsocus salpinx	X		
Ectopsocus strauchi		X	
Ectopsocus titschacki		X	
Peripsocus reductus			X
Peripsocus phaeopterus			X
Heterocaecilius sp.			X
Trichopsocus dalii		X	
Cuneopalpus cyanops			X
Elipsocus moebiusi			X
Elipsocus pumilis (= westwoodi)			X
Propsocus pulchripennis			X
Mesopsocus immunis			X
Hemipsocus chloroticus		X	
Myopsocus eatoni		X	
Myopsocus minutus		X	
Loensia fasciata		X	
Amphigerontia contaminata			X

The constituents of the Southeastern Subtropical Pattern would appear to be species which have expanded out of the Tropics but which have, for the most part, been unable to adapt to climates of the study area beyond the subtropical climate of the southeastern coastal plain. The status of a species with respect to its relationship to tropical populations is probably a question of the amount of time that the species has existed in the area. Species which are at the northern end of a tropical range are presumably the most recent invaders. Those which have speciated apart from their tropical close relatives have presumably been in the area longer. We cannot, however, assume that the remaining (12) species have been in the area long enough to have lost their tropical roots. It may be that we have simply not yet discovered them.

(3) The Boreal and Holarctic Pattern. To this pattern belong a small array of species which tend to be northern in distribution within the study area and/or also occur in the Palearctic Region. Seventeen species (Table II), i.e., 7.3% of the native species show this pattern. Of these, seven species are boreal Nearctic and ten are Holarctic. Within the seven boreal Nearctic species, *Caecilius hyperboreus* is closely related to the holarctic *C. flavidus*; *C. boreus* is closely related to the holarctic *C. burmeisteri*. *Lachesilla arnae* and *L. nubiloides*, both species of the andra group, are probably rather close to each other, and both are rather close to the eastern Palearctic species *L. kerzhneri* Roesler. Thus, for these four species, a holarctic ancestry seems possible. For the remaining three, little can be said. *Caecilius graminis* belongs to a small species group which includes several Palearctic species; *Xanthocaecilius quillayute* and *Lachesilla albertina* are both probably derived from Meso-American ancestral lines.

(4) The Southwestern and Rocky Mountains Pattern. Thirty-five species, i.e., 15.0 % of the native fauna, show a pattern of distribution based on the southern and central Rocky Mountains and surrounding arid lands (Table II). Of these, ten species are largely Mexican, reaching their northern limit in the Southern Rockies (species designated by M in Table II), while *Cerastipsocus trifasciatus* continues northeastward. Six others have their closest affinities in the Mexican or South American highlands (species designated by S in Table II). Five have their closest relationships in Pacific Coastal species (species designated by P in Table II). Two are closest to eastern North American species (species designated by E in Table II); two others are closest to boreal species (species designated by B in Table II). *Asiopsocus sonorensis*, with range restricted to southern Arizona, southwestern New Mexico, and northern Sonora (Mexico) shows a relict distribution. Its closest relatives are in Spain and Mongolia. *Liposcelis formicaria* may prove to be part of the Boreal and Holarctic pattern when its distribution is better understood.

Table II. List of native species and their distribution patterns.

Species	Pattern 1	Pattern 2	Pattern 3	Pattern 4	Pattern 5	Pattern 6
Thylacella cubana		X (N)[1]				
Soa flaviterminata		X (N)				
Nepticulomima sp.		X				
Proentomum personatum		X (N)				
Echmepteryx hageni	X					
Echmepteryx intermedia		X (N)				
Echmepteryx youngi		X (C)				
Echmepteryx madagascariensis		X (N)				
Neolepolepis caribensis		X				
Neolepolepis occidentalis	X					
Pteroxanium kelloggi					X	
Rhyopsocus bentonae		X (C)				
Rhyopsocus micropterus					X (SW)	
Rhyopsocus eclipticus		X (C)				
Rhyopsocus texanus		X (N)				
Cerobasis guestfalica	X				X (EU)	
Leptnotus reticulatus				X		
Myrmecodipnella aptera					X	
Psyllipsocus oculatus		X (N)				
Psyllipsocus ramburii						X
Speleketor flocki				X (P)		
Speleketor irwini					X (SW)	
Speleketor pictus					X (SW)	
Pachytroctes neoleonensis				X (M)		
Tapinella maculata		X (N)				
Sphaeropsocus sp.					X	
Belaphotroctes allieni				X (S)		
Belaphotroctes badonneli		X (C)				
Belaphotroctes ghesquierei		X (N)				
Belaphotroctes hermosus		X				
Belaphotroctes simberloffi		X				
Embidopsocus bousemani	X					
Embidopsocus laticeps		X (C)				
Embidopsocus mexicanus		X (N)				
Embidopsocus citrensis		X				
Embidopsocus needhami	X					
Embidopsocus femoralis		X (N)				
Liposcelis brunnea				X		
Liposcelis deltachi				X		
Liposcelis hirsutoides		X				
Liposcelis nasa		X (N)				
Liposcelis ornata		X (N)				
Liposcelis pallens	X					
Liposcelis pallida				X (P)		
Liposcelis villosa					X (SW)	
Liposcelis nigra	X					
Liposcelis lacinia				X		
Liposcelis silvarum					X (EU)	
Liposcelis tricellata					X	
Liposcelis formicaria				X		
Liposcelis bostrychophila						X
Liposcelis prenolepidis					X	
Lithoseopsis hellmani		X (N)				
Lithoseopsis hystrix		X (N)				
Epipsocus sp.		X (C)				
Bertkauia crosbyana	X				X	
Bertkauia lepicidinaria	X					
Loneura sp.				X (M)		
Asiopsocus sonorensis				X		
Notiopsocus sp.		X				
Pronotiopsocus sp.		X (C)				
Polypsocus corruptus	X				X	
Teliapsocus contermnus						X

Table II. continued.

Species	Pattern 1	Pattern 2	Pattern 3	Pattern 4	Pattern 5	Pattern 6
Caecilius africanus		X (N)				
Caecilius antillanus		X (N)				
Caecilius casarum		X (N)				
Caecilius insularum		X (N)				
Caecilius indicator		X (C)				
Caecilius confluens	X					
Caecilius graminis			X			
Caecilius nadleri	X					
Caecilius totonacus					X (M)	
Caecilius boreus			X			
Caecilius burmeisteri						X (EU)
Caecilius caloclypeus		X (C)				
Caecilius croesus		X (N)				
Caecilius flavidus			X			
Caecilius hyperboreus			X			
Caecilius lochloosae		X				
Caecilius manteri	X					
Caecilius maritimus					X	
Caecilius micanopi		X (C)				
Caecilius perplexus				X		
Caecilius pinicola	X					
Caecilius tamiami		X				
Caecilius posticus	X					
Caecilius incoloratus		X (C)				
Caecilius juniperorum		X (C)				
Caecilius subflavus		X (C)				
Xanthocaecilius quillayute			X			
Xanthocaecilius sommermanae	X					
Anomopsocus amabilis	X					
Nanolachesilla chelata		X				
Nanolachesilla hirundo		X (C)				
Prolachesilla terricola				X (S)		
Lachesilla andra	X					
Lachesilla arnae			X			
Lachesilla dona					X	
Lachesilla kola				X (S)		
Lachesilla nubilis	X					
Lachesilla nubiloides			X			
Lachesilla punctata				X (B)		
Lachesilla centralis					X	
Lachesilla perezi		X (N)				
Lachesilla albertina			X			
Lachesilla corona	X					
Lachesilla anna	X					
Lachesilla bottimeri		X				
Lachesilla chapmani		X (C)				
Lachesilla contraforcepeta	X					
Lachesilla forcepeta	X					
Lachesilla gracilis		X (N)				
Lachesilla kathrynae		X (N)				
Lachesilla major	X					
Lachesilla penta		X (N)				
Lachesilla sulcata		X (N)				
Lachesilla aethiopica		X (N)				
Lachesilla pacifica					X	
Lachesilla pallida	X					
Lachesilla pedicularia						X
Lachesilla rena				X (M)		
Lachesilla tectorum		X (N)				
Lachesilla riegeli		X (N)				
Lachesilla tropica		X (N)				
Lachesilla ultima		X (C)				
Lachesilla arida				X (B)		

Table II. continued.

Species	Pattern 1	Pattern 2	Pattern 3	Pattern 4	Pattern 5	Pattern 6
Lachesilla chiricahua				X		
Lachesilla jeanae				X (S)		
Lachesilla rita		X (N)				
Lachesilla rufa	X					
Lachesilla yakima					X (SW)	
Lachesilla texcocana				X (M)		
Ectopsocopsis cryptomeriae	X					
Ectopsocus briggsi					X (EU)	
Ectopsocus californicus					X	
Ectopsocus maindroni		X (N)				
Ectopsocus meridionalis	X					
Ectopsocus thibaudi		X (N)				
Ectopsocus vachoni						X
Kaestneriella fumosa				X (S)		
Kaestneriella tenebrosa				X (S)		
Peripsocus madidus	X					
Peripsocus pauliani		X (N)				
Peripsocus subfasciatus			X			
Peripsocus alachuae		X				
Peripsocus alboguttatus			X			
Peripsocus maculosus	X					
Peripsocus madescens	X					
Peripsocus minimus						X
Peripsocus potosi		X (N)				
Peripsocus stagnivagus	X					
Archipsocopsis frater		X (N)				
Archipsocopsis parvula		X (N)				
Archipsocus floridanus		X (N)				
Archipsocus gurneyi		X (N)				
Archipsocus nomas		X (N)				
Pseudocaecilius citricola		X (N)				
Pseudocaecilius tahitiensis		X (N)				
Trichopsocus acuminatus				X (EU)		
Aaroniella achrysa		X (N)				
Aaroniella badonneli	X					
Aaroniella maculosa	X					
Philotarsus kwakiutl					X	
Philotarsus picicornis			X			
Elipsocus abdominalis				X (EU)		
Elipsocus guentheri				X (P)		
Elipsocus hyalinus				X (EU)		
Elipsocus obscurus				X (SW)		
Neplomorpha perpsocoides		X (C)				
Palmicola aphrodite		X (C)				
Palmicola solitaria		X (C)				
Reutereila helvimacula			X			
Mesopsocus laticeps			X			
Mesopsocus unipunctatus			X			
Hemipsocus pretiosus		X (N)				
Lichenomima coloradensis				X (E)		
Lichenomima lugens	X					
Lichenomima sparsa	X					
Myopsocus antillarus		X (N)				
Atropsocus atratus	X					
Hyalopsocus floridanus	X					
Hyalopsocus striatus	X					
Psocus crosbyi					X	
Psocus leidyi	X					
Camelopsocus buctrianus					X (SW)	
Camelopsocus hiemalis					X (SW)	
Camelopsocus monticolus				X (M)		
Camelopsocus similis				X (M)		
Camelopsocus tucsonensis				X (P)		

Table II. continued.

Species	Pattern 1	Pattern 2	Pattern 3	Pattern 4	Pattern 5	Pattern 6
Indiopsocus bisignatus	X					
Indiopsocus ceterus		X (N)				
Indiopsocus coquilletti					X	
Indiopsocus trifumatus	X					
Indiopsocus insulanus		X (C)				
Indiopsocus texanus		X (N)				
Loensia conspersa				X		
Loensia maculosa				X (E)		
Loensia moesta	X					
Ptycta lineata		X (N)				
Ptycta polluta	X					
Steleops elegans	X					
Steleops lichenatus	X					
Trichadenotecnum alexanderae	X					
Trichad. castum	X					
Trichad. innuptum			X			
Trichad. merum	X					
Trichad. circularoides		X (N)				
Trichad. desolatum				X (M)		
Trichad. majus			X			
Trichad. pardus		X (N)				
Trichad. slossonae	X					
Trichad. quaestum	X					
Metylophorus barretti				X (M)		
Metylophorus novaescotiae	X					
Metylophorus purus	X					
Cerastipsocus trifasciatus	X			X (M)		
Cerastipsocus venosus	X					
Amphigerontia bifasciata			X			
Amphigerontia infernicola				X		
Amphigerontia montivaga	X					
Amphigerontia petiolata	X					
Blaste cockerelli				X		
Blaste garciorum		X (N)				
Blaste longipennis				X (P)		
Blaste opposita	X					
Blaste oregona					X (SW)	
Blaste osceola		X (C)				
Blaste persimilis		X				
Blaste posticata		X (N)				
Blaste quieta	X					
Blaste subapterous						X (SW)
Blaste subquieta	X					
Blastopsocus lithinus	X					
Blastopsocus semistriatus	X					
Blastopsocus variabilis	X					

[1]B – close relatives boreal.
C – close relatives in American Tropics.
E – close relatives in eastern North America.
Eur– species shared with Europe.
M – most of range in Mexico.
N – at northern end of range extending into Tropics.
P – close relatives in Pacific Coastal Pattern.
S – close relatives in Mexican or South American highlands.
Sw– close relatives in Southwestern and Rocky Mountain Pattern.

Of the other eight constituent species of this pattern, nothing can be said at present about affinities.

(5) The Pacific Coast Pattern. Thirty-two species, i.e. 13.7% of the native fauna show a Pacific Coastal distribution pattern. Of these, ten show affinity with species of the Southwestern and Rocky Mountain Pattern (species designated by SW in Table II). Seven species are shared with Europe (species designated by EU in Table II), but because they do not occur inland across the North American continent, they are not regarded as constituents of the Boreal and Holarctic Pattern. One of these, *Cerobasis guestfalica*, also occurs on the eastern coastal plain. Two species, *Bertkauia crosbyana* and *Polypsocus corruptus*, are also constituents of the Eastern Deciduous Forest Pattern. Both occur spottily across southern Canada, but because of their wide north-south distributions in the east and their Meso-American affinities, they are not regarded as constituents of the Boreal Pattern. *Caecilius maritimus* was probably derived by speciation from the holarctic *C. flavidus*. *Lachesilla dona*, although it ranges far to the south, may have boreal affinities. *Blaste oregona* and *B. subapterous* have apparently originated by speciation in the mountains of California and Oregon of a complex which extends eastward into the Rocky Mountains and eastern United States. Three species, *Pteroxanium kelloggi*, *Ectopsocus briggsi* (also in Europe), and *E. californicus*, also occur in the South Temperate Zone, either in southern South America, or in the Australian Region, or both. The affinities of these three suggest that they may have originated in the Australian Region. Their mode of dispersal is not understood. *Sphaeropsocus* sp. is apparently a relict. The genus is otherwise known from the Baltic amber. *Liposcelis silvarum*, known in the study area only from a few counties in Oregon, may be relictual there, although it is widespread in the Palearctic Region.

(6) Other patterns. Two species exhibit patterns of distribution which do not fit any of those described above. *Teliapsocus conterminus* has a distribution divided into an eastern and a western component. The eastern component is a fairly typical Eastern Deciduous Forest distribution except that it extends farther south in Florida and shows a large central gap including the states of Illinois, Indiana, and Ohio. The western component includes the Pacific Coast from northern California to southern Alaska, and inland to the Rocky Mountains. Its absence in the central grasslands suggests a humidity-limited distribution. Its absence in Illinois, Indiana, and Ohio, but presence in southern Missouri, Michigan, and northern Minnesota, suggests some additional limiting factor acting either now or in the recent past.

Ectopsocus vachoni occurs spottily across the southern portion of the study area, from southern Georgia west to southern California. It has an extensive distribution in Mexico and Central America. Its range

In the study area appears to be simply the northern limit for a species which is adapted to both semiarid and more moist conditions.

Four species are widely distributed in the study area with numerous out-door stations and are also commonly associated with human commerce. These are *Lepinotus reticulatus*, *Psyllipsocus ramburii*, *Liposcelis bostrychophila*, and *Lachesilla pedicularia*. It is possible that all were introduced into the study area, but their large out-door ranges suggest that they may have arrived by other means. *Lepinotus reticulatus* is known to be phoretic on birds (Mockford, 1967).

Literature Cited

Aaron, S. F. 1883. Description of new Psocidae in the Collection of the American Entomological Society. Trans. Am. Entomol. Soc. 11: 37-40, pl. IX.

Aaron, S. F. 1886. On some new Psocidae. Proc. Acad. Nat. Sci. Phila. 1886: 13-18, pl. I.

Allen, R. W. 1973. The biology of *Thysanosoma actinioides* (Cestoda: Anoplocephalidae) a parasite of domestic and wild ruminants. New Mex. St. Univ. Ag. Expt. Sta. Bull. 604: 1-69.

Ashmead, W. H. 1879. On a new *Psocus*. Canad. Entomol. 11: 228-229.

Ashmead, W. H. 1894. Notes on cotton insects found in Mississippi. Insect Life 7: 25-29, 240-247 (psocids pp. 28-29).

Badonnel, A. 1931. Contribution a l'étude de la faune du Mozambique, voyage de M. P. Lesne (1928-1929). 4e note. -- Copéognathes. Ann. Sci. Nat., Ser. Bot., Zool. 14: 229-260.

Badonnel, A. 1935a. Psocoptères nouveaux d'Afrique et d'Arabie. Rev. Fr. Entomol. 2: 76-82.

Badonnel, A. 1935b. Psocoptères de France (5e note). Liste d'espèces nouvelles ou peu connues avec indication de quelques synonymies. Bull. Soc. Entomol. Fr. 40: 199-203.

Badonnel, A. 1936. Psocoptères de France (7e note) Espèces nouvelles ou peu connues, et description de deux espèces inédites. Bull. Soc. Entomol. Fr. 41: 24-29.

Badonnel, A. 1938. Psocoptères de France (9e note). Diagnoses preliminaires et nouvelles captures. Bull. Soc. Entomol. Fr. 43: 17-22.

Badonnel, A. 1943. Faune de France 42. Psocoptères. P. Lechevalier et Fils, Paris, pp. 1-164.

Badonnel, A. 1944. Contribution a l'étude de Psocoptères de l'Atlantide. Rev. Fr. Entomol. 11: 47-60.

Badonnel, A. 1945. Contribution a l'étude des Psocoptères du Maroc, voyage de L. Berland et M. Vachon (1939). Rev. Fr. Entomol. 12: 31-50.

Badonnel, A. 1948. Psocoptères du Congo belge (2e note). Rev. Zool. Bot. Afr. 40: 266-322.

Badonnel, A. 1949a. Psocoptères de la Cote d'Ivoire. Mission Paulian-Delamare (1945). Rev. Fr. Entomol. 16: 20-46.

Badonnel, A. 1949b. Psocoptères du Congo Belge (3e note). Bull. Inst. Roy. Sci. Nat. Belg. 25: 1-64.

Badonnel, A. 1951. Ordre des Psocoptères. In Grassé, P. (ed.). Traité de Zoologie. Paris. 17 vols (Psocoptera: vol. 10, fasc. 2, pp. 1301-1340).

Badonnel, A. 1955. Psocoptères de l'Angola. Publ. Cult. Cia. Diamant. Angola. 26: 1-267.

Badonnel, A. 1962. Psocoptères. In Biologie de l'Amerique Australe. Vol. I. Études sur la Faune du Sol. Eds. Centre Nat. Recherche Sci., Paris (Psocoptera, pp. 185-229).

Badonnel, A. 1963. Psocoptères terricoles, lapidicoles, et corticoles du Chili. Biologie de l'Amerique Australe. Vol. II. Etudes sur la faune du sol. Eds. Centre Nat. Recherche Sci., Paris (Psocoptera, pp. 291-338).

Badonnel, A. 1966. Sur le genre *Archipsocus* Hagen (Psocoptera, Archipsocidae). Bull. Mus. natl. Hist. Nat., Paris. 2e Ser. 38: 409-415.

Badonnel, A. 1967. Faune de Madagascar XXIII. Insectes Psocoptères. O.R.S.T.O.M. and C.N.R.S., Paris, 237 pp.

Badonnel, A. 1968. Trois espèces Americaines inédites de *Liposcelis* (Psocoptera, Liposcelidae). Bull. Soc. Zool. Fr. 93: 535-544, Pls. I, II.

Badonnel, A. 1969. Psocoptères de l'Angola et de pays voisins, avec révision de types africains d'Enderlein (1902) et de Ribaga (1911). Publ. Cult. Cia Diamant. Angola. 79: 1-152.

Badonnel, A. 1971a. *Liposcelis* (Psocoptera, Liposcelidae) de l'Ile de Chypre. Bull. Mus. natl. Hist. Nat., Paris 2e Ser., 42: 1212-1223.

Badonnel, A. 1971b. *Embidopsocus thorntoni* (Psocoptera: Liposcelidae) nouvelle espèce de l'Archipel des Galapagos. Nouv. Rev. Entomol. I: 325-327.

Badonnel, A. 1971c. Psocoptères édaphiques du Chili (3e note) (Insects). Bull. Mus. natl. Hist. Nat., Paris, 3e ser., 1: 1-38.

Badonnel, A. 1971d. *Sphaeropsocopsis reisi* n. sp., premier représentant africain connu de la famille des Sphaeropsocidae (Psocoptera, Nanopsocetae), avec compléments a la faune des Psocoptères angolais. Publ. Cult. cia. Diamant. Angola. 84: 13-28.

Badonnel, A. 1972. Espèces brésiliennes de la sous-famille des Embidopsocinae (Psocoptera: Liposcelidae). Bull. Mus. natl. Hist. Nat., Paris. 3e Ser., 87: 1097-1139.

Badonnel, A. 1976. *Archipsocus etiennei* n. sp. (Psocoptera, Archipsocidae) de l'Ile de La Réunion. Nouv. Rev. Entomol. 6: 3-8.

Badonnel, A. 1978. Compléments a l'etude des Archipsocidae du Brésil (Insecta, Psocoptera). Rev. Brasil. Biol. 38: 177-186.

Badonnel, A. 1979. *Ectopsocus thibaudi* n. sp., du département de la Guadeloupe, avec remarques sur le group hirsutus (Psocoptera Ectopsocidae). Bull. Soc. Entomol. Fr. 84: 52-57.

Badonnel, A. 1980. Compléments a la description de *Mesopsocus dubosquei* (Psocoptera, Mesopsocidae), avec analyse du complex "unipunctatus". Revue Fr. Entomol., (N.S.) 2: 99-106.

Badonnel, A. 1986a. Psocoptères du Sénégal. III. Liposcelidae (Psocoptera). Nouv. Rev. Entomol. 3: 69-76.

Badonnel, A. 1986b. Description du mâle de *Peripsocus potosi* Mockford, un curieux cas de gynandromorphisme. Rev. Fr. Entomol. 8: 97-99.

Badonnel, A. 1986c. Psocoptères (Insecta) de la bordure pacifique de l'Etat de Jalisco, Mexique. Revue suisse Zool. 93: 693-723.

Badonnel, A. 1986d. Psocopteres de Colombie (Insecta, Psocoptera) Missions écologiques du Professeur Sturm (1956a, 1978). Spixiana. 9: 179-223.

Badonnel, A. and A. N. García Aldrete. 1980. *Lachesilla nuptialis* n. sp. Espece-Soeur de *Lachesilla aethiopica* (Enderlein) (Psocoptera: Lachesillidae). Folia Entomol. Mex. 44: 5-18.

Ball, A. 1926. Les Psocidae de Belgique. Ann. et Bull. Soc. Entomol. Belg. 66: 331-349, Pls. I-III.

Ball, A. 1943. Contribution a l'étude des Psocoptères, III.---*Ectopsocus* de Congo belge, avec une remarque sur le rapport I.O./D. Bull. Mus. Roy. Hist. Nat. Belg. 19: 1-28.

Banks, N. 1892. A synopsis, catalogue, and bibliography of the Neuropteroid insects of temperate North America. Trans. Am. Entomol. Soc. 19: 327-372.

Banks, N. 1897. Psocidae. *In* Call, R. E. Some notes on the flora and fauna of Mammoth Cave, KY. Amer. Nat. 31: 377-392 (Psocid p. 382, pl. X).

Banks, N. 1900a. A new genus of Atropidae. Entomol. News 11: 431-432.

Banks, N. 1900b. New genera and species of Nearctic neuropteroid insects. Trans. Am. Entomol. Soc. 26: 239-259.

Banks, N. 1900c. Two new species of Troctes. Entomol. News 11: 559-560.

Banks, N. 1903a. Neuropteroid insects from Arizona. Proc. Entomol. Soc. Wash. 5: 237-245.

Banks, N. 1903b. Some new neuropteroid insects. J. N. Y. Entomol. Soc. 11: 236-243.

Banks, N. 1904a. Neuropteroid insects from New Mexico. Trans. Am. Entomol. Soc. 30: 97-110, Pl. I.

Banks, N. 1904b. A list of Neuropteroid insects, exlusive of Odonata, from the vicinity of Washington, D.C. Proc. Entomol. Soc. Wash. 6: 201-217.

Banks, N. 1905. Descriptions of new nearctic neuropteroid insects. Trans. Am. Entomol. Soc. 32: 1-20, Pls. I, II.

Banks, N. 1907a. New Trichoptera and Psocidae. Journ. N. Y. Entomol. Soc. 15: 162-166.

Banks, N. 1907b. Catalogue of the neuropteroid insects (except Odonata) of the United States. Am. Entomol. Soc. (separate publication), 53 pp.

Banks, N. 1908. Neuropteroid insects -- notes and descriptions. Trans. Am. Entomol. Soc. 34: 255-266.

Banks, N. 1913. Neuropteroid insects from Brazil. Psyche 20: 83-89.

Banks, N. 1914. New neuropteroid insects, native and exotic. Proc. Acad. Nat. Sci. Phila. 66: 608-618, pl. XXVIII.

Banks, N. 1918. New neuropteroid insects. Bull. Mus. Comp. Zool. 62: 3-5, pls. 1-2.

Banks, N. 1920. New neuropteroid insects. Bull. Mus. Comp. Zool. 64: 299-362, pls. 1-7.

Banks, N. 1924. Descriptions of new neuropteroid insects. Bull. Mus. Comp. Zool. 65: 421-544, pls. 1-4.

Banks, N. 1930a. Some new neuropteroid insects. Psyche 37: 183-191, pl. 9.

Banks, N. 1930b. New neuropteroid insects from the United States. Psyche 37: 223-232, pl. 12.

Banks, N. 1931. On some Psocidae from the Hawaiian Islands. Proc. Haw. Entomol. Soc. 7: 437-441, pls. VII-IX.

Banks, N. 1938. New native neuropteroid insects. Psyche 45: 72-78.

Banks, N. 1941. New neuropteroid insects from the Antilles. Mem. Soc. Cubana Hist. Nat. 15: 385-402, pls. 43-45.

Baz, A. 1988. Psocópteros de Azores: nuevas citas, descripciones y sinonimias. Bol. Soc. Portug. Entomol. 93: 1-15.

Baz, A. Unpublished. Los Psocópteros (Insecta: Psocoptera) del sistema Iberico Meridional. Doctoral thesis, Facultad de Ciencias, Univ. de Alcala de Henares. 230 pp.

Bertkau, P. 1883. Ueber einen auffallenden Geschlechtsdimorphismus bei Psociden nebst Beschreibung einiger neuer Gattungen und Arten. Arch. Naturgesch. 49: 97-101, pl. 1.

Betts, M. M. 1956. A list of insects taken by tit mice in the Forest of Dean (Gloucestershire). Entomol. Mon. Mag. 92: 68-71.

Betz, B. W. 1983a. The biology of *Trichadenotecnum alexanderae* Sommerman (Psocoptera: Psocidae). III. Analysis of mating behavior. Psyche, 90: 97-117.

Betz, B. W. 1983b. Systematics of the *Trichadenotecnum alexanderae* species complex (Psocoptera: Psocidae) based on an investigation of modes of reproduction and morphology. Canad. Entomol. 115: 1329-1354.

Betz, B. W. 1983c. The biology of *Trichadenotecnum alexanderae* Sommerman (Psocoptera: Psocidae). II. Duration of biparental and parthenogenetic reproductive abilities. J. Kans. Entomol. Soc. 56: 420-426.

Bigot, L., G. Bonin, and M. Roux. 1983. Variations spatio-temporelles entre la communauté des Coléoptères et Psocoptères frondicoles et la végétation dans le massif de la Sainte Baume (Provence). Ecol. Medit. 9: 173-191.

Billberg, G. J. 1820. Enumeratio insectorum in Museo Billberg. Holm. (Psocids Part 4: 94).

Boring, A. M. 1913. The odd chromosome in *Cerastipsocus venosus*. Biol. Bull., Woods Hole, 24: 125-132.

Brauer, F. 1857. Neuroptera austriaca; die im Erzhergothum Oesterreich bis jetzt angefunden Neuropteren. Wien. XXIII+ 80 pp, 5 pls.

Braun, E. L. 1972. Deciduous forests of eastern North America. Hafner Co., New York, XIV + 596 pp, map.

Broadhead, E. 1947a. New species of *Liposcelis* Motschoulsky (Corrodentia, Liposcelidae) in England. Trans. R. Entomol. Soc. Lond. 98: 41-58, pl. 1.

Broadhead, E. 1947b. The life-history of *Embidopsocus enderleini* (Ribaga) (Corrodentia, Liposcelidae). Entomol. Mon. Mag. 83: 200-203.

Broadhead, E. 1950. A revision of the genus *Liposcelis* Motschulsky with notes on the position of this genus in the order Corrodentia and on the variability of ten *Liposcelis* species. Trans. R. Entomol. Soc. Lond. 101: 335-388, Pls. I-III.

Broadhead, E. 1952. A comparative study of the mating behavior of eight *Liposcelis* species. Trans. IX Internat. Congr. Entomol. (1951): 380-382.

Broadhead, E. 1958a. Some records of animals preying upon psocids. Entomol. Mon. Mag. 94: 68-69.

Broadhead, E. 1958b. The psocid fauna of larch trees in northern England -- an ecological study of mixed species populations exploiting a common resource. J. Anim. Ecol. 27: 217-263.

Broadhead, E. 1971. A new species of *Liposcelis* (Psocoptera, Liposcelidae) from North America with records of another species. J. Nat. Hist. 5: 263-270.

Broadhead, E. and B. M. Hobby. 1944. Studies on a species of *Liposcelis* (Corrodentia, Liposcelidae) occurring in stored products in Britain.--Part I. Entomol. Mon. Mag. 80: 45-59.

Broadhead, E. and A. M. Richards. 1982. The Psocoptera of East Africa -- a taxonomic and ecological survey. Biol. J. Linn. Soc. 17: 137-216.

Broadhead, E. and A. J. Wapshere. 1966. *Mesopsocus* populations on larch in England -- the distribution and dynamics of two closely-related coexisting species of Psocoptera sharing the same food resource. Ecol. Monog. 36: 327-388.

Burmeister, H. 1839. Handbuch der Entomologie. Berlin. Vol. 2. (Psocina pp. 772-782).

Carpenter, F. M. 1926. Fossil insects from Kansas. Bull. Mus. Comp. Zool. 67. Copeognatha pp. 440-441, pls. 2, 3.

Carpenter, F. M. 1932. The lower Permian insects of Kansas. Part 5. Psocoptera and additions to the Homoptera. Am. J. Sci., Fifth Ser., 29: Psocoptera pp. 1-17.

Carpenter, F. M. 1933. The lower Permian insects of Kansas. 6: Delopteridae, Protelytroptera, Plecoptera and a new collection of Protodonata, Megasecoptera, Homoptera, and Psocoptera. Proc. Amer. Acad. Arts. Sci. 68: Psocoptera pp. 443-461, figs. 13-16.

Carpenter, F. M. 1939. The lower Permian insects of Kansas. Part 8. Additional Megasecoptera, Protodonata, Odonata, Homoptera, Psocoptera, Protelytroptera, Plectoptera, and Protoperlaria. Proc. Am. Acad. Arts, Sci. 73: Psocoptera pp. 54-57, figs. 3, 5, 6.

Chapman, P. J. 1930. Corrodentia of the United States of America: I. Suborder Isotecnomera. J. N. Y. Entomol. Soc. 38: 219-290, 319-383, pls. XII-XXI.

Chapman, P. J. and A. M. Nadler. 1928. *In* Leonard, M. D. A list of insects of New York. Mem. Cornell Ag. Exp. Sta. 101, 1121 pp. (Corrodentia, pp. 60-63).

Comstock, J. H. 1924. An introduction to entomology. Ed. 1. Comstock Publ. Co., Ithaca. xix + 1044 pp.

Corbett, G. H. and E. Hargreaves. 1915. *Vulturops floridensis*, a new member of the psocid subfamily Vulturopinae from the United States. Psyche 22: 142-143, pl. XI.

Curran, C. H. 1925. Descriptions of two insects found in imported foodstuffs. Canad. Entomol. 57: 292-293.

Curtis, J. 1837. British Entomology. London. Vol. IV. Hymenoptera, Part II. Neuroptera. Trichoptera. (Psocids pt. 14: 648-651).

Dalman, J. W. 1823. Analecta Entomologica. Holm. (psocids p. 98).

Danks, L. 1950. New species of psocid [in Russian] Psocopteres, *Philotarsus badonnell* Danks, sp. n.. National Museum of Natural Science, Riga LSSR, Bulletin No. 1: 2 pp.

Danks, L. 1955. Psocoptera of the Batum and Sachi Botanical Gardens [in Russian]. Entomol. Obozr. 34: 180-184.

Edwards, B. A. B. 1950. Tasmanian Psocoptera with descriptions of new species. Pap. and Proc. R. Soc. Tasm., 1949: 93-134.

Eertmoed, G. E. 1966. The life history of *Peripsocus quadrifasciatus* (Psocoptera: Peripsocidae). J. Kans. Entomol. Soc. 39: 54-65.

Eertmoed, G. E. 1973. The phenetic relationships of the Epipsocetae (Psocoptera): the higher taxa and the species of two new families. Trans. Am. Entomol. Soc. 99: 373-414.

Eertmoed, G. E. 1978. Embryonic diapause in the psocid *Peripsocus quadrifasciatus*: photoperiod, temperature, ontogeny, and geographic variation. Physiol. Entomol. 3: 197-206.

Eertmoed, G. E. and E. Eertmoed. 1983. A collection of Psocoptera from Voyageurs National Park, Minnesota. Great Lakes Entomol. 16: 123-126.

Enderlein, G. 1900. Die Psocidenfauna Perus. Zool. Jahrb., Abt. f. Syst., Geogr. u. Biol. Thiere. 14: 133-139, pls. 8, 9.

Enderlein, G. 1901. Neue deutsche und exotische Psociden, sowie Bemerkungen zur Systematik. Zool. Jahrb., Abt. f. Syst., Geogr. u. Biol. Thiere. 14: 537-548, pl. 35.

Enderlein, G. 1902. Psociden aus Deutsch-Ostafrika. Mitt. Zool. Mus. Berlin 2: 7-15, pl. 5.

Enderlein, G. 1903a. Neue Copeognathen aus Kamerun. Zool. Jahrb., Abt. f. Syst., Geogr. u. Biol. Thiere. 19: 1-8, pl. 1.

Enderlein, G. 1903b. Die Copeognathen des indo-australischen Faunengebietes. Ann. hist-nat., Mus. Natl. Hung. 1: 179-344, pls. III-XIV.

Enderlein, G. 1903c. Über die Morphologie, Gruppierung und systematische Stellung der Corrodentien. Zool. Anz. 26: 423-437.

Enderlein, G. 1903d. Ein neuer Copeognathentypus, zugleich ein neuer deutscher Wohnungsschädling. Zool. Anz. 27: 76.

Enderlein, G. 1903e. Über die Stellung von Leptella Reut. und Reuterella nov. gen., die Vertreter zweier neuer europäischer Copeognathensubfamilien. Zool. Anz. 27: 131-134.

Enderlein, G. 1903f. Zur Kenntnis amerikanischer Psociden. Zool. Jb., Abt. f. Syst., Geogr. U. Biol. Thiere. 18: 351-362, pl. 17.

Enderlein, G. 1903g. Zur Kenntniss europäischer Psociden. Zool. Jb., Abt. f. Syst., Geogr. u. Biol. Thiere. 18: 365-382, pl. 19.

Enderlein, G. 1904. Die von Herrn Prof. Dr. Friedr. Dahl in Bismark-Archipel gesammelten Copeognathen, nebst Bemerkungen über die physiologische Bedeutung des Stigmasackes. Zool. Jahrb. Abt. f. Syst., Geogr. u. Biol. Thiere. 20: 105-112, pl. 7.

Enderlein, G. 1905. Morphologie, Systematik und Biologie der Atropiden und Troctiden, sowie Zusammenstellung aller bisher bekannten recenten und fossilen Formen. Results of the Swedish Zoological Expedition to Egypt and the White Nile 1901 under the direction of L. A. Jagerskiöld. No. 18: 58 pp, 4 pls. Upsala.

Enderlein, G. 1906a. Zur Kenntnis der Copeognathen-Fauna West-preussens. Ber. Westpreuss. Bot.-Zool. Ver. 28: 71-88.

Enderlein, G. 1906b. Die Copeognathen-Fauna Japans. Zool. Jb., Abt. f. Syst., Geogr. u. Biol. Thiere. 23: 243-256, pls. 10-11.

Enderlein, G. 1906c. Die australischen Copeognathen. Zool. Jb. Abt. f. Syst., Geogr. u. Biol. Thiere. 23: 401-412, pl. 23.

Enderlein, G. 1906d. Zehn neue aussereuropäische Copeognathen. Stett. entomol. Zeit. 67: 306-316.

Enderlein, G. 1906e. Einige Notizen zur Kenntnis der Copeognathen Nordamerikas. Stett entomol. Zeit. 67: 317-320.

Enderlein, G. 1906f. The Scaly-winged Copeognatha. Spolia Zeylan. 4: 39-122, pls. A-E.

Enderlein, G. 1907a. *Troctes entomophilus*, ein neuer Insekten-liebhaber aus Columbien. Stett. entomol. Zeit. 68: 34-36.

Enderlein, G. 1907b. Neue Beiträge zur Kenntnis der Copeognathen Japans. Stett. entomol. Zeit. 68: 90-106.

Enderlein, G. 1908. Die Copeognathenfauna der Insel Formosa. Zool. Anz. 33: 759-779.

Enderlein, G. 1909a. Neue Gattungen und Arten nordamerikanischen Copeognathen. Bull. Lab. Zool. gen. e agr., Portici 3: 329-339.

Enderlein, G. 1909b. Biospeologica XI. Copeognathen (Erste Reihe). Arch. Zool. Exp. Gen. 5e Ser. 1: 533-539, pl. XVIII.

Enderlein, G. 1909c. Neue Gattungen und Arten von Copeognathen aus Transvaal sowie aus der Ohaus'schen Ausbeute aus Ecuador. Stett. entomol. Zeit. 70: 266-273.

Enderlein, G. 1910. Eine Dekade neuer Copeognathengattungen. Sitz. Ges. naturf. Freunde, Berlin 1910: 63-78.

Enderlein, G. 1911. Die fossilen Copeognathen und ihre Phylogenie. Palaeontographica 58: 278-360, pls. XXI-XXVII.

Enderlein, G. 1912. Über einige hervorragende neue Copeognathen-Gattungen. Zool. Anz. 39: 298-306.

Enderlein, G. 1913. Beiträge zur Kenntnis der Copeognathen I, II. Zool. Anz. 41: 354-360.

Enderlein, G. 1915. Copeognatha. Collections Zoologiques du Baron Edm. de Selys Longchamps. Catalogue Systematique et Descriptif. Fasc. 3: 55pp., pls. 1-5. Brussels.

Enderlein, G. 1922. A scaly-winged psocid, new to science, discovered in Britain. Entomol. Mon. Mag. 58: 101-104.

Enderlein, G. 1924. Copeognathen. *In* Damph, A. Zur Kenntnis der estländischen Moorfauna (II). Sitzb. Ges. naturf. Fr. Berlin 31: 34-37.

Enderlein, G. 1925. Beiträge zur Kenntnis der Copeognathen X. Konowia 4: 97-108.

Enderlein, G. 1927. Ordnung Flechtlinge, Copeognatha. In Brohmer, P. et al. Die Tierwelt Mitteleuropas. IV Band: Insekten, I. Teil, Lief 2: 16 pp.

Enderlein, G. 1929. Entomologica Canaria II. Zool. Anz. 84: 221-225.

Enderlein, G. 1931. Die Copeognathen-Fauna der Seychellen. Trans. Linn. Soc. Lond. 19: 207-239, pls. 14-16.

Essig, E. O. 1940. A seed-infesting psocid new to North America. J. Econ. Entomol. 33: 946.

Fabricius, J. C. 1775. Systema entomologiae, sistens insectorum classes, ordines, genera, species, adiectis synonymis, locis, descriptionibus, observationibus. Flensberg & Leipzig, 832 pp.

Fabricius, J. C. 1787. Mantissa insectorum sistens eorum species nuper detectas adiectis characteribus, genericis, diferentiis specificis, emendationibus, observationibus. Copenhagen. Vol. I, i-xx + 348 pp., Vol. II, 382 pp.

Fabricius, J. C. 1793. Entomologia systematica emendata et aucta, secundum classes, ordines, genera, species, adiectis, synonymis, locis, observationibus, descriptionibus. Copenhagen. 4 vols. (psocids, Vol. II: 81-87).

Fabricius, J. C. 1798. Supplementum Entomologia systematicae. Copenhagen.

Galil, B. 1984. On a collection of Psocoptera from the Azores. Bocagiana 71: 1-9.

García Aldrete, A. N. 1972. Tres nuevas especies del género *Lachesilla* (Psocoptera: Lachesillidae). Rev. Soc. Mex. Hist. Nat. 33: 123-129, Figs. 1-21.

García Aldrete, A. N. 1974. A classification above species level of the genus *Lachesilla* Westwood (Psocoptera: Lachesillidae). Folia Entomol. Mex. 27: 1-88.

García Aldrete, A. N. 1975. Nuevas especies Americanas del género *Lachesilla* (Psocoptera: Lachesillidae). Nouv. Rev. Entomol. 5: 217-220.

García Aldrete, A. N. 1982. The species group "riegeli" of the genus *Lachesilla* (Psocoptera: Lachesillidae). Diagnoses, records, and descriptions of new species. Zool. Anz. 209: 196-210.

García Aldrete, A. N. 1983. The *Lachesilla centralis* complex: descriptions and records of constituent species (Insecta: Psocoptera: Lachesillidae). Folia Entomol. Mex. 55: 13-29.

García Aldrete, A. N. 1984a. Troglomorpha (Psocoptera) de Nuevo León, México. An. Inst. Biol. Univ. Nat. Autón. México 55, Ser. Zoología (1): 103-122.

García Aldrete, A. N. 1984b. Psocoptera (Insecta) de nidos de rata (Neotoma floridana smalli Sherman) en Cayo Largo, Florida. Rev. Biol. Trop. 32: 299-302.

García Aldrete, A. N. 1984c. Abundancia relativa de especies de psócidos (Insecta: Psocoptera) en nidos de la ardilla gris (*Sciurus carolinensis* Gmelin), en Tallahassee, Florida, Estados Unidos de América. An. Inst. Biol. Univ. Nal. Autón. México 55, Ser. Zool. (2): 39-44.

García Aldrete, A. N. 1985. The species of *Lachesilla* in the group "texcocana" (Psocoptera: Lachesillidae). Descriptions, records, and relationships. Folia Entomol. Mex. 65: 37-62.

García Aldrete, A. N. 1986a. The species group "patzunensis" of the genus *Lachesilla* (Psocoptera: Lachesillidae). An. Inst. Biol. Univ. Nal. Autón. Méx. 56, Ser. Zool. (1): 53-72.

García Aldrete, A. N. 1986b. Descripciones y registros de especies Mexicanas de *Pachytroctes* (Psocoptera: Pachytroctidae). Folia Entomol. Mex. 69: 5-17.

García Aldrete, A. N. 1987. Las Especies Mexicanas de *Rhyopsocus* (Psocoptera: Psoquillidae). Folia Entomol. Mex. 71: 5-15.

García Aldrete, A. N. 1988a. Species of *Lachesilla* (Psocoptera: Lachesillidae) from the coast of the Mexican state of Jalisco. Folia Entomol. Mex. 77: 33-61.

García Aldrete, A. N. 1988b. Especies de Psócidos (Psocoptera), en nidos de aves en México. An. Inst. Biol. Univ. Nal. Autón. México 58, Ser. Zool. (2): 507-524.

García Aldrete, A. N. 1990a. Ch. 34. Insecta: Psocoptera, in Dindal, D. L. (ed.). Soil Biology Guide, pp. 1033-1052. Wiley & Sons, Inc.

García Aldrete, A. N. 1990b. Sistemática de las especies de *Lachesilla* en el grupo rufa. Distribución geográfica y afinidades (Insecta: Psocoptera: Lachesillidae). Anales Inst. Biol. Univ. Nac. Autón. México, Ser. Zool., 61: 13-97.

García Aldrete, A. N. In press. A new species of *Lachesilla* from Canada, and a list of Canadian *Lachesilla* (Psocoptera: Lachesillidae). Folia Entomol. Mex.

de Geer, C. 1778. Mémoires pour servir à l'histoire des Insectes. Stockholm. 7 vols. (psocids vol. 7, pp. 41-48, 68, 869, pl. 4, figs. 1-4).

Glinyanaya, Y. I. 1975. The importance of day length in regulating the seasonal cycles and diapause in some Psocoptera. Entomol. Rev. 54: 10-13.

Günther, K. K. 1968. Staubläuse (Psocoptera) aus der Mongolei. Mitt. Zool. Mus. Berlin 44: 125-141.

Günther, K. K. 1974a. Staubläuse, Psocoptera. Die Tierwelt Deutschlands. 61. Teil. 314 pp. Jena.

Günther, K. K. 1974b. Psocoptera of the Mongolian Peoples Republic [in Russian]. Insects of Mongolia. Leningrad. No. 2: 34-50.

Günther, K. K. 1980. Beiträge zur Kenntnis der Psocoptera -- Fauna Mazedoniens. Acta Mus. Mac. Scient. Nat. 16: 1-32.

Gurney, A. B. 1939. Nomenclatorial notes on Corrodentia, with descriptions of two new species of *Archipsocus*. J. Wash. Acad. Sci. 29: 501-515.

Gurney, A. B. 1943. A synopsis of the psocids of the tribe Psyllipsocini, including the description of an unusual new genus from Arizona (Corrodentia: Empheriidae: Empheriinae). Ann. Entomol. Soc. Am. 36: 195-220.

Gurney, A. B. 1949. Distributional and synonymic notes on psocids common to Europe and North America, with remarks on the distribution of Holarctic insects (Corrodentia). J. Wash. Acad. Sci. 39: 56-65.

Gurney, A. B. 1950. Psocids likely to be encountered by pest control operators. Pest Control Tech., Entomol. Sec. 1950: 131-163.

Hagen, H. A. 1858. Synopsis der Neuroptera Ceylons. Verh. Zool.-Bot. Ver., Wien. 8: 471-489 (psocids pp. 473-475).

Hagen, H. A. 1861. Synopsis of the Neuroptera of North America; with a list of South American species. Smithson. Misc. Coll. 4: xx + 358 pp. (psocids pp. 7-14, 302).

Hagen, H. A. 1865. Synopsis of the Psocina without ocelli. Entomol. Mon. Mag. 2: 121-124.

Hagen, H. A. 1866. Psocinorum et Embidinorum synopsis synonymica. Verh. zool.-bot. Ges. Wien 16: 1-22 (psocids pp. 1-20).

Hagen, H. A. 1876. Pseudo-Neuroptera. *In* Kidder, J. H. Contributions to the natural history of Kerguelen Island. Bull. U. S. Nat. Mus. 1: 1-122 (psocids pp. 52-57).

Hagen, H. A. 1882. Ueber Psociden in Bernstein. Stett. entomol. Zeit. 43: 217-237, 265-300, 524-526.

Hagen, H. A. 1883. Beiträge zur Monographie der Psociden. Familie Atropina. Stett. entomol. Zeit. 44: 285-332.

Harris, T. W. 1869. Entomological correspondence of Thaddeus William Harris, M. D. Edited by S. H. Scudder. Occ. Pap. Boston Soc. Nat. Hist. 1: x/vii + 375 pp, 4 pls. (psocids pp. 327-333).

Harrison, J. W. H. 1916. Notes on Psocoptera. The Entomologist 49: 134-135.

Heyden, G. H. 1850. Zwei neue deutsche Neuropteren-Gattungen. Stett. entomol. Zeit. 11: 83-85.

Heymons, R. 1909. Ein neuer *Troctes* als Schädling in Buchwelzengrütze. Dtsch. Entomol. Zeit. 1909: 452-455.

Hickman, V. V. 1934. A contribution to the study of Tasmanian Copeognatha. Pap. and Proc. R. Soc. Tasmania 1933: 77-89.

Hicks, E. A. 1937. Check list and bibliography of the occurrence of insects in birds' nests. Ames, Ia., 681 pp. (Corrodentia: pp. 158-164).

Hong, Y.-C. 1983. Middle Jurassic fossil insects in North China. [In Chinese with English descriptions of new taxa]. Geological Publishing House. Beijing. pp. 1-223, Psocoptera pp. 74-79, 188-191.

Illiger, J. C. W. 1798. Kugelann Verzeichniss der Käfer Preussens, ausgearbeitet von Illiger, mit einer Vorrede von Hellwig und angelängten Versuch einer natürlichen Ordnung und Gattungsfolge der Insecten. Halle. xlii + 510 pp.

Jacobson, G. G. and W. L. Bianchi. 1904. The Orthoptera and Pseudo-neuroptera of the Russian nation and bordering lands [in Russian]. Petersburg. (psocids pp. 482-496).

Jentsch, S. 1938. Beiträge zur Kenntnis der Überordnung Psocoidea 3. Zur Copeognathenfauna Nordwestfalens. Abh. Prov. Westf. Mus. Naturk. 9: 3-42.

Jentsch, S. 1939. Beiträge zur Kenntnis der Überordnung Psocoidea. 8. Die Gattung *Ectopsocus* (Psocoptera). Zool. Jahrb. Abt. f. Syst., Geogr. u. Biol. Thiere 73: 111-128.

Jostes, R. F., Jr. 1975. A method for determining the chromosome numbers of parthenogenetic psocids (Insecta: Psocoptera). Cytologia 40: 553-555.

Karny, H. H. 1926. On some tropical Copeognatha, especially from the Fiji Islands. Bull. Entomol. Res. 16: 285-290.

Kimmins, D. E. 1941. Notes on British Psocoptera. -- I. *Elipsocus hyalinus* (Steph.), and its allies. Ann. Mag. Nat. Hist., Ser. 11, 7: 520-530.

Kislow, C. J. and R. W. Matthews. 1977. Nesting behavior of *Rhopalum atlanticum* Bohart (Hymenoptera: Sphecidae: Crabro-ninae). J. Georgia Entomol. Soc. 12: 85-89.

Klier, E. 1956. Zur Konstruktionsmorphologie des männlichen Geschlechtsapparates der Psocopteren. Zool. Jb., Abt. f. Anat. u. Ontog. Tiere. 75: 207-286.

Knulle, W. and R. R. Spadafora. 1969. Water vapor sorption and humidity relationships in Liposcelis (Insecta: Psocoptera). J. Stored Prod. Res. 5: 49-55.

Kolbe, H. J. 1880. Monographie der deutschen Psociden mit besonderer Berücksichtigung der Fauna Westfalens. Jahresb. zool. Sek. Westf. Prov.-Ver. Wiss. Kunst. 1879-80: 73-142, pls. 1-4.

Kolbe, H. J. 1881. Psocidologische Berichtigungen. Entomol. Nachr. 7: 254-256.

Kolbe, H. J. 1882a. Das phylogenetische Alter der europäischen Psocidengruppen. Jber. westf. ProvVer. Wiss u. Kunst. 10: 18-27.

Kolbe, H. J. 1882b. Neue Psociden der palaärktischen Region. Entomol. Nachr. 8: 207-212.

Kolbe, H. J. 1883. Neue Psociden des Königl. Zoologischen Museums zu Berlin. Stett. entomol. Zeit. 44: 65-87.

Kolbe, H. J. 1884. Der Entwickelungsgang der Psociden im Individuum und in der Zeit. Berlin entomol. Zeit. 28: 35-38.

Kolbe, H. J. 1885. Zur Kenntnis der Psociden-Fauna Madagaskars. Berl. entomol. Zeit. 29: 183-192, pl. 4-B.

Kolbe, H. J. 1888a. *Troctes silvarum* n. sp., eine im Freien lebende Verwandte der Staublaus. Entomol. Nachr. 14: 1-2.

Kolbe, H. J. 1888b. Psocidae. In Rostock, M. Neuroptera germanica. Jber. Ver. Naturk. Zwickau 1887: 1-198, pls. i-x.

Lacroix, J. L. 1915. Psocides nouveaux [Nevr.]. Bull. Soc. Entomol. Fr. 1915: 192-195.

Latreille, P. A. 1794. Découverte de nids de *Termes*, et *Psocus* décrits. Bull. Soc. Philom. 1: 84-85.

Latreille, P. A. 1796. Précis des characteres génériques des insectes disposés dans un ordre naturel. Paris (psocids p. 99).

Latreille, P. A. 1799. Le genre *Psocus*. In Coquebert, A. J. Illustrata Iconographica Insectorum, quae in Musaeis parisinis observavit et in lucem edidit Joh. Christ. Fabricius, praemissis eiusdem descriptionibus. Paris (psocids: part I: pp. 8-14).

Leach, W. 1815. The Edinburgh Encyclopaedia. Edinburgh (psocids vol. 9, p. 139).

Lee, S. S. and I. W. B. Thornton. 1967. The family Pseudocaeciliidae (Psocoptera) -- a reappraisal based on the discovery of new Oriental and Pacific species. Pacific Insects Monogr. 16: 116 pp.

Lienhard, C. 1977. Die Psocopteren des Schweizerischen Nationalparks und seiner Umgebung (Insecta: Psocoptera). Resultats des Recherches Sci., Park Natl. Suisse 14: 415-551.

Lienhard, C. 1981. Neue und interessante Psocopteren aus Griechenland, Spanien und Portugal. Dtsch. entomol. Zeit. 28: 147-163.

Lienhard, C. 1983. Description d'un nouveau psoque italien et remarques sur la position systematique de *Psocus morio* Latreille. Bull. Soc. Entomol. Ital. 115: 9-14.

Lienhard, C. 1984. Études préliminaires pour une faune des Psocoptères de la région ouest-palearctique. I. Le genre *Cerobasis* Kolbe, 1882 (Psocoptera: Trollidae). Revue suisse Zool. 91: 747-764.

Lienhard, C. 1985. Vorarbeiten zu einer Psocopteren-Fauna der Westpaläarktis. II. Die europäischen Arten der Gattung *Elipsocus* Hagen, 1866 (Psocoptera: Elipsocidae). Bull. Soc. Entomol. Suisse. 58: 113-127.

Lienhard, C. 1986a. Beitrag zur Kenntnis der Psocopteren-Fauna Ungarns (Insecta). Annls. Hist-Nat. Mus. natn. Hung. 78: 73-78.

Lienhard, C. 1986b. Études préliminaires pour une faune des Psocoptères de la région ouest-palearctique. III. Contribution 'a la connaissance de la famille des Psocidae (Insecta: Psocoptera). Rev. Suisse Zool. 93: 297-328.

Lienhard, C. 1989. Zwei interessante europäische *Lachesilla*-Arten (Psocoptera: Lachesillidae). Bull. Soc. Entomol. Suisse. 62: 307-314.

Lienhard, C. 1990a. Revision of the western Palaearctic species of *Liposcelis* Motschulsky (Psocoptera: Liposcelididae). Zool. Jb., Abt. f. Syst. Geogr. u. Biol. Tiere 117: 117-174.

Lienhard, C. 1990b. New records and synonymies in western Palaerctic Psocoptera. Dtsch. ent. Z., N.F. 37: 205-212.

Linnaeus, C. 1758. Systema Naturae per regna tria naturae secundum classes, ordines, genera, species, cum characteribus, differentis, synonymis, locis. Ed. dec. reformata. Holm. 2 vols.

Linnaeus, C. 1761. Fauna Suecica sistens Animalia Sueciae regni. Stockholm.

Linnaeus, C. 1768. Systema Naturae, etc., ed. 13.

Loens, H. 1890. Zur Psocidenfauna Westfalens. Stett. entomol. Zeit. 51: 5-8.

Machado-Allison, C. E. and N. Papavero. 1962. Um novo gênero e uma nova especie de Corrodentia do Brasil: Lenkoella neotropica (Reuterellinae, Elipsocidae). Papéis Avulsos do Depto. de Zool., Sec. Agr., Sao Paulo, Brasil. 15: 311-315.

McLachlan, R. 1866. New genera and species of Psocidae. Trans. R. Entomol. Soc. Lond. Ser. 3. 5: 345-352.

McLachlan, R. 1867. A monograph of the British Psocidae. Entomol. Mon. Mag. 3: 177-181, 194-197, 226-231, 241-245, 270-276, pl. II.

McLachlan, R. 1869. Description of a new species of Psocidae (*Caecilius atricornis*) inhabiting Britain. Entomol. Mon. Mag. 5: 196.

McLachlan, R. 1877. Description d'un psocide nouveau de la Belgique. Comptes rendus des Seances Soc. Entomol. Belg. 20: liv-lv.

McLachlan, R. 1880. Notes on the entomology of Portugal. II. Pseudo-Neuroptera (in part) & Neuroptera-Planipennia. Entomol. Mon. Mag. 17:103-108.

McLachlan, R. 1881. Editorial comment. Zool. Record 1880: 211.

McLachlan, R. 1883. Remarks on certain Psocidae, chiefly British. Entomol. Mon. Mag. 19: 181-185.

McLachlan, R. 1899. *Ectopsocus briggsi,* a new genus and species of Psocidae found in England. Entomol. Mon. Mag. Ser. 2, 10: 277-278.

Meinander, M., O. Halkka, and V. Söderlund. 1974. Chromosomal evolution in the Psocoptera. Notulae Entomol. 54: 81-84.

Mockford, E. L. 1950. The Psocoptera of Indiana. Proc. Ind. Acad. Sci. 60: 192-204.

Mockford, E. L. 1951. On two North American Philotarsids (Psocoptera). Psyche 58: 102-107.

Mockford, E. L. 1952. Additional notes on Indiana Psocoptera. Proc. Ind. Acad. Sci. 62: 198-199.

Mockford, E. L. 1953. Three new species of *Archipsocus* from Florida (Psocoptera: Archipsocidae). Fla. Entomol. 36: 113-124.

Mockford, E. L. 1955a. Studies on the reuterelline psocids. Proc. Entomol. Soc. Wash. 57:97-108.

Mockford, E. L. 1955b. Notes on some eastern North American psocids with descriptions of the two new species. Amer. Midl. Nat. 53: 436-441.

Mockford, E. L. 1957a. Life history studies on some Florida insects of the genus *Archipsocus* (Psocoptera). Bull. Fla. St. Mus. 1: 253-274.

Mockford, E. L. 1957b. A new species of *Archipsocus* from Florida (Psocoptera: Archipsocidae). Fla. Entomol. 40: 33-34.

Mockford, E. L. 1959. The *Ectopsocus briggsi* complex in the Americas. Proc. Entomol. Soc. Wash. 61: 260-266.

Mockford, E. L. 1961. An annotated list of the Psocoptera of the Flint-Chattanhoochee-Apalachicola region of Georgia, Florida, and Alabama. Fla. Entomol. 44: 129-140.

Mockford, E. L. 1963. The species of Embidopsocinae of the United States (Psocoptera: Liposcelidae). Ann. Entomol. Soc. Amer. 56: 25-37.

Mockford, E. L. 1965a. The genus *Caecilius* (Psocoptera: Caeciliidae). Part I. Species groups and the North American species of the flavidus group. Trans. Am. Entomol. Soc. 91: 121-166.

Mockford, E. L. 1965b. Some South African Psocoptera from termite nests. Entomol. News 76: 169-176.

Mockford, E. L. 1965c. A new genus of hump-backed psocids from Mexico and southwestern United States (Psocoptera: Psocidae). Folia Entomol. Mex. No. 11: 1-11, Table I, pls. I, II.

Mockford, E. L. 1965d. Polymorphism in the Psocoptera: a review. Proc. N. C. Branch, Entomol. Soc. Am. 20: 82-86.

Mockford, E. L. 1966. The genus *Caecilius* (Psocoptera: Caeciliidae). Part II. Revision of the species groups, and the North American species of the fasciatus, confluens, and africanus groups. Trans. Am. Entomol. Soc. 92: 133-172, pls. 9-15.

Mockford, E. L. 1967. Some Psocoptera from plumage of birds. Proc. Entomol. Soc. Wash. 69: 307-309.

Mockford, E. L. 1969a. The genus *Caecilius* (Psocoptera: Caeciliidae). Part III. The North American species of the alcinus, caligonus, and subflavus groups. Trans. Am. Entomol. Soc. 95: 77-151.

Mockford, E. L. 1969b. Fossil insects of the Order Psocoptera from Tertiary amber of Chiapas, Mexico. J. Paleo. 43: 1267-1273.

Mockford, E. L. 1971a. Psocoptera from sleeping nests of the dusky-footed wood rat in southern California. Pan-Pac. Entomol. 47: 127-140.

Mockford, E. L. 1971b. *Peripsocus* species of the alboguttatus group (Psocoptera: Peripsocidae). J. N. Y. Entomol. Soc. 79: 89-115.

Mockford, E. L. 1971c. Parthenogenesis in psocids (Insecta: Psocoptera). Am. Zoologist 11: 327-339.

Mockford, E. L. 1972. New species, records, and synonymy of Florida Belaphotroctes (Psocoptera: Liposcelidae). Fla. Entomol. 55: 153-163.

Mockford, E. L. 1974a. Records and descriptions of Cuban Psocoptera. Entomol. Amer. 48: 103-215.

Mockford, E. L. 1974b. The Echmepteryx hageni complex (Psocoptera: Lepidopsocidae) in Florida. Fla. Entomol. 57: 255-267.

Mockford, E. L. 1977. Morphological characters of the Florida species of *Archipsocus* with closed phallosome (Psocoptera: Archipsocidae). Fla. Entomol. 60: 41-48.

Mockford, E. L. 1978a. A generic classification of Family Amphipsocidae (Psocoptera: Caecilietae). Trans. Amer. Entomol. Soc. 104: 139-190.

Mockford, E. L. 1978b. New species, records, and key to Texas Liposcelidae (Psocoptera). Proc. Entomol. Soc. Wash. 80: 556-574.

Mockford, E. L. 1979a. Diagnoses, distribution, and comparative life history notes on *Aaroniella maculosa* (Aaron) and *A. eertmoedi* n. sp. (Psocoptera: Philotarsidae). Great Lakes Entomol. 12: 35-44.

Mockford, E. L. 1979b. Chapter 33. Psocoptera. *In* Danks, H. V. (ed.), Canada and its insect fauna. Mem. Entomol. Soc. Canad. 108. 573 pp. (Psocoptera pp. 324-326).

Mockford, E. L. 1980. Identification of *Elipsocus* species of western North America with descriptions of two new species (Psocoptera: Elipsocidae). Pan-Pac. Entomol. 56: 241-259.

Mockford, E. L. 1982. Redescription of the type species of *Myopsocus*, *M. unduosus* (Hagen), and resulting nomenclatural changes in genera and species of Myopsocidae (Psocoptera). Psyche 89: 211-220.

Mockford, E. L. 1983. Systematics of Asiopsocidae (Psocoptera) including *Pronotiopsocus amazonicus* n. gen. n. sp. Fla. Entomol. 66: 241-249.

Mockford, E. L. 1984a. Two new-species of *Speleketor* from southern California with comments on the taxonomic position of the genus (Psocoptera: Prionoglaridae). Sw. Nat. 29: 169-179.

Mockford, E. L. 1984b. A systematic study of the genus *Camelopsocus* with descriptions of three new species (Psocoptera: Psocidae). Pan-Pac. Entomol. 60: 193-212.

Mockford, E. L. 1984c. Systematics of the *Blaste posticata* complex with descriptions of three new species (Psocoptera: Psocidae). Fla. Entomol. 67: 548-566.

Mockford, E. L. 1985a. Systematics of the genus *Nadleria* (Psocoptera: Lachesillidae) with description of a new species and hypotheses on evolution of male external genitalia in the family Lachesillidae. Ann. Entomol. Soc. Am. 78: 94-100.

Mockford, E. L. 1985b. Forms, distribution, and relationships of *'Psocus' coquilletti* Banks (Psocoptera: Psocidae). Sw. Nat. 30: 13-22.

Mockford, E. L. 1987a. Order Psocoptera. *In* Stehr, F. (ed.). Immature Insects. Kendall/Hunt Co., Dubuque, Ia. xiv + 754 pp. (Psocoptera Ch. 21, pp. 196-214).

Mockford, E. L. 1987b. Systematics of North American and Greater Antillean species of *Embidopsocus* (Psocoptera: Liposcelidae). Ann. Entomol. Soc. Am. 80: 849-864.

Mockford, E. L. 1989. *Xanthocaecilius* (Psocoptera: Caeciliidae), a new genus from the Western Hemisphere: I. Description, species complexes, and species of the *quillayute* and *granulosus* complexes. Trans. Am. Entomol. Soc. 114: 265-294.

Mockford, E. L. 1991. Psocids (Psocoptera). Chapter 22. *In*, Gorham, J. R. (ed.) Insect and mite pests in food an illustrated Key. U. S. Dept. Agr. and U. S. Dept. Health and Hum. Svces., Agr. Handbook No. 655, Vol. 2, psocids pp. 371-402.

Mockford, E. L. In press. New species and records of Psocoptera (Insecta) from Roraima State, Brazil. Acta Amazon.

Mockford, E. L. and A. N. García Aldrete. 1974. Two new synonymies and a new name in North American *Lachesilla*. Pan-Pac. Entomol. 50: 235-237.

Mockford, E. L. and A. N. García Aldrete. 1976. A new species and notes on the taxonomic position of *Asiopsocus* Günther (Psocoptera). Sw. Nat. 21: 335-346.

Mockford, E. L. and A. N. García Aldrete. 1991. *Rhyopsocus texanus* (Psocoptera: Psoquillidae): its synonymy, forms, and distribution. Entomol. News 102: 133-136.

Mockford, E. L. and A. B. Gurney. 1956. A review of the psocids, or book-lice and bark-lice, of Texas (Psocoptera). J. Wash. Acad. Sci. 46: 353-368.

Mockford, E. L. and D. M. Sullivan. 1986. Systematics of the graphocaecilline psocids with a proposed higher classification of the Family Lachesillidae (Psocoptera). Trans. Am. Entomol. Soc. 112: 1-80.

Mockford, E. L. and D. M. Sullivan. 1990. *Kaestneriella* Roesler (Psocoptera: Peripsocidae): new and little known species from the southwestern United States and Mexico and a revised species key. Pan-Pac. Entomol. 66: 281-291.

Mockford, E. L. and S. K. Wong. 1969. The genus *Kaestneriella* (Psocoptera: Peripsocidae). J. N. Y. Entomol. Soc. 77: 221-249.

Motschulsky, V. 1851. Enumérations des nouvelles espèces des Coléoptères rapportés par M. Victor Motschulsky de son dernier voyage. Bull. Soc. imp. Nat. Moscow 24: 479-511 (psocids pp. 510-511).

Motschulsky, V. 1852. Études entomologiques. Helsingfors. (Psocids pp. 19-20).

Müller, O. F. 1764. Fauna insectorum Friedrichsdalina sive methodica descriptio insectorum agri Friedrichdalensis, etc. Copenhagen. i-xxiv + 96 pp.

Müller, O. F. 1776. Zoologiae Danicae prodromus seu animalium Daniae et Norvegiae indigenarum characteres, nomina, et synonymia imprimis popularium. Copenhagen. 32 + 282 pp.

Navás, L. 1908. Neurópteros nuevos de América. Broteria 10: 196-198.

Navás, L. 1909. Neurópteros nuevos de la fauna ibérica. Act. Mem. Prim. Congr. Natur. Espan. 1: 143-158.

Navás, L. 1912. Neurópteros nuevos de América. Broteria ser. zool. 10: 194-202, Figs. 1-7.

Navás, L. 1913. Cuatro pequeñas colecciones de Neurópteros de la Península Ibérica. Bol. Soc. Aragon Cienc. Nat. 12: 85-89.

Navás, L. 1916. Neurópteros nuevos de España. Rev. Acad. Cienc. Madrid 14: 593-601.

Navás, L. 1927. Comunicaciones entomológicas 8. Socópteros del Museo de Hamburgo. Rev. Acad. Cienc. Zaragoza 11: 37-52.

Navás, L. 1932. Insectos de la Argentina. Rev. Acad. Cienc. Zaragoza 16: 87-120.

Navás, L. 1934. Insectes de la India. Rev. Acad. Cienc. Zaragoza 17: 29-48.

New, T. R. 1969. Observations on the biology of Psocoptera found in leaf litter in southern England. Trans. Soc. Brit. Entomol. 18: 171-180.

New, T. R. 1970. The relative abundance of some British Psocoptera on different species of trees. J. Anim. Ecol. 39: 521-540.

New, T. R. 1972. Some Brazilian Psocoptera from bird nests. Entomologist, 105: 153-160.

New, T. R. 1974. Psocoptera. Handbooks for the identification of British insects. London. pp. 1-102.

New, T. R. 1987. Biology of the Psocoptera. Oriental Insects 21: 1-109.

New, T. R. and C. C. Loan. 1972. Psocoptera collected near Belleville, Ontario. Proc. Entomol. Soc. Ontario 102: 16-22.

New, T. R. and A. M. Nadler. 1970. A North American record of *Cuneopalpus cyanops* (Rostock) (Psocoptera). Entomologist 103: 44.

Nitzch, C. L. 1821. Ueber die Eingeweide der Bücherlaus (*Psocus pulsatorius*) und über das Verfahren bei der Zergliederung sehr kleiner Insekten. Germar's Mag. Entomol. 4: 276-290, pl. 2.

Obr, S. 1948. A la connaissance des Psocoptères de Moravie (Tchécoslovaquie). [in Czech]. Pub. Fac. Sci. Univ. Masaryk 360: 1-108.

Packard, A. S. 1870. New or rare American Neuroptera, Thysanoptera and Myriapoda. Proc. Boston Soc. Nat. Hist. 13: 405-409.

Palmgren, P. 1932. Zur Biologie von *Regulus r. regulus* (L.) und *Parus atricapillus borealis* Selys. Eine vergleichendikologische Untersuchung. Acta Zool. Fenn. 14: 1-113.

Pearman, J. V. 1924. A new species of *Caecilius* (Psocoptera). Entomol. Mon. Mag. 60: 58-61.

Pearman, J. V. 1925. Additions to the British psocid fauna. Entomol. Mon. Mag. 61: 124-129.

Pearman, J. V. 1928a. Some Psocoptera from the New Hebrides. Entomol. Mon. Mag. 64: 133-137.

Pearman, J. V. 1928b. Biological observations on British Psocoptera. Entomol. Mon. Mag. 64: 209-218.

Pearman, J. V. 1928c. Biological observations on British Psocoptera. II. Hatching and ecdysis. Entomol. Mon. Mag. 64: 239-243.

Pearman, J. V. 1928d. Biological observations on British Psocoptera. III. Sex behavior. Entomol. Mon. Mag. 64: 263-268.

Pearman, J. V. 1929. New species of Psocoptera from warehouses. Entomol. Mon. Mag. 65: 104-109.

Pearman, J. V. 1931a. A new species of *Lepinotus* (Psocoptera). Entomol. Mon. Mag. 67: 47-50.

Pearman, J. V. 1931b. More Psocoptera from warehouses. Entomol. Mon. Mag. 67: 95-98.

Pearman, J. V. 1932. Notes on the genus *Psocus* with special reference to the British species. Entomol. Mon. Mag. 68: 193-204.

Pearman, J. V. 1933. A new species of *Terracaecilius* (Psocoptera). Entomol. Mon. Mag. 69: 81-83.

Pearman, J. V. 1934. New and little known African Psocoptera. Stylops 3: 121-132.

Pearman, J. V. 1935. Notes on some dimorphic psocids. Entomol. Mon. Mag. 71: 82-85.

Pearman, J. V. 1936a. The taxonomy of the Psocoptera: preliminary sketch. Proc. R. Entomol. Soc. London. (B) 5: 58-62.

Pearman, J. V. 1936b. Two new psocids from Ceylon. Ceylon J. Sci. (B) 20: 1-7, pl. I.

Pearman, J. V. 1942. Third note on Psocoptera from warehouses. Entomol. Mon. Mag. 78: 289-292.

Pearman, J. V. 1946. A specific characterization of *Liposcelis divinatorius* (Mueller) and *mendax* n. sp. (Psocoptera). Entomologist 79: 235-244.

Pearman, J. V. 1951. Additional species of British Psocoptera. Entomol. Mon. Mag. 87: 84-89.

Pearman, J. V. 1960. Some African Psocoptera found on rats. Entomologist 93: 246-250.

Perkins, R. C. L. 1899. Fauna Hawaiiensis. Cambridge. (Psocidae: vol. 2 (2): pp. 77-87).

Provancher, L. 1876. Petite faune entomologique du Canada. Les Neuroptères. Nat. Canad. 8: 177-187 (psocids pp. 181-187).

Rambur, M. P. 1842. Histoire naturelle des insectes. Néuroptères. Paris. xvii + 534 pp., 12 pls. (psocids: pp. 317-324.).

Rapp, W. F. 1961. Corrodentia in cliff swallow nests. Entomol. News. 72: 195.

Reuter, O. M. 1894. Corrodentia Fennica. I. Psocidae. Acta Soc. Faun. Flor. Fenn. 9: 1-47, Figs. 1-10.

Reuter, O. M. 1899. Antechningar om Finska psocider. Act. Soc. Fauna Flora Fenn. 17: 1-7.

Reuter, O. M. 1904. Neue Beiträge zur Kenntnis der Copeognathen Finnlands. Acta Soc. Fauna Flora Fenn. 26: 1-26, pls. I-III.

Ribaga, C. 1899. Descrizione di un nuovo genere e di una nuova specie di Psocidi trovata in Italia. Riv. Pat. veg. 8: 156-158, pl. vii.

Ribaga, C. 1904. Sul genere *Ectopsocus* McL. e descrizione di una nuova varieta dell. *E. briggsi* McL. Redia 1: 294-298.

Ribaga, C. 1905. Descrizione di nuovi Copeognati. Redia 2: 99-110, pls. IX-X.

Ribaga, C. 1907. Copeognati nuovi. Redia 4: 1-9, pl. IV.

Ribaga, C. 1911. Nuovi Copeognati Sudafricani. Redia 7: 156-171.

Roesler, R. 1935. Zur Kenntnis der Mecklenburgischen Fauna VII. 22. Die Copeognathen Mecklenburgs. Arch. Ver. f. Naturges. Mecklenburg. 9: 18-30.

Roesler, R. 1939. Beiträge zur Kenntnis der Copeognathenfauna Deutschlands. Zool. Anz. 125: 157-176.

Roesler, R. 1940. Neue Copeognathen. Arb. morph. tax. Entomol. 7: 236-244.

Roesler, R. 1943. Über einige Copeognathengenera. Stett. entomol. Zeit. 104: 1-14.

Roesler, R. 1944. Die Gattungen der Copeognathen. Stett entomol. Zeit. 105: 117-166.

Roesler, R. 1954. Neue Gattungen und Arten der deutschen Psocopterenfauna. Beitr. Entomol. 4: 559-574.

Rostock, M. 1876. Psocidenjagd im Hause. Entomol. Nachr. 2: 190-192.

Rostock, M. 1878. Die Ephemeriden und Psociden Sachsens mit Berücksichtigung des meistens übrigen deutschen Arten. Jber. Ver. Naturk. Zwickau 1877-1878: 76-100.

Rudolph, D. 1982a. Occurrence, properties, and biological implications of the active uptake of water vapour from the atmosphere in Psocoptera. J. Insect Physiol. 28: 111-121.

Rudolph, D. 1982b. Site, process, and mechanism of active uptake of water vapor from the atmosphere in the Psocoptera. J. Insect Physiol. 28: 205-212.

Schmidt, C. M. (unpubl.). The taxonomy and phylogeny of the *Trichadenotecnum alexanderae* species complex (order Psocoptera). Doctoral dissertation, Illinois State University.

Schmidt, D. J. (unpubl.). Seasonality of psocids (class: Insecta, Order: Psocoptera) at the Archbold Biological Station, Lake Placid, Florida. Master of Science Thesis, Illinois State University. vl + 85 pp.

Schneider, H. 1955. Vergleichende Untersuchungen über Parthenogenese und Entwicklungsrhythmen bei einheimischen Psocopteren. Biol. Zentralb. 74: 273-310.

Schrank, F. 1781. Enumeratio insectorum Austriae indigenorum. Vindelicor. ix + 548 pp.

Scopoli, J. A. 1763. Entomologia carniolica exhibens insecta carnioliae indigena et distributa in ordines, genera, species, varietates, methodo Linneana. Vindobona. 28 + 421 pp., 43 pls.

Selys-Longchamps, E. de. 1872. Notes on two new genera of Psocidae. Entomol. Mon. Mag. 9: 145-146.

Shipley, A. E. 1904. The orders of Insects. Zool. Anz. 27: 259-262.

Smithers, C. N. 1958. A new genus and species of domestic psocid (Psocoptera) from Southern Rhodesia. J. Entomol. Soc. S. Afr. 21: 113-116.

Smithers, C. N. 1963. The Elipsocidae (Psocoptera) of Australia. Pac. Insects 5: 885-898.

Smithers, C. N. 1967. A catalogue of the Psocoptera of the world. Austr. Zool. 14: 3-145.

Smithers, C. N. 1969. The Psocoptera of New Zealand. Rec. Canterbury Mus. 8: 259-344.

Smithers, C. N. 1970. Redefinition of *Teliapsocus* Chapman, *Zelandopsocus* Tillyard, and *Cladioneura* Enderlein (Psocoptera). Proc. R. Entomol. Soc. Lond. (B) 39: 79-84.

Smithers, C. N. 1972. The classification and phylogeny of the Psocoptera. Austr. Mus. Mem. 14: 349 pp.

Smithers, C. N. 1978. A new species and new records of Psocoptera (Insecta) from Ireland. Ir. Nat. J. 19: 141-148.

Smithers, C. N. 1983. A reappraisal of Clematostigma Enderlein with notes on related genera (Psocoptera: Psocidae). Aust. Entomol. Mag. 9: 71-79.

Sommerman, K. M. 1943a. Description and bionomics of *Caecilius manteri* n. sp. (Corrodentia). Proc. Entomol. Soc. Wash. 45: 29-39.

Sommerman, K. M. 1943b. Bionomics of *Lachesilla nubilis* (Aaron) (Corrodentia, Caeciliidae). Canad. Entomol. 75: 99-105.

Sommerman, K. M. 1943c. Bionomics of *Ectopsocus pumilis* (Banks) (Corrodentia, Caeciliidae). Psyche 50: 53-64.

Sommerman, K. M. 1944. Bionomics of *Amapsocus amabilis* (Walsh) (Corrodentia, Psocidae). Ann. Entomol. Soc. Am. 37: 359-364.

Sommerman, K. M. 1946. A revision of the genus *Lachesilla* north of Mexico. Ann. Entomol. Soc. Am. 39: 627-661.

Sommerman, K. M. 1948. Two new Nearctic psocids of the genus *Trichadenotecnum* with a nomenclatural note on a third species. Proc. Entomol. Soc. Wash. 50: 165-173.

Sommerman, K. M. 1956a. Two new species of *Rhyopsocus* (Psocoptera) from the U.S.A., with notes on the bionomics of one household species. J. Wash. Acad. Sci. 46: 145-149.

Sommerman, K. M. 1956b. Parasitization of nymphal and adult psocids (Psocoptera). Proc. Entomol. Soc. Wash. 58: 149-152.

Sommerman, K. M. 1957. Three new species of *Liposcelis* (= *Troctes*) (Psocoptera) from Texas. Proc. Entomol. Soc. Wash. 59: 125-127.

Spieksma, F. T. and C. Smits. 1975. Some ecological and biological aspects of the booklouse *Liposcelis bostrychophilus* Badonnel 1931 (Psocoptera). Netherlands J. Zool. 25: 219-230.

Stephens, F. 1829. A systematic catalogue of British insects: being an attempt to arrange all the hitherto discovered indigenous insects in accordance with their natural affinities. London xx + 388 pp. (psocids pp. 312-314).

Stephens, F. 1836. Illustrations of British entomology. London. (psocids part 6, pp. 115-129).

Tetens, H. 1891. Zur Kenntnis der deutschen Psociden. Entomol. Nachr. 17: 369-384.

Thornton, I. W. B. 1961. The Trichadenotecnum group (Psocoptera: Psocidae) in Hong Kong with descriptions of new species. Trans. R. Entomol. Soc. Lond. 113: 1-24.

Thornton, I. W. B. and E. Broadhead. 1954. The British species of *Elipsocus* Hagen (Corrodentia, Mesopsocidae). J. Soc. Brit. Entomol. 5: 47-64.

Thornton, I. W. B., S. S. Lee, and W. D. Chui. 1972. Insects of Micronesia. Psocoptera. B. P. Bishop Museum. Insects of Micronesia 8: 45-144.

Thornton, I. W. B. and S. K. Wong. 1968. The Peripsocid fauna (Psocoptera) of the Oriental Region and the Pacific. Pacific Insects Monogr. 19: 1-158.

Tillyard, R. J. 1923. A monograph of the Psocoptera, or Copeognatha, of New Zealand. Trans. N. Z. Inst. 54: 170-196.

Tillyard, R. J. 1926. Kansas permian insects. Part 8. The Order Copeognatha. Am. J. Sci. 11: 315-349.

Tillyard, R. J. 1928. Kansas permian insects. Part 12. The Family Delopteridae, with a discussion of its ordinal position. Am. J. Sci. 16: 469-484.

Tillyard, R. J. 1935. Upper permian insects of New South Wales III. The Order Copeognatha. Proc. Linn. Soc. N. S. W. 60: 265-279.

Tillyard, R. J. 1937. Kansas permian insects. Am. J. Sci. 33: 81-110.

Townsend, C. H. T. 1912. Vulturopinae, a new subfamily of the Psocidae; type *Vulturops* gen. nov. (Platyp., Corrod.). Entomol. News 23: 266-269.

Turner, B. D. 1974. The population dynamics of tropical arboreal Psocoptera (Insecta) on two species of conifers in the Blue Mountains, Jamaica. J. Anim. Ecol. 43: 323-337.

Turner, B. D. 1975. The Psocoptera of Jamaica. Trans. R. Entomol. Soc. Lond. 126: 533-609.

Turner, B. D. 1979. Psocids as prey for Mascarene Swiftlets. Entomol. Mon. Mag. 113: 210.

Turner, B. D. 1987. Forming a clearer view of *L. bostrychophilus*. Environmental Health 95: 9-13.

Turner, B. D. and E. Broadhead. 1974. The diversity and distribution of psocid populations on *Mangifera indica* L. in Jamaica and their relationship to altitude and micro-epiphyte diversity. J. Anim. Ecol. 43: 173-190.

Turner, B. D. and H. Maude-Roxby. 1987. A survey of the incidence of booklice in the domestic kitchen environment. BCPC. Monograph No. 37. Stored Products Pest Control, pg. 275.

Turner, B. D. and H. Maude-Roxby. 1988. Starvation survival of the stored product pest *Liposcelis bostrychophilus* Badonnel (Psocoptera, Liposcelidae). J. Stored Prod. Res. 24: 23-28.

Verrill, A. E. 1902. The Bermuda Islands, their scenery, climate, production, physiography, natural history, and geology. Trans. Conn. Acad. Arts Sci. 11: 413-957, pls. LXV-CIV.

Vishniakova, V. N. 1975. Psocoptera in late-Cretaceous insect-bearing resins from the Taimyr. Entomol. Rev. 54: 63-75.

Vishniakova, V. N. 1976. O Reliktovikh Cenoedakh (Insecta, Psocoptera) Mesozoickoi fauny. Paleont. J. Akad. Nauk., U.S.S.R. 2: 76-94.

Vishniakova, V. N. 1981. New Paleozoic and Mesozoic Lophioneuridae (Thripida, Lophioneuridae). [In Russian] Trud. Paleo. Inst., Akad. nauk, U.S.S.R. 183: 43-63.

Walker, F. 1853. List of the specimens of neuropterous insects in the collection of the British Museum. London. Part III. (Termitidae-Ephemeridae), pp. 477-585. (psocids, pp. 477-501).

Walsh, B. D. 1862. List of the Pseudoneuroptera of Illinois contained in the cabinet of the writer, with descriptions of over forty new species, and notes on their structural affinities. Proc. Acad. Nat. Sci. Philad. 14: 361-402. (psocids pp. 361-362).

Walsh, B. D. 1863. Notes by Benj. D. Walsh. Proc. Entomol. Soc. Philad. 2: 182-186.

Weber, S. E. 1906. The possible dissemination of tubercle bacilli by insects. N. Y. Med. J. 84: 884-888.

Weber, S. E. 1907. A new genus of Atropidae. Entomol. News 18: 189-194.

Westwood, J. O. 1840. Synopsis of the genera of British insects. London, 158 pp.

Westwood, J. O. 1841. Observations on the structural characters of the deathwatch with the description of a new British genus belonging to the family Psocidae. Ann. Mag. Nat. Hist. 6: 480.

Wlodarczyk, J. 1963. Psocoptera of some bird nests. [In Polish]. Polsk. Akad. Nauk., Inst. Zool., Fragmenta Faunistica 10: 361-366.

Wlodarczyk, J. and J. Martini. 1969. An attempt at analysis of the occupation of birds nests by Psocoptera. [In Polish with English summary]. Ekologia Polska, Ser. B. 15: 323-336.

Wolda, H. and E. Broadhead. 1985. Seasonality of Psocoptera in two tropical forests in Panama. J. Anim. Ecol. 54: 519-530.

Wong, S. K. and I. W. B. Thornton. 1966. Chromosome numbers of some psocid genera. Nature 211: 214-215.

Zetterstedt, J. W. 1840. Insecta Lapponica descripta. Leipzig. (psocids: Sectio 5: 1053-1056).

Zimmerman, E. C. 1948. Insects of Hawaii. Honolulu. Vol. 2, pp. 217-252.

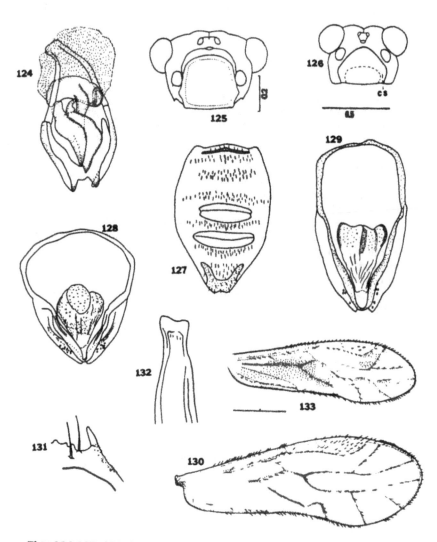

Figs. 124-133. 124 - *Ectopsocus strauchi*, phallosome; 125 - *Atopsocus* sp., front view of cleared head capsule showing narrow inner clypeal shelf; 126 - *Xanthocaecilius* sp., front view of cleared head capsule showing wide inner clypeal shelf (c.s.); 127 - *Caecilius* sp., ventral view of abdomen showing two ventral vesicles; 128 - *Caecilius subflavus*, phallosome; 129 - *Xanthocaecilius sommermanae*, phallosome; 130 - *X. sommermanae*, forewing; 131 - *Caecilius insularum*, right, front edge of labrum with stylet; 132 - *Caecilius casarum*, lacinial tip; 133 - *Caecilius posticus* male, forewing (lack of one M branch is anomalous).

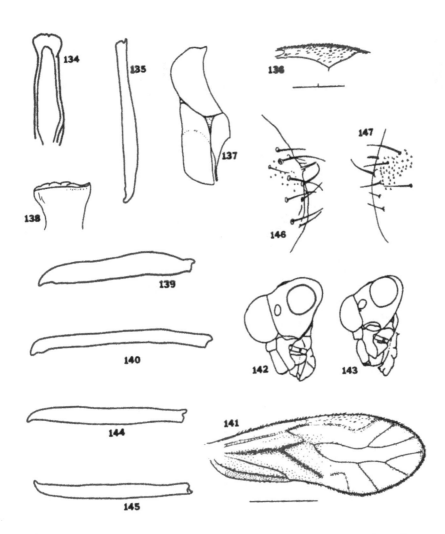

Figs. 134-147. *Caecilius* spp. 134 - *C. flavidus*, lacinial tip; 135 - *C. flavidus* male, fore tibia; 136 - *C. indicator* male, pterostigma with spur vein; 137 - *C. jamaicensis*, mesepisternum with precoxal suture meeting episternal suture; 138 - *C. africanus*, lacinial tip; 149 - *C. micanopi* male, fore tibia; 140 - *C. caloclypeus* male, fore tibia; 141 - *C. hyperboreus*, forewing; 142 - *C. perplexus* female, head in side view; 143 - *C. pinicola* female, head in side view; 144 - *C. croesus* male fore tibia; 145 - *C. burmeisteri* male, fore tibia; 146 - *C. boreus* male, edge of paraproct with papillar field; 147 - *C. croesus* male, edge of paraproct with papillar field.

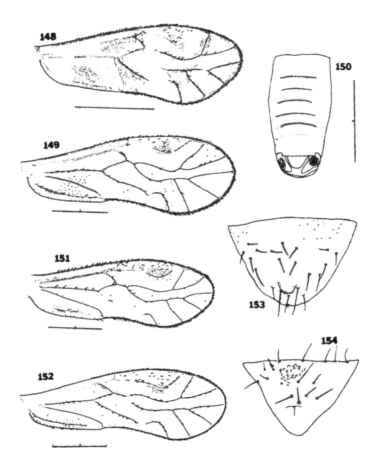

Figs. 148-154. *Caecilius* spp. 148 - *C. distinctus*, forewing; 149 - *C. nadleri* male, forewing; 150 - *C. totonacus* male, abdomen dorsally showing segmental sclerotic strips; 151 - *C. juniperorum*, forewing; 152 - *C. subflavus* male, forewing; 153 - *C. confluens* male, epiproct; 154 - *C. gonostigma* male, epiproct.

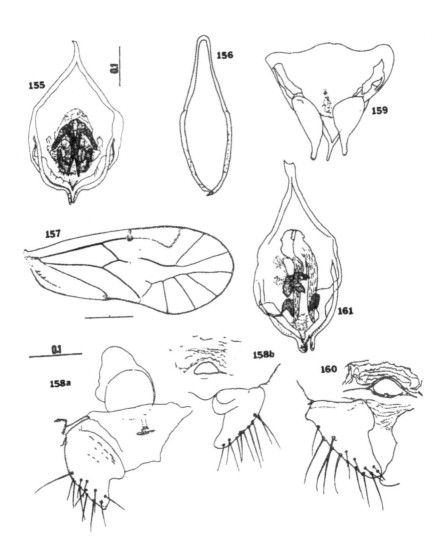

Figs. 155-161. Lachesillids. 155 - *Nanolachesilla hirundo*, phallosome; 156 - *Lachesilla texcocana*, phallosome; 157 - *Anomopsocus amabilis* female, forewing; 158 - *Nanolachesilla chelata*, ovipositor valvulae, ninth sternum, and spermatheca; 156 - *Nanolachesilla hirundo*, ovipositor valvulae and ninth sternum; 159 - *Prolachesilla terricola*, phallosome; 160 - *P. terricola*, ovipositor valvulae and ninth sternum; 161 - *N. chelata*, phallosome.

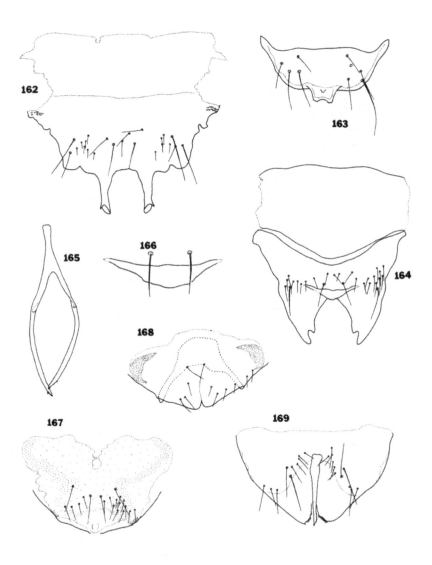

Figs. 162-169. *Lachesilla* spp. 162 - *L. texcocana*, hypandrium; 163 - *L. texcocana* male, epiproct; 164 - *L. centralis*, hypandrium; 165 - *L. cintalapa*, phallosome; 166 - *L. perezi*, median distal sclerite of hypandrium; 167 - *L. centralis*, subgenital plate; 168 - *L. fuscipalpis*, subgenital plate; 169 - *L. corona*, subgenital plate.

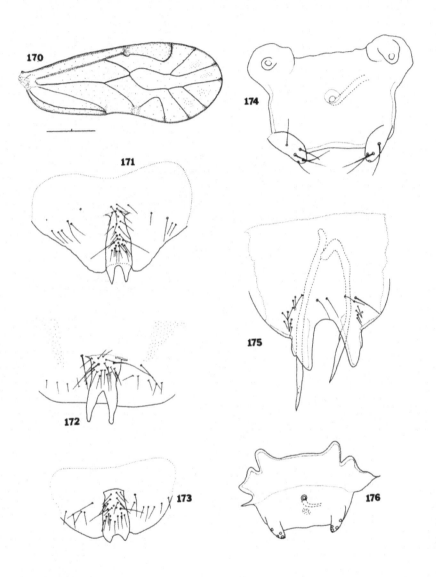

Figs. 170-176. *Lachesilla* spp. 170 - *L. nita* female, forewing; 171 - *L. jeanae*, subgenital plate; 172 - *L. rufa*, subgenital plate; 173 - *L. yakima*, subgenital plate; 174 - *L. andra*, ninth sternum and ovipositor valvulae; 175 - *L. kola*, hypandrium and phallosome; 176 - *L. kola*, ninth sternum and ovipositor valvulae.

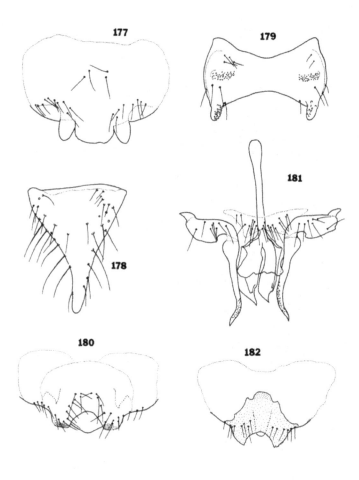

Figs. 177-182. *Lachesilla* spp. 177 - *L. punctata*, subgenital plate; 178 - *L. punctata* male, epiproct; 179 - *L. dona* male, epiproct; 180 - *L. nubiloides*, subgenital plate; 181 - *L. nubiloides*, hypandrium and phallosome; 182 - *L. dona*, subgenital plate.

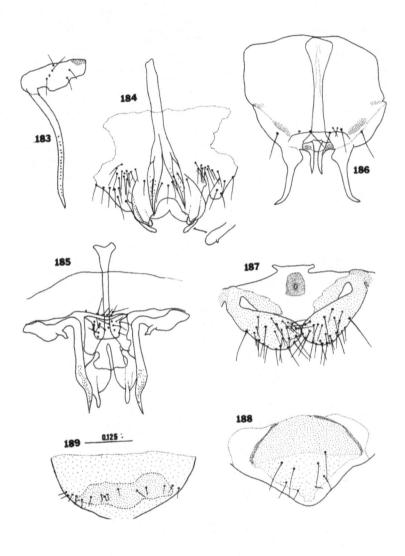

Figs. 183-189. *Lachesilla* spp. 183 - *L. dona*, clasper; 184 - *L. arnae*, hypandrium and phallosome; 185 - *L. nubilis*, hypandrium and phallosome; 186 - *L. riegeli*, hypandrium and phallosome; 187 - *L. riegeli*, ninth sternum and ovipositor valvulae; 188 - *L. tropica*, subgenital plate; 189 - *L. ultima*, subgenital plate.

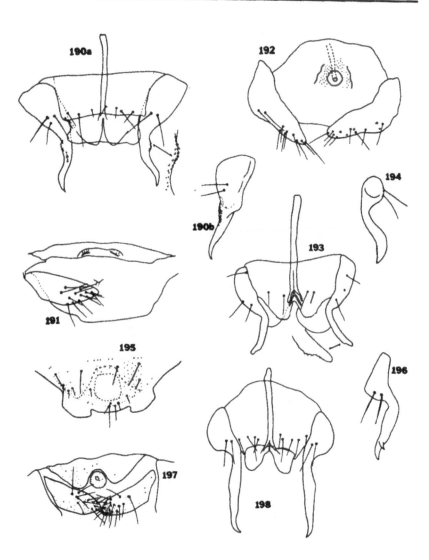

Figs. 190-198. *Lachesilla* spp. 190a - *L. major*, hypandrium and phallosome; 190b - *L. forcepeta*, clasper; 191 - *L. forcepeta*, ninth sternum and left ovipositor valvula; 192 - *L. pedicularia*, ninth sternum and ovipositor valvulae; 193 - *L. contraforcepeta*, hypandrium and phallosome; 194 - *L. alpejia*, claspers; 195 - *L. alpejia*, subgenital plate; 196 - *L. denticulata*, clasper; 197 - *L. denticulata*, ninth sternum and ovipositor valvulae; 198 - *L. bottimeri*, hypandrium and phallosome.

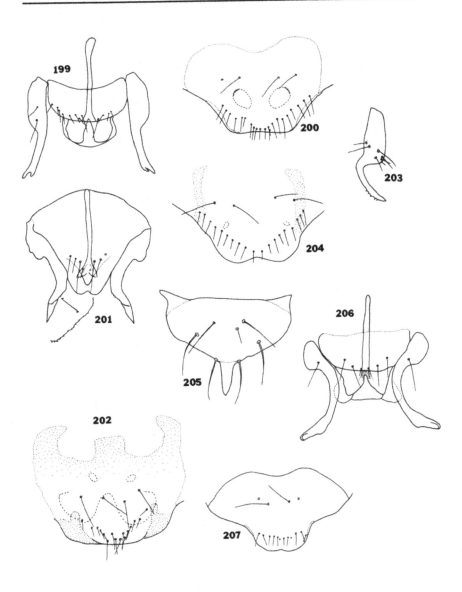

Figs. 199-207. - *Lachesilla* spp. 199 - *L. anna*, hypandrium and phallosome; 200 - *L. anna*, subgenital plate; 201 - *L. penta*, hypandrium and phallosome; 202 - *L. penta*, subgenital plate; 203 - *L. gracilis*, clasper; 204 - *L. gracilis*, subgenital plate; 205 - *L. kathrynae* male, epiproct; 206 - *L. kathrynae*, hypandrium and phallosome; 207 - *L. kathrynae*, subgenital plate.

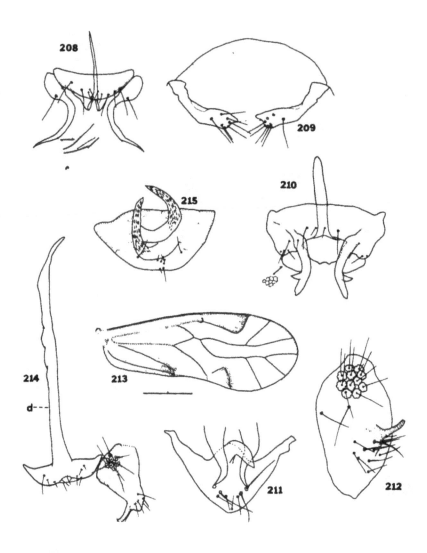

Figs. 208-215. *Lachesilla* spp. 208 - *L. chapmani*, hypandrium and phallosome; 209 - *L. aethiopica*, ninth sternum and ovipositor valvulae; 210 - *L. quercus*, hypandrium and phallosome; 211 - *L. pedicularia* male, clunial processes and epiproct; 212 - *L. pallida* male, paraproct; 213 - *L. pallida* male, forewing; 214 - *L. rena* male, epiproct and paraproct; 215 - *L. pacifica* male, epiproct.

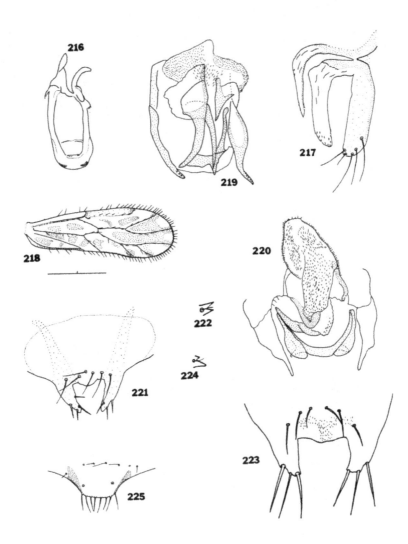

Figs. 216-225. Ectopsocids. 216 - *Ectopsocopsis cryptomeriae* male, clunial frame; 217 - *Ectopsocus pumilis*, ovipositor valvulae; 218 - *Ectopsocus strauchi*, forewing; 219 - *Ectopsocus thibaudi*, phallosome; 220 - *Ectopsocus briggsi*, phallosome; 221 - *E. briggsi*, subgenital plate; 222 - *E. briggsi* female, spine of paraproctal margin; 223 - *Ectopsocus californicus*, subgenital plate; 224 - *Ectopsocus meridionalis* female, spine of paraproctal margin; 225 - *Ectopsocus richardsi*, subgenital plate.

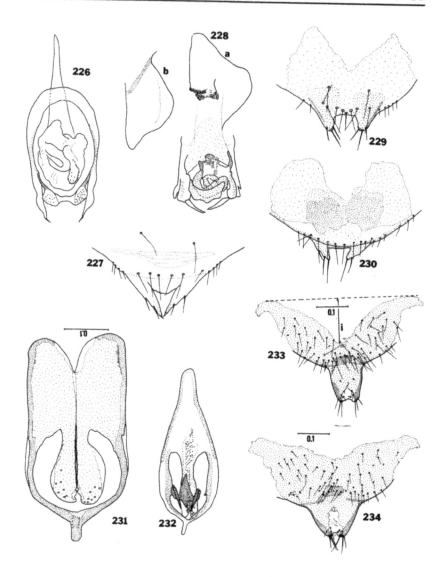

Figs. 226-234. Fig. 226 - *Ectopsocus pumilis*, phallosome; 227 - *E. pumilis*, subgenital plate; 228 - *E. salpinx*, phallosome (a) and anterior process (b); 229 - *Ectopsocus maindroni*, subgenital plate; 230 - *Ectopsocus titschacki*, subgenital plate; 231 - *Kaestneriella fumosa*, phallosome; 232 - *Peripsocus reductus*, phallosome; 233 - *Kaestneriella tenebrosa*, subgenital plate showing procedure for measuring indentation; 234 - *K. fumosa*, subgenital plate.

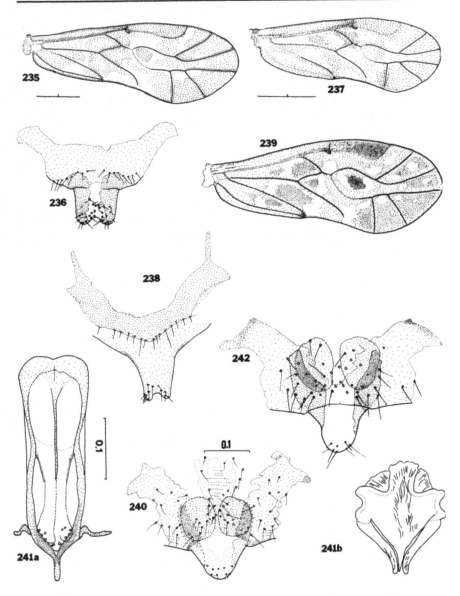

Figs. 235-242. *Peripsocus* spp. 235 - *P. subfasciatus* female, forewing; 236 - *P. subfasciatus*, subgenital plate; 237 - *P. pauliani* female, forewing; 238 - *P. pauliani*, subgenital plate; 239 - *P. potosi* female, forewing; 240 - *P. minimus*, subgenital plate; 241a - *P. minimus*, phallosome; 241b - *P. minimus*, endophallus; 242 - *P. alboguttatus*, subgenital plate.

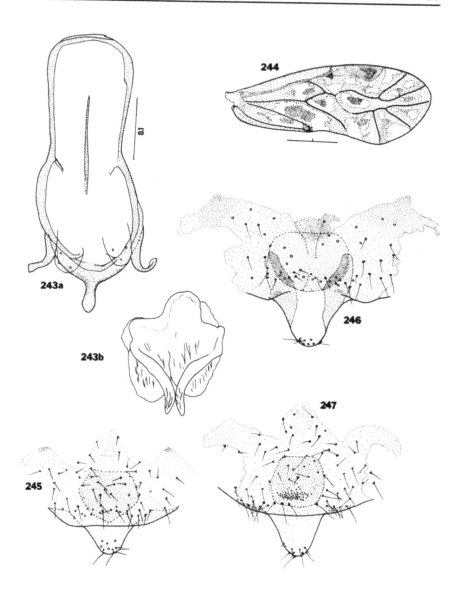

Figs. 243-247. *Peripsocus* spp. 243a - *P. madescens* phallosome; 243b - *P. madescens*, endophallus; 244 - *P. alachuae* female, forewing; 245 - *P. alachuae*, subgenital plate; 246 - *P. maculosus*, subgenital plate; 247 - *P. madescens*, subgenital plate.

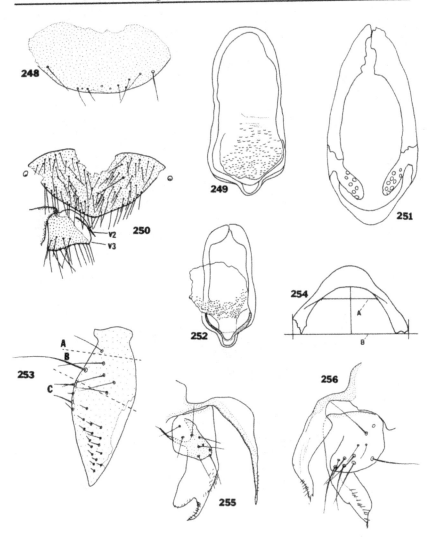

Figs. 248-256. 248 - *Archipsocopsis frater*, subgenital plate; 249 - *A. frater*, phallosome; 250 - *Archipsocus floridanus*, subgenital plate and ovipositor valvulae; 251 - *A. floridanus*, phallosome; 252 - *Archipsocopsis parvulus*, phallosome; 253 - *A. floridanus* female paraproct showing setal zones; 254 - *A. floridanus*, aedeagal arch showing lines measured for calculating index of internal roundness (line A is one-fourth distance from inner "top" of arch to basal line B); 255 - *Heterocaecilius* sp. , ovipositor valvulae; 256 - *Pseudocaecilius citricola*, ovipositor valvulae.

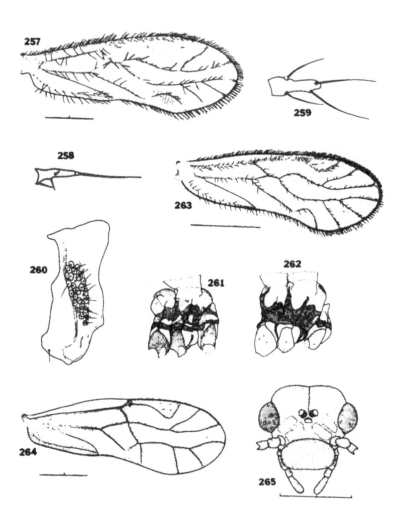

Figs. 257-265. 257 - *Pseudocaecilius tahitiensis* female, forewing; 258 - *Aaroniella maculosa*, terminal flagellomere; 259 - *Philotarsus kwakiutl*, terminal flagellomere; 260 - *P. kwakiutl* male, paraproct; 261 - *Aaroniella badonneli* female, thoracic pleura; 262 - *A. maculosa* female, thoracic pleura; 263 - *Philotarsus picicornis* female, forewing; 264 - *Propsocus pulchripennis* male, forewing; 265 - *Elipsocus hyalinus* female, head markings.

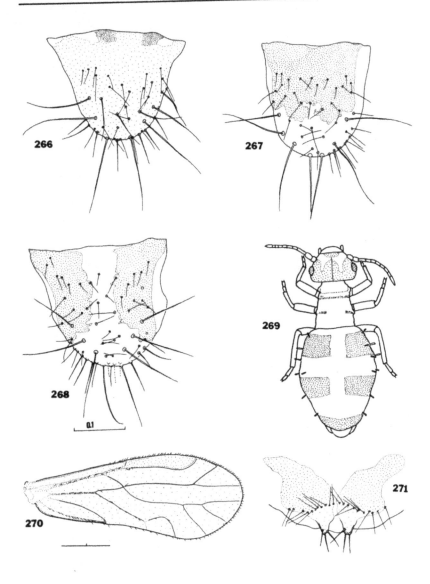

Figs. 266-271. Elipsocids. 266 - *Elipsocus abdominalis* female, epiproct (showing color pattern); 267 - *Elipsocus moebiusi* female, epiproct; 268 - *Elipsocus obscurus* female, epiproct; 269 - *Nepiomorpha peripsocoides*, apterous female, habitus; 270 - *Reuterella helvimacula* male, forewing; 271 - *R. helvimacula*, subgenital plate.

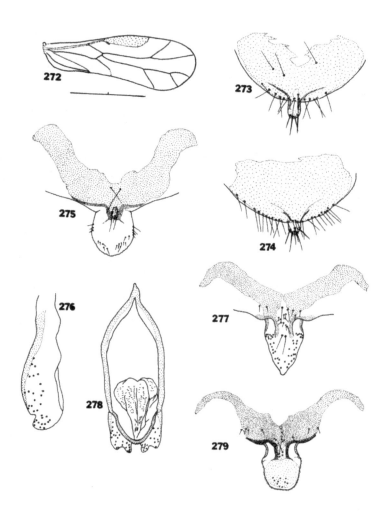

Figs. 272-279. 272 - *Palmicola aphrodite* male, forewing; 273 - *P. aphrodite*, subgenital plate; 274 - *Palmicola solitaria*, subgenital plate; 275 - *Mesopsocus laticeps*, subgenital plate; 276 - *Mesopsocus immunis*, external paramere; 277 - *M. immunis*, subgenital plate; 278 - *Mesopsocus unipunctatus*, phallosome; 279 - *M. unipunctatus*, subgenital plate.

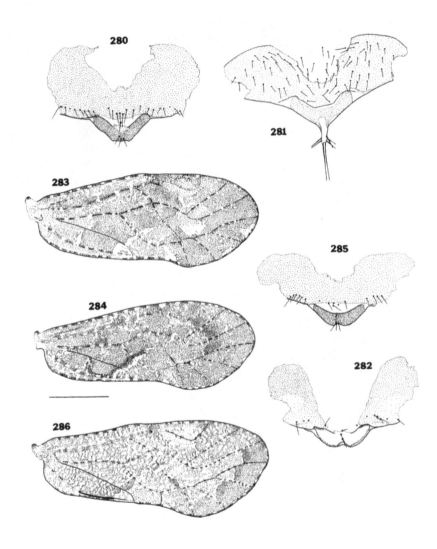

Figs. 280-286. Myopsocids. 280 - *Lichenomima lugens*, subgenital plate; 281 - *Myopsocus antillanus*, subgenital plate; 282 - *Lichenomima coloradensis*, subgenital plate; 283 - *L. coloradensis* female, forewing; 284 - *L. lugens* female, forewing; 285 - *Lichenomima sparsa*, subgenital plate; 286 - *L. sparsa* female, forewing.

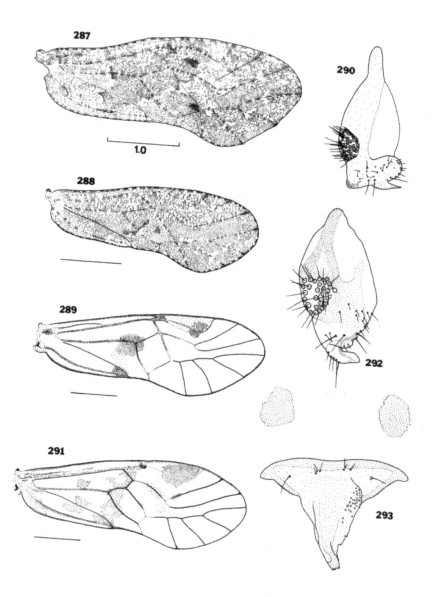

Figs. 287-293. 287 - *Myopsocus antillanus* female, forewing; 288 - *Myopsocus eatoni* male, forewing; 289 - *Psocus leidyi* female, forewing; 290 - *P. leidyi* male, paraproct; 291 - *Amphigerontia bifasciata* female, forewing; 292 - *A. bifasciata* male, paraproct; 293 - *Indiopsocus texanus*, hypandrium.

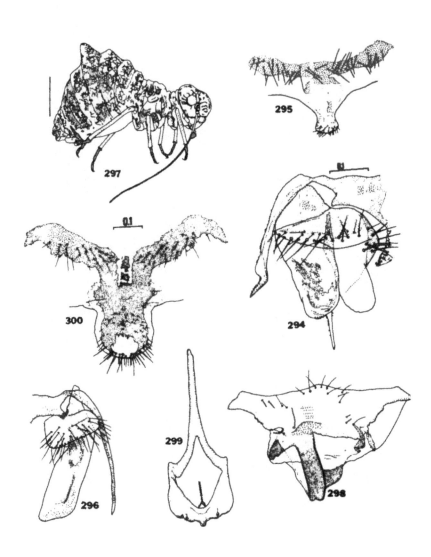

Figs. 294-300. Psocidae. 294 - *Camelopsocus bactrianus*, ovipositor valvulae; 295 - *Cerastipsocus venosus*, subgenital plate; 296 - *Metylophorus novaescotiae*, ovipositor valvulae; 297 - *C. bactrianus* female, habitus; 298 - *C. bactrianus*, hypandrium; 299 - *C. bactrianus*, phallosome; 300 - *C. bactrianus*, subgenital plate.

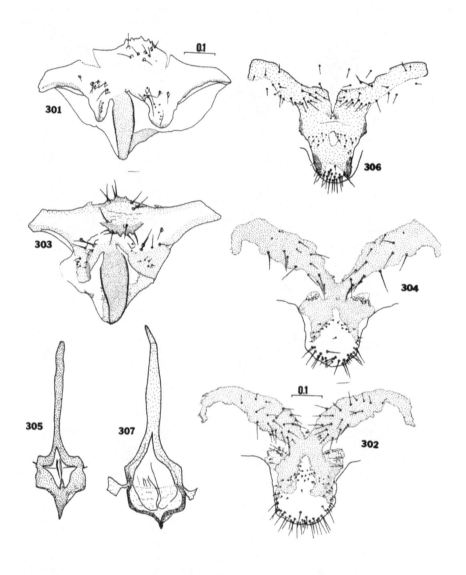

Figs. 301-307. *Camelopsocus* spp. 301 - *C. tucsonensis*, hypandrium; 302 - *C. tucsonensis*, subgenital plate; 303 - *C. hiemalis*, hypandrium; 304 - *C. hiemalis*, subgenital plate; 305 - *C. similis* phallosome; 306 - *C. similis*, subgenital plate; 307 - *C. monticolus*, phallosome.

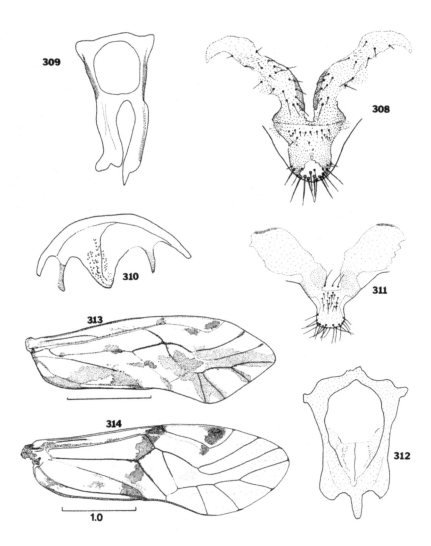

Figs. 308-314. Psocidae. 308 - *Camelopsocus monticolus*, subgenital plate; 309 - *Indiopsocus campestris*, phallosome; 310 - *I. campestris*, hypandrium; 311 - *I. campestris*, subgenital plate; 312 - *I. coquillettii*, phallosome; 313. *I. coquillettii* macropterous female, forewing; 314 - *I. ceterus*, forewing.

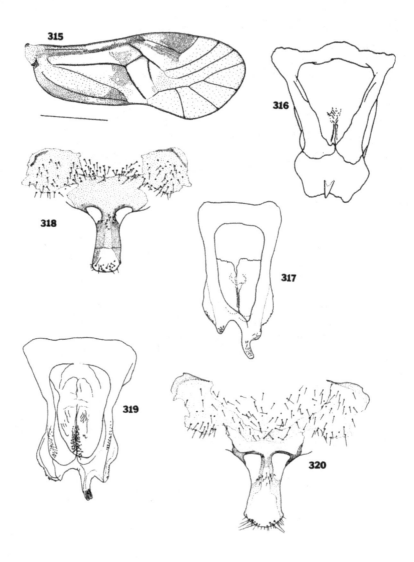

Figs. 315-320. *Indiopsocus* spp. 315 - *I. bisignatus* male, forewing; 316 - *I. bisignatus*, phallosome; 317 - *I. ceterus*, phallosome; 318 - *I. ceterus*, subgenital plate; 319 - *I. texanus*, phallosome; 320 - *I. texanus*, subgenital plate.

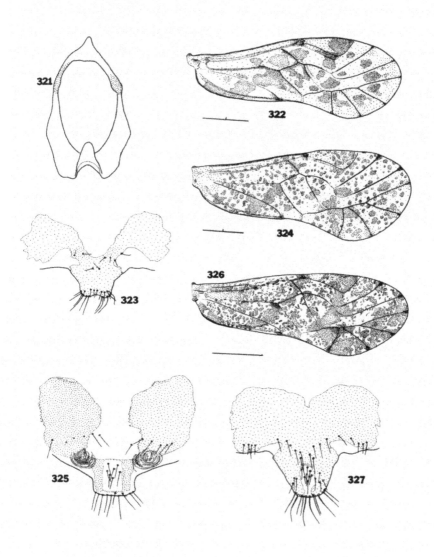

Figs. 321-327. *Trichadenotecnum* spp. 321 - *T. desolatum*, phallosome; 322 - *T. circularoides* female, forewing; 323 - *T. circularoides*, subgenital plate; 324 - *T. pardus* female, forewing; 325 - *T. pardus*, subgenital plate; 326 - *T. desolatum* female, forewing; 327 - *T. desolatum*, subgenital plate.

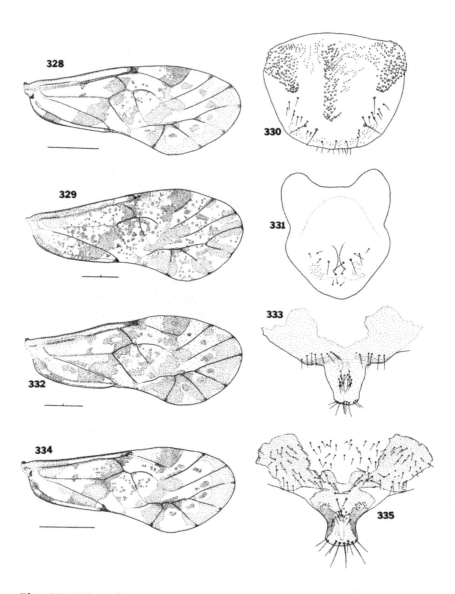

Figs. 328-335. *Trichadenotecnum* spp. 328 - *T. majus* female, forewing; 329 - *T. alexanderae* female, forewing; 330 - *T. majus* male, epiproct; 331 - *T. slossonae* female, forewing; 333 - *T. slossonae*, subgenital plate; 334 - *T. quaesitum* female, forewing; 335 - *T. quaesitum*, subgenital plate.

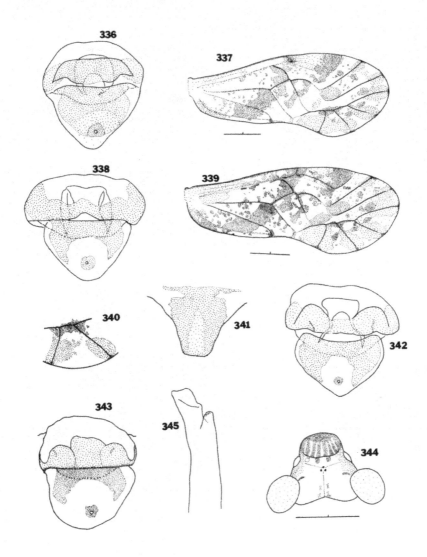

Figs. 336-345. *Trichadenotecnum* and *Steleops* spp. 336 - *T. innuptum* female, ninth sternum; 337 - *T. innuptum* female, forewing; 338 - *T. castum* female, ninth sternum; 339 - *T. castum* female, forewing; 340 - *T. merum* female, cell M3 and surroundings of forewing; 341 - *T. merum*, egg guide showing pigment pattern; 342 - *T. merum* female, ninth sternum; 343 - *T. alexanderae* female, ninth sternum; 344 - *Steleops elegans* female, head; 345 - *S. elegans* female, lacinial tip.

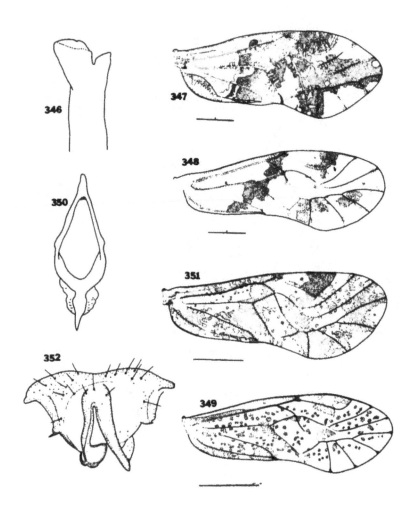

Figs. 346-352. *Steleops* and *Loensia* spp. 346 - *L. moesta* female, lacinial tip; 347 - *S. lichenatus* female, forewing; 348 - *S. elegans* female, forewing; 349 - *L. conspersa* female, forewing; 350 - *L. conspersa*, phallosome; 351 - *L. fasciata* female, forewing; 352 - *L. fasciata*, hypandrium (re-drawn from Badonnel, 1943).

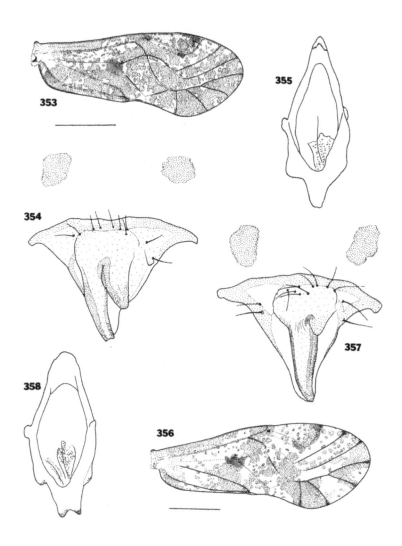

Figs. 353-358. *Loensia* spp. 353 - *L. moesta* female, forewing; 354 - *L. moesta*, hypandrium; 355 - *L. moesta*, phallosome; 356 - *L. maculosa* female, forewing; 357 - *L. maculosa*, hypandrium; 358 - *L. maculosa*, phallosome.

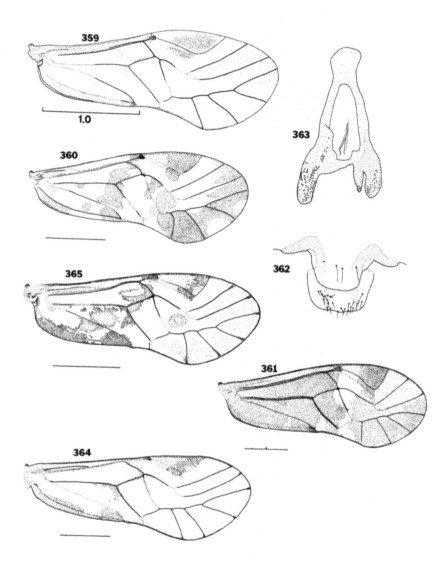

Figs. 359-365. Psocidae. 359 - *Ptycta lineata* male, forewing; 360 - *Ptycta polluta* female, forewing; 361 - *Atropsocus atratus* female, forewing; 362 - *A. atratus* male, clunial shelf and epiproct; 363 - *Hyalopsocus striatus*, phallosome; 364 - *H. striatus* male, forewing; 365 - *Hyalopsocus floridanus* female, forewing.

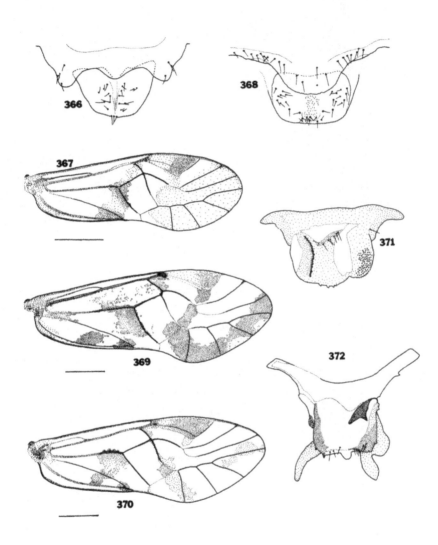

Figs. 366-372. *Psocus* and *Metylophorus* spp. 366 - *P. leidyi* male, clunial shelf and epiproct; 367 - *P. crosbyi* female, forewing; 368 - *P. crosbyi* male, clunial shelf and epiproct; 369 - *M. barretti* female, forewing; 370 - *M. novaescotiae* female, forewing; 371 - *M. novaescotiae*, hypandrium; 372 - *M. purus*, hypandrium.

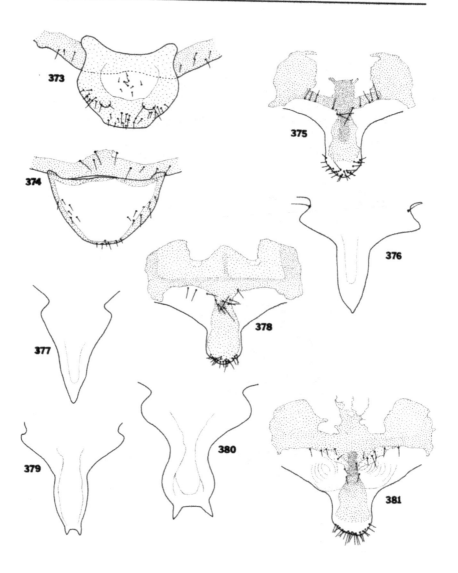

Figs. 373-381. *Amphigerontia* and *Blaste* spp. 373 - *A. bifasciata* male, epiproct and edge of clunium; 374 - *B. quieta* male, epiproct and edge of clunium; 275 - *A. bifasciata*, subgenital plate; 376 - *A. bifasciata*, distal end of hypandrium; 377 - *A. petiolata*, distal end of hypandrium; 378 - *A. petiolata*, subgenital plate; 379 - *A. montivaga*, distal end of hypandrium; 380 - *A. contaminata*, distal end of hypandrium; 381 - *A. contaminata*, subgenital plate.

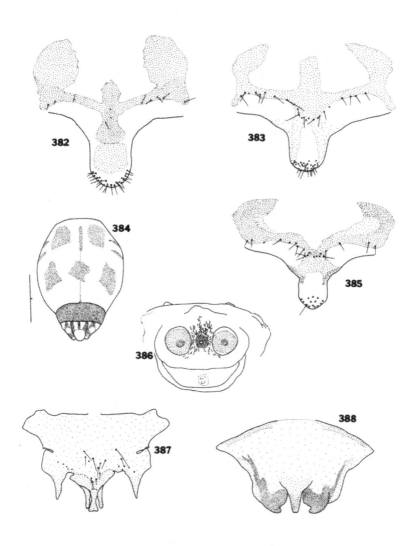

Figs. 382-388. *Amphigerontia, Blaste,* and *Blastopsocus* spp. 382 - *A. montivaga,* subgenital plate; 383 - *Blastopsocus lithinus,* subgenital plate; 384 - *Blastopsocus variabilis,* abdomen showing color pattern; 385 - *B. variabilis,* subgenital plate; 386 - *Blaste cockerelli* female, ninth sternum; 387 - *Blaste posticata,* distal segment of hypandrium; 388 - *Blaste subquieta,* distal segment of hypandrium.

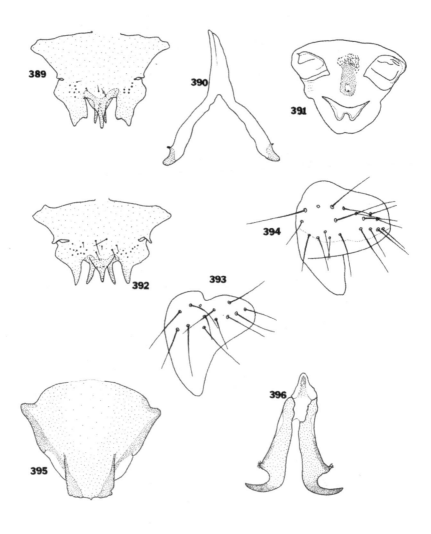

Figs. 389-396. *Blaste* spp. 389 - *B. garciorum,* distal segment of hypandrium; 390 - *B. garciorum,* phallosome; 391 - *B. garciorum,* female ninth sternum; 392 - *B. osceola,* third valvula; 394 - *B. posticata,* third valvula; 395 - *B. persimilis,* distal segment of hypandrium; 396 - *B. persimilis,* phallosome.

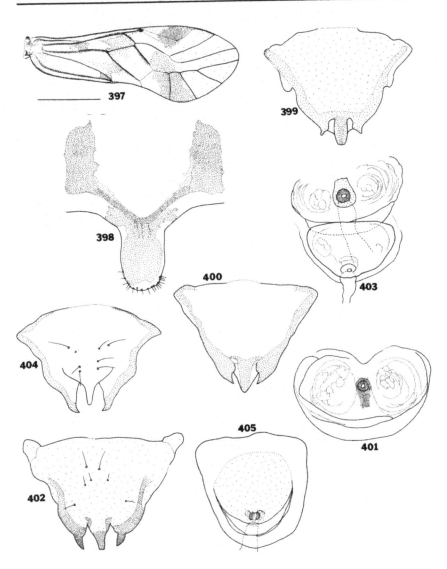

Figs. 397-405. *Blaste* spp. 397 - *B. opposita* female, forewing; 398 - *B. opposita*, subgenital plate; 399 - *B. quieta*, distal segment of hypandrium; 400 - *B. subapterous*, distal segment of hypandrium; 401 - *B. subquieta* female, ninth sternum; 402 - *B. longipennis*, distal segment of hypandrium; 403 - *B. longipennis* female, ninth sternum; 404 - *B. oregona*, distal segment of hypandrium; 405 - *B. oregona* female, ninth sternum.

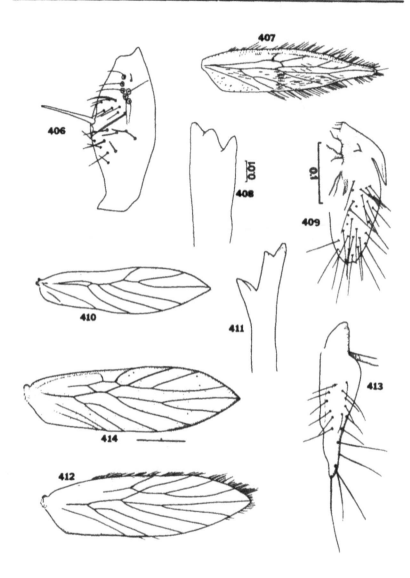

Figs. 406-414. Lepidopsocids. 406 - *Nepticulomima* sp. female, paraproct; 407 - *Thylacella cubana* female, forewing; 408 - *T. cubana* female, lacinial tip; 409 - *T. cubana*, ovipositor valvulae; 410 - *Proentomum personatum* female, hindwing; 411 - *Nepticulomima* sp. female, lacinial tip; 412 - *N.* sp. female, forewing; 413 - *N.* sp. , ovipositor valvulae; 414 - *P. personatum* female, forewing.

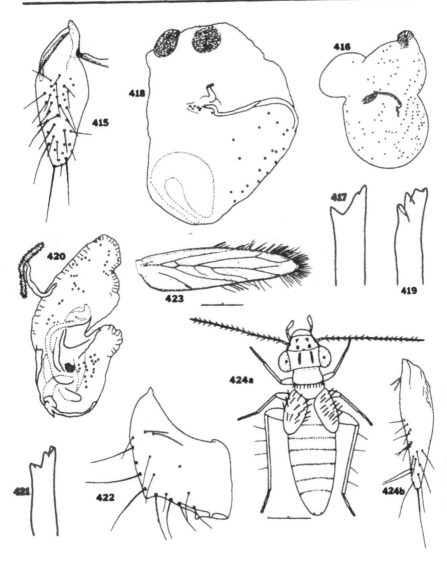

Figs. 415-424. Lepidopsocids. 415 - *P. personatum*, ovipositor valvulae; 416 - *Echmepteryx hageni*, spermathecal sac; 417 - *Echmepteryx madagascariensis*, lacinial tip; 418 - *E. madagascariensis*, spermathecal sac; 419 - *Echmepteryx falco*, lacinial tip; 420 - *E. falco*, spermathecal sac; 421 - *Neolepolepis occidentalis*, lacinial tip; 422 - *N. occidentalis*, fore coxa; 423 - *N. occidentalis*, macropterous female, forewing; 424a - *N. occidentalis*, brachypter, habitus; 424b - *N. occidentalis*, ovipositor valvulae.

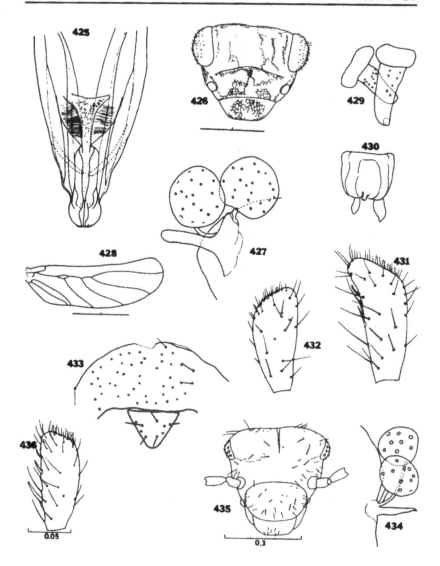

Figs. 425-436. 425 - *Neolepolepis occidentalis*, distal end of phallosome; 426 - *Pteroxonium kelloggi*, facial markings; 417 - *Rhyopsocus eclipticus*, accessory bodies and beak of spermatheca; 428 - *Balliella ealensis* female, hindwing; 429 - *B. ealensis*, spermathecal accessory bodies; 430 - *Psoquilla marginepunctata*, spermapore sclerite; 431 - *Rhyopsocus bentonae*, mx4; 432 - *Rhyopsocus disparilis*, mx4; 433 - *R. disparilis*, accessory bodies and beak of spermatheca; 435 - *Rhyopsocus micropterus*, front view of head; 436 - *R. micropterus*, mx4.

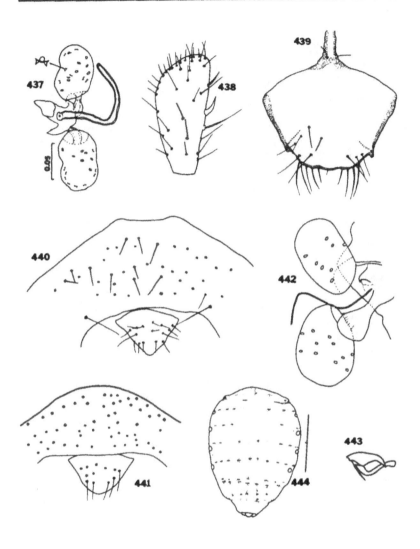

Figs. 437-444. *Rhyopsocus* and *Cerobasis* spp. 437 - *R. micropterus*, accessory bodies and beak of spermatheca; 438 - *R. eclipticus*, mx4; 439 - *R. eclipticus*, hypandrium; 440 - *R. eclipticus* female, clunium and epiproct; 441 - *R. texanus* female, clunium and epiproct; 442 - *R. texanus*, accessory bodies and beak of spermatheca; 443 - *C. annulata*, pretarsal claw; 444 - *C. annulata* female, abdomen showing color pattern.

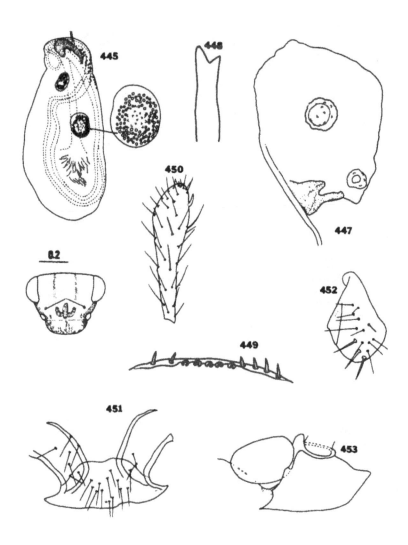

Figs. 445-453. 445 - *Cerobasis annulata*, spermathecal sac; 446 - *Cerobasis guestfalica*, facial markings; 447 - *C. guestfalica*, spermathecal sac; 448 - *Lepinotus reticulatus*, lacinial tip; 449 - *Psyllipsocus oculatus*, distal inner labral sensilla; 450 - *Dorypteryx pallida*, mx4; 451 - *D. pallida*, hypandrium and phallosome; 452 - *D. pallida*, ovipositor valvulae; 453 - *Dorypteryx domestica*, structures around opening of spermathecal sac.

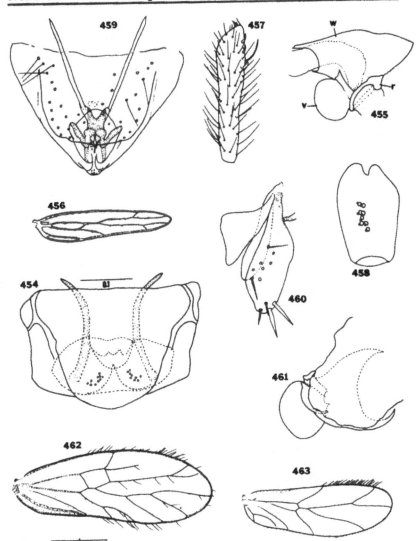

Figs. 454-463. Psyllipsocids. 454 - *Dorypteryx domestica*, hypandrium and phallosome (re-drawn from Lienhard, 1977); 455 - *Dorypteryx pallida*, structures around opening of spermathecal sac (v - vesicle, w - wing, r - ring); 456 - *Pseudodorypteryx mexicana*, forewing; 457 - *P. mexicana*, mx4; 458 - *P. mexicana*, pedicel with microspades organ; 459 - *P. mexicana*, hypandrium and phallosome; 460 - *P. mexicana*, ovipositor valvulae; 461 - *P. mexicana*, structures around opening of spermathecal sac; 462 - *Psocatropos microps*, macropterous female, forewing; 463 - *P. microps*, macropterous female, hindwing.

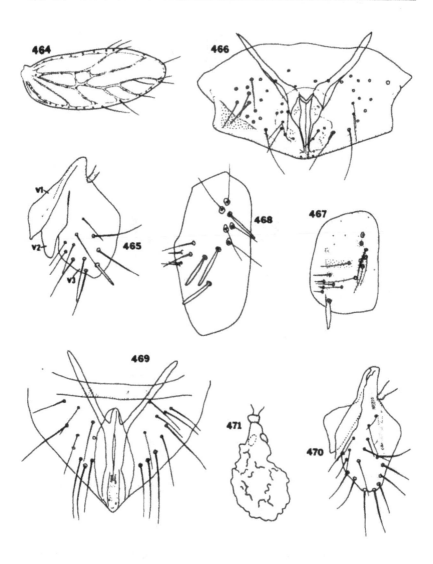

Figs. 464-471. Psyllipsocids. 464 - *Psocatropos microps*, brachypterous female, forewing; 465 - *P. microps*, ovipositor valvulae (labelled); 466 - *P. microps*, hypandrium and phallosome; 467 - *P. microps* male, paraproct; 468 - *P. microps* female, paraproct; 469 - *Psyllipsocus oculatus*, hypandrium and phallosome; 470 - *P. oculatus*, ovipositor valvulae; 471 - *Psyllipsocus ramburti*, spermathecal sac.

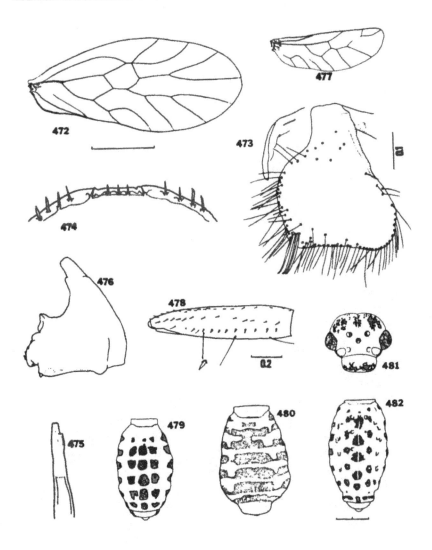

Figs. 472-482. *Speleketor* spp. 472 - *S. irwini*, forewing; 473 - *S. flocki*, ovipositor valvulae; 474 - *S. irwini*, distal inner labral sensilla; 475 - *S. irwini*, lacinial tip; 476 - *S. flocki*, mandible; 477 - *S. irwini*, hindwing; 478 - *S. irwini*, fore femur; 479 - *S. flocki*, dorsal abdominal color pattern; 480 - *S. irwini*, dorsal abdominal color pattern; 481 - *S. pictus*, facial markings; 482 - *S. pictus*, dorsal abdominal color pattern.

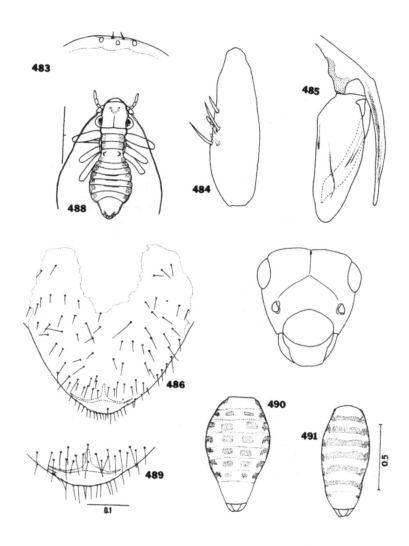

Figs. 483-491. Pachytroctids. 483 - *Tapinella maculata*, distal inner labral sensilla; 484 - *T. maculata*, mx4 showing claviform sensilla; 485 - *T. maculata*, ovipositor valvulae; 486 - *Nanopsocus oceanicus*, subgenital plate; 487 - *Pachytroctes neoleonensis*, head in front view; 488 - *T. maculata* male, habitus; 489 - *T. maculata*, subgenital plate; 490 and 491 - *T. maculata* female, abdominal color pattern.

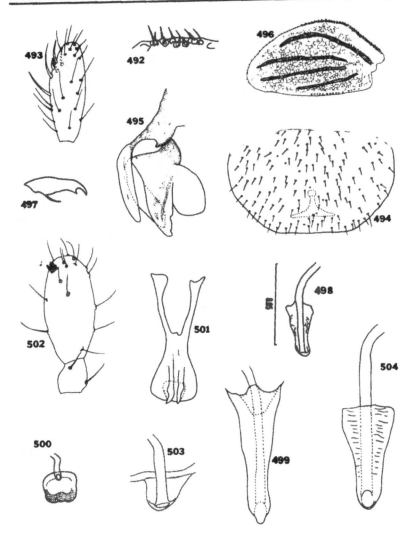

Figs. 492-504. *Sphaeropsocopsis* and *Belaphotroctes* spp. 492 - *S. argentinus*, distal inner labral sensilla; 493 - *S. argentinus*, mx4; 494 - *S. argentinus*, subgenital plate; 495 - *S. argentinus*, ovipositor valvulae; 496 - *S. argentinus*, forewing; 497 - *B. alleni*, pretarsal claw; 498 - *B. alleni*, spermapore sclerite; 499 - *B. badonneli*, spermapore sclerite; 500 - *B. ghesquierei*, spermapore sclerite; 501 - *B. ghesquierei*, middle piece of phallosome; 502 - *B. hermosus* female, mx3 and mx4; 503 - *B. hermosus*, spermapore sclerite; 504 - *B. simberloffi*, spermapore sclerite.

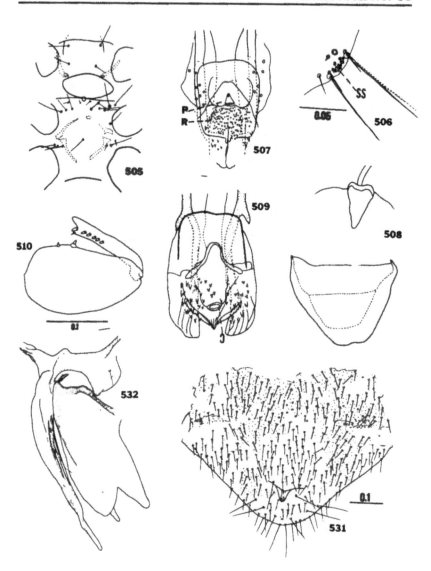

Figs. 505-510, 531, 532. *Embidopsocus* and *Lithoseopsis* spp. 505 - *E. citrensis* female, thoracic sterna; 506 - *E. laticeps* male, edge of tg 10 with field of trichoid sensilla(ss); 507 - *E. bousemani*, endophallus (R = radula, P = end of penis); 508 - *E. needhami*, spermapore sclerite and u-shaped sclerite; 509 - *E. needhami*, median distal region of phallosome; 510 - *E. femoralis* female, fore femur and tibia; 531 - *L. hystrix*, subgenital plate; 532 - *L. hystrix*, ovipositor valvulae.

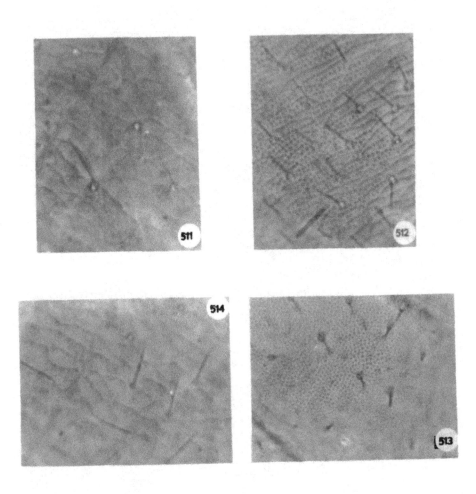

Figs. 511-514. *Liposcelis* spp. (females), sculpture of integument. 511 - *L. brunnea*, vertex; 512 - *L. fusciceps*, vertex; 513 - *L. fusciceps*, abdominal terga; 514 - *L. pallens*, vertex.

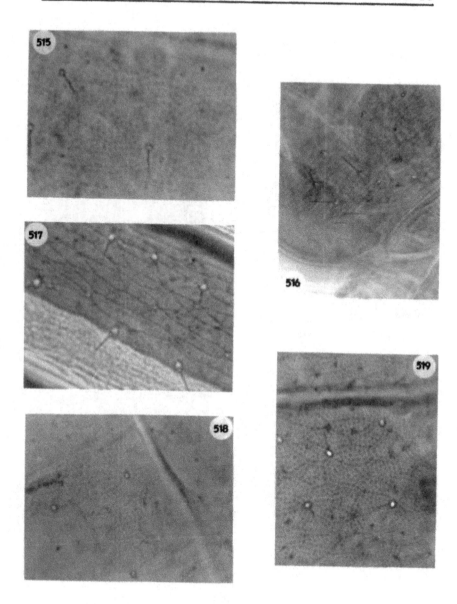

Figs. 515-519. *Liposcelis* spp. (females), sculpture of integument. 515 - *L. pallens*, abdominal terga; 516 - *L. villosa*, vertex; 517 - *L. nigra*, abdominal terga; 518 - *L. rufa*, abdominal terga; 519 - *L. silvarum*, abdominal terga.

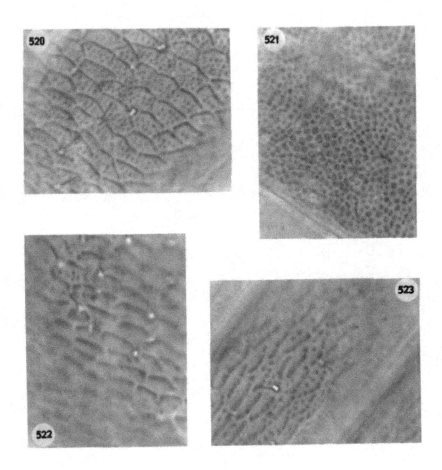

Figs. 520-523. *Liposcelis spp.* (females), sculpture of integument. 520 - *L. triocellata*, vertex; 521 - *L. triocellata*, abdominal terga; 522 - *L. formicaria*, vertex; 523 - *L. formicaria*, abdominal terga.

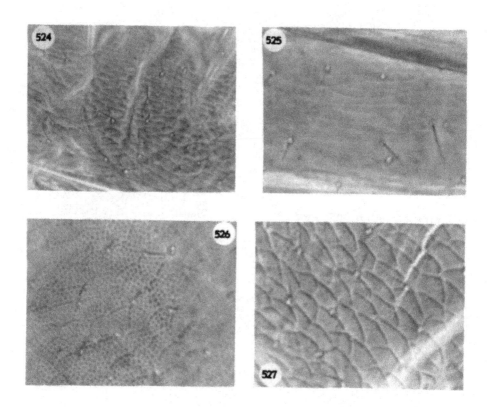

Figs. 524-527. *Liposcelis* spp. (females), sculpture of integument. 524 - *L. mendax*, vertex; 525 - *L. mendax*, abdominal terga; 526 - *L. bostrychophila*, vertex; 527 - *L. corrodens*, vertex.

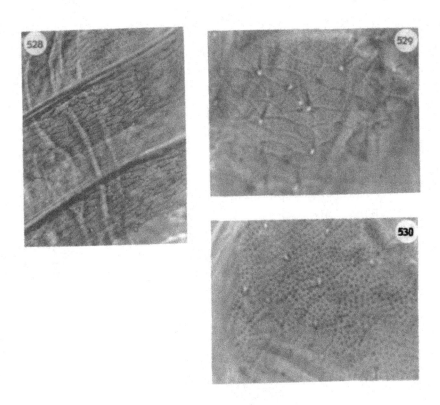

Figs. 528-530. *Liposcelis* spp. (females), sculpture of integument. 528 - *L. corrodens*, abdominal terga; 529 - *L. paeta*, vertex; 530 - *L. prenolepidis*, vertex.

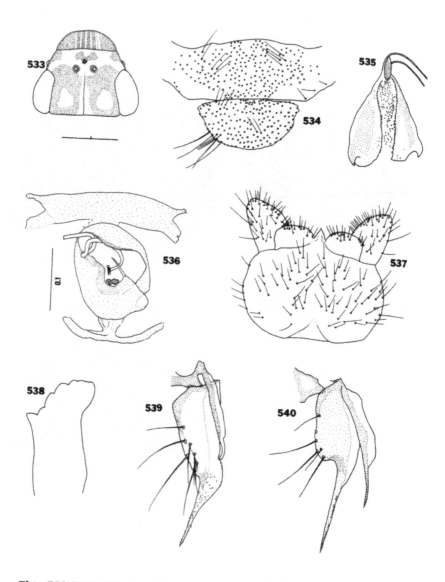

Figs. 533-540. 533 - *Stimulopalpus japonicus* female, head color pattern; 534 - *Lithoseopsis hellmani*, clunium and setal plate; 535 - *L. hellmani*, sclerites flanking opening of spermathecal sac; 536 - *L. hystrix*, spermapore sclerite; 537 - labium of typical psocomorph (*Epipsocus* sp.); 538 - *Epipsocus* sp. , lacinial tip; 539 - *Epipsocus* sp. , ovipositor valvulae; 540 - *Bertkauia crosbyana*, ovipositor valvulae.

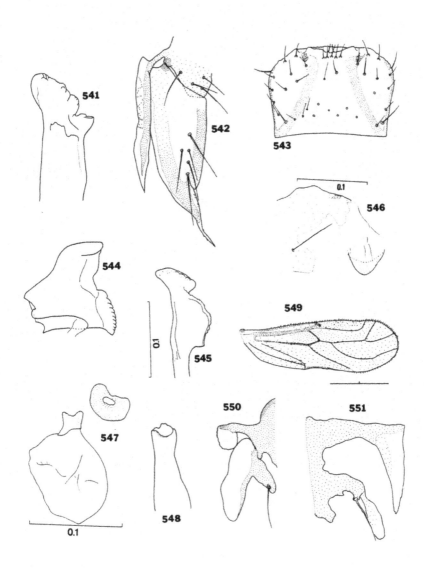

Figs. 541-551. *Loneura* sp. , lacinial tip; 542 - *Loneura* sp. , ovipositor valvulae; 543 - *Astopsocus sonorensis*, lacinial tip; 546 - *A. sonorensis*, ovipositor valvulae; 547 - *A. sonorensis*, spermatheca; 548 - *Nottopsocus* sp. , lacinial tip; 549 - *N.* sp. , forewing; 550 - *N.* sp. , ovipositor valvulae; 551 - *Pronottopsocus* sp. , ovipositor valvulae.

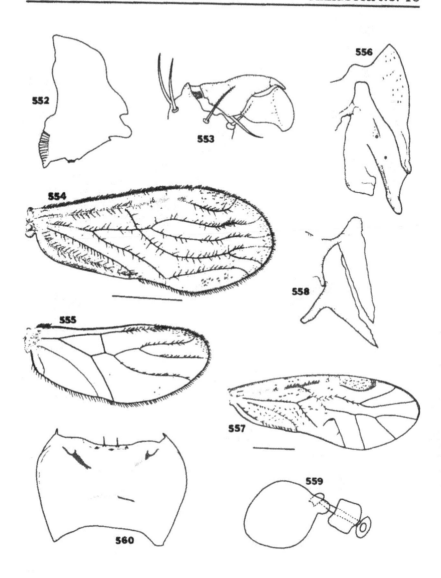

Figs. 552-560. 552 - Caecilioid mandible (*Caecilius insularus*); 553 - Caecilioid pretarsal claw (*Teliapsocus conterminus*); 554 - *Polypsocus corruptus* female, forewing; 555 - *P. corruptus* female, hindwing; 556 - *T. conterminus*, ovipositor valvulae; 557 - *T. conterminus* female, forewing; 558 - *Graphopsocus cruciatus*, ovipositor valvulae; 559 - *G. cruciatus*, spermatheca; 560 - *G. cruciatus*, labrum with stylets.

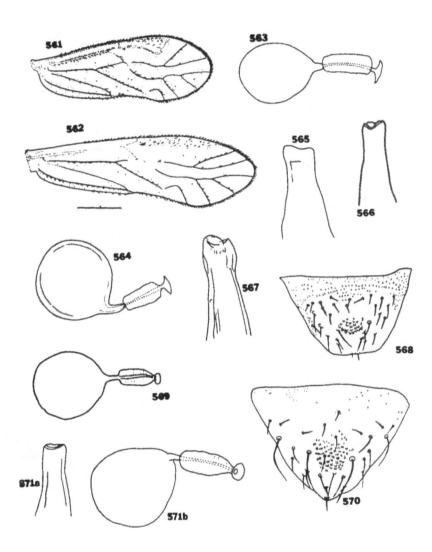

Figs. 561-571. *Caecilius* spp. 561 - *C. africanus* female, forewing; 562 - *C. casarum* female, forewing; 563 - *C. antillanus*, spermatheca; 564 - *C. insularum*, spermatheca; 565 - *C. antillanus*, lacinial tip; 566 - *C. insularum*, lacinial tip; 567 - *C. indicator*, lacinial tip; 568 - *C. indicator* male, epiproct; 569 - *C. indicator*, spermatheca; 570 - *C. graminis* male, paraproctal papillar field; 571 - *C. confluens*, spermatheca.

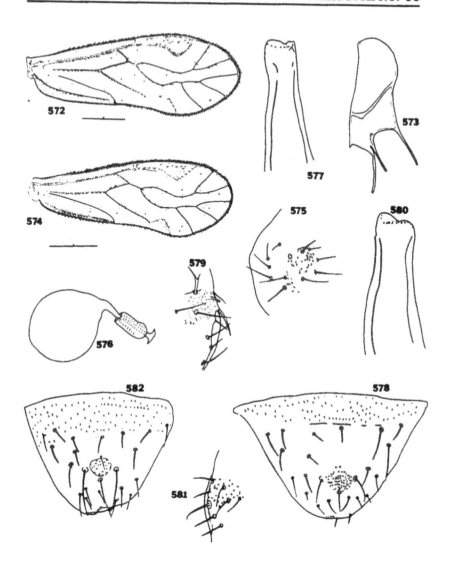

Figs. 572-582. *Caecilius* spp. 572 - *C. confluens* female, forewing; 573 - *C. confluens*, mesopleuron; 574 - *C. gonostigma* male, forewing; 575 - *C. gonostigma* male, paraproctal papillar field; 576 - *C. nadleri*, spermatheca; 577 - *C. nadleri*, lacinial tip; 578 - *C. nadleri* male, epiproctal papillar field; 579 - *C. nadleri* male paraproctal papillar field; 580 - *C. totonacus*, lacinial tip; 581 - *C. totonacus* male, paraproctal papillar field; 582 - *C. totonacus* male, epiproctal papillar field.

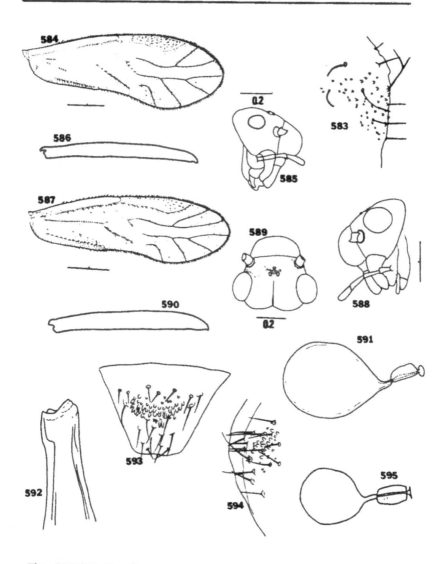

Figs. 583-595. *Caecilius* spp. 583 - *C. burmeisteri* male, paraproctal papillar field; 584 - *C. flavidus* female, forewing; 585 - *C. lochloosae* female, side view of head; 586 - *C. lochloosae* male, fore tibia; 587 - *C. micanopi* female, forewing; 588 - *C. micanopi* female, side view of head; 589 - *C. pinicola* female, head, dorsal view; 590 - *C. tamiami* male, fore tibia; 591 - *C. posticus*, spermatheca; 592 - *C. subflavus*, lacinial tip; 593 - *C. subflavus* male, paraproctal papillar field; 595 - *C. subflavus*, spermatheca.

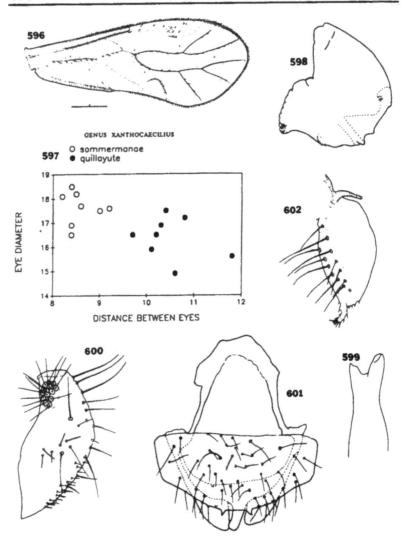

Figs. 596-602. 596 - *Xanthocaecilius quillayute* male, forewing; 597 - scatter diagrams of distance between eyes (IO) plotted against transverse eye diameter (d) for males of two species of *Xanthocaecilius* (units = mm measured on camera lucida drawings at 73x magnification); 598 - *Elipsocus obscurus*, mandible; 599 - *Lachesilla punctata*, lacinial tip; 600 - *Anomopsocus amabilis*, hypandrium and phallosome; 602 - *A. amabilis* ovipositor valvulae.

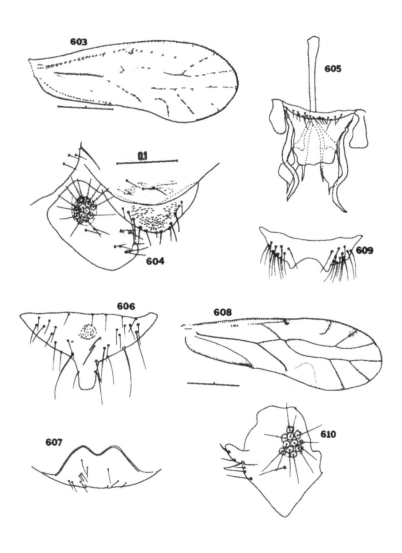

Figs. 603-610. Lachesillids. 603 - *Nanolachesilla hirundo* female, forewing; 604 - *Nanolachesilla chelata* male, epiproct and left paraproct; 605 - *Lachesilla andra*, hypandrium and phallosome; 606 - *L. andra* male, epiproct; 607 - *L. andra*, subgenital plate; 608 - *Lachesilla arnae* male, forewing; 609 - *L. arnae* male, epiproct; 610 - *L. arnae* male, paraproct.

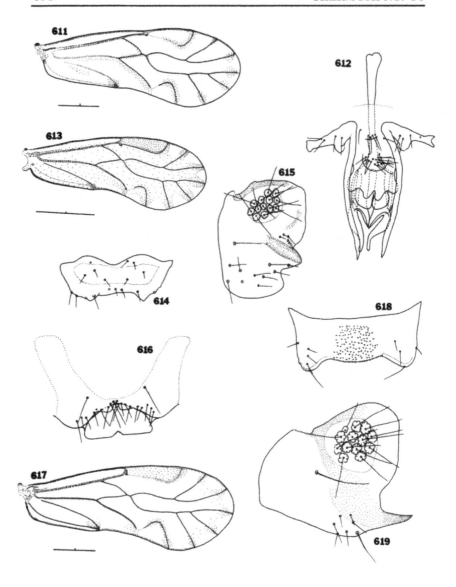

Figs. 611-619. *Lachesilla* spp. 611 - *L. dona* female, forewing; 612 - *L. dona*, hypandrium and phallosome; 613 - *L. kola* female, forewing; 614 - *L. kola* male, epiproct; 615 - *L. kola* male paraproct; 616 - *L. kola*, subgenital plate; 617 - *L. nubilis* male, forewing; 618 - *L. nubilis* male, epiproct; 619 - *L. nubilis* male, paraproct.

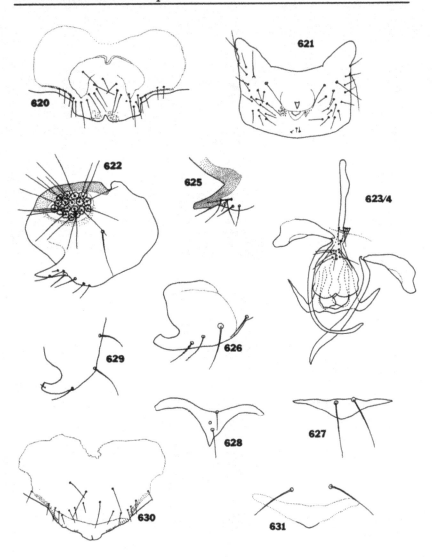

Figs. 620-631. *Lachesilla* spp. 620 - *L. nubilis*, subgenital plate; 621 - *L. nubiloides* male, epiproct; 622 - *L. nubiloides* male paraproct; 623/624 - *L. punctata*, hypandrium and phallosome; 625 - *L. punctata* male, paraproctal process; 627 - *L. centralis*, median distal sclerite of hypandrium; 628 - *L. cintalapa*, median distal sclerite of hypandrium; 629 - *L. cintalapa* male, paraproctal process; 630 - *L. cintalapa*, subgenital plate; 631 - *L. perezi*, median distal sclerite of hypandrium.

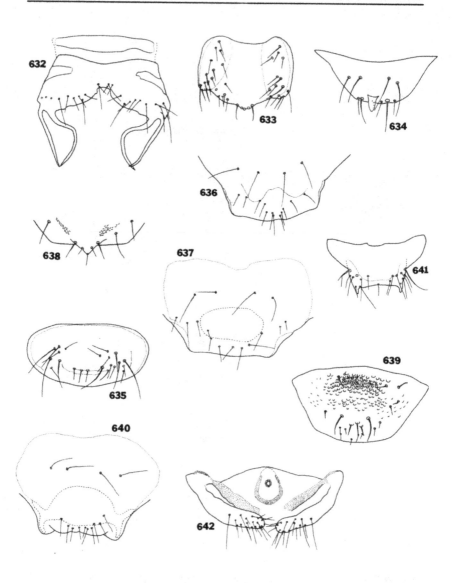

Figs. 632-642. *Lachesilla* spp. 632 - *L. corona*, hypandrium; 633 - *L. bottimeri* male, epiproct; 634 - *L. chapmani* male, epiproct; 635 - *L. denticulata* male, epiproct; 636 - *L. denticulata*, subgenital plate; 637 - *L. forcepeta*, subgenital plate; 638 - *L. gracilis* male, distal half of epiproct (re-drawn from Garcia Aldrete, 1988a); 639 - *L. major* male, epiproct; 640 - *L. major*, subgenital plate; 641 - *L. penta* male, epiproct; 642 - *L. penta*, ninth sternum and ovipositor valvulae.

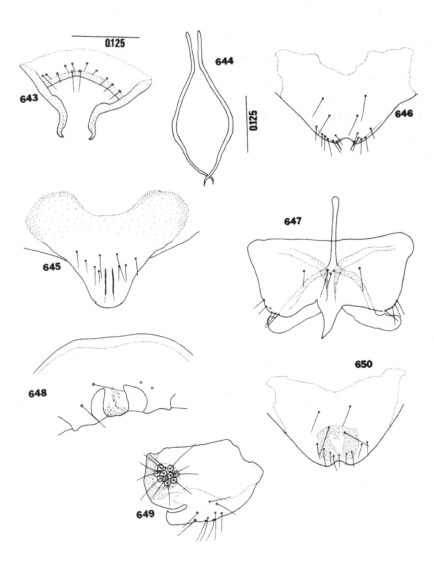

Figs. 643-650. *Lachesilla* spp. 643 - *L. fuscipalpis*, hypandrium (re-drawn from García Aldrete, 1974); 644 - *L. fuscipalpis*, phallosome (re-drawn from García Aldrete, 1974); 645 - *L. sulcata*, subgenital plate; 646 - *L. aethiopica*, subgenital plate; 647 - *L. pacifica*, hypandrium and phallosome; 648 - *L. pacifica* male, clunium; 649 - *L. pacifica* male, paraproct; 650 - *L. pacifica*, subgenital plate.

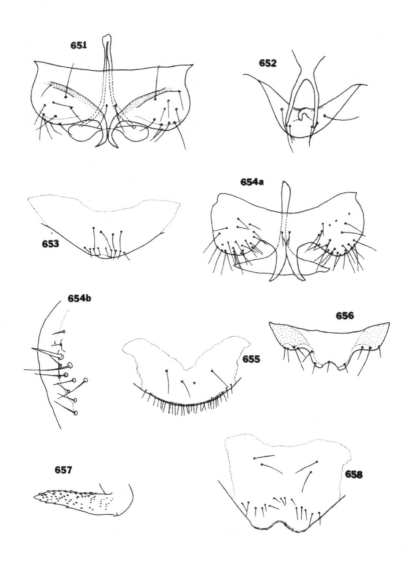

Figs. 651-658. *Lachesilla* spp. 651 - *L. pallida*, hypandrium and phallosome; 652 - *L. pallida* male, clunial processes and epiproct; 653 - *L. pallida*, subgenital plate; 654a - *L. pedicularia*, hypandrium and phallosome; 654b - free margin of male paraproct; 655 - *L. pedicularia*, subgenital plate; 656 - *L. quercus* male, epiproct; 657 - *L. quercus*, subgenital plate.

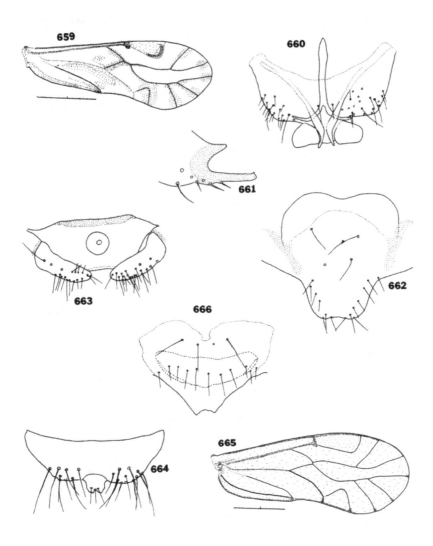

Figs. 659-666. *Lachesilla* spp. 659 - *L. rena* male, forewing; 660 - *L. rena*, hypandrium and phallosome; 661 - *L. rena* male, paraproctal process; 662 - *L. rena*, subgenital plate; 663 - *L. tectorum*, ninth sternum and ovipositor valvulae; 664 - *L. riegeli* male, epiproct; 665 - *L. riegeli*, forewing; 666 - *L. riegeli*, subgenital plate.

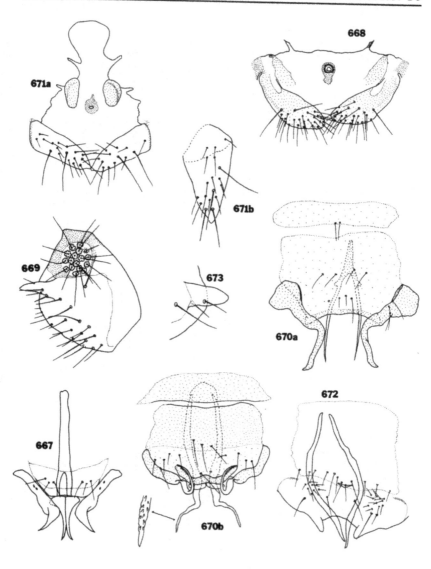

Figs. 667-673. *Lachesilla* spp. 667 - *L. tropica*, hypandrium and phallosome; 668 - *L. tropica*, ninth sternum and ovipositor valvulae; 669 - *L. rufa* male, paraproct; 670a - *L. rufa*, hypandrium and phallosome; 670b - *L. jeanae*, hypandrium and phallosome; 671a - *L. rufa*, ninth sternum and ovipositor valvulae; 671b - *L. jeanae*, ovipositor valvula; 672 - *L. nita*, hypandrium and phallosome; 673 - *L. nita* male, paraproctal process.

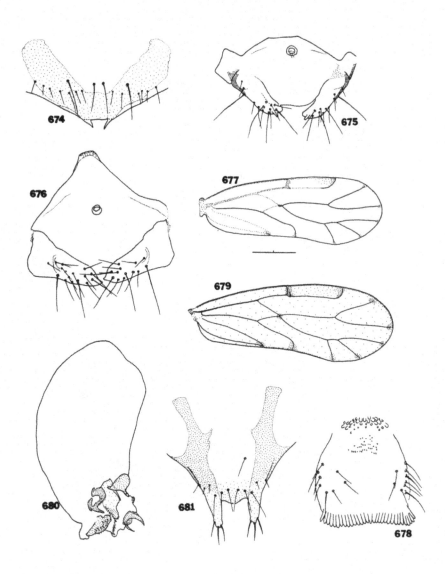

Figs. 674-681. *Lachesilla* and *Ectopsocus* spp. 674 - *L. nita*, subgenital plate; 675 - *L. nita*, ninth sternum and ovipositor valvulae; 676 - *L. yakima*, ninth sternum and ovipositor valvulae; 677 - *E. briggsi* male, clunial comb; 679 - *E. californicus* male, forewing; 680 - *E. maindroni*, phallosome; 681 - *E. meridionalis*, subgenital plate.

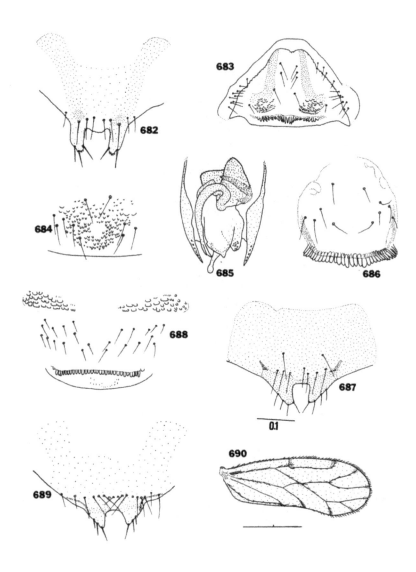

Figs. 682-690. *Ectopsocus* spp. 682 - *E. petersi*, subgenital plate; 683 - *E. pumilis* male, clunial comb; 684 - *E. richardsi* male, clunial comb; 685 - *E. richardsi*, phallosome; 686 - *E. salpinx* male, clunial comb; 687 - *E. salpinx*, subgenital plate; 688 - *E. strauchi* male, clunial comb; 689 - *E. strauchi*, subgenital plate; 690 - *E. thibaudi*, forewing.

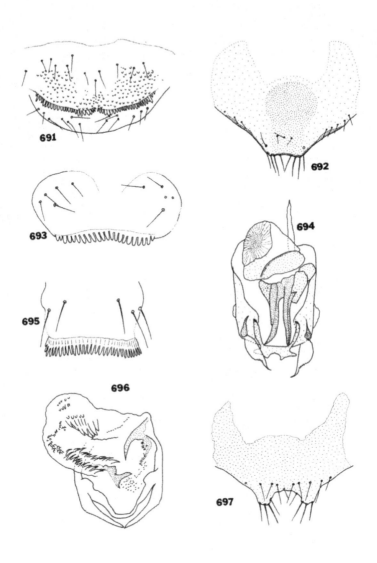

Figs. 691-697. *Ectopsocus* spp. 691 - *E. thibaudi* male, clunial comb; 692 - *E. thibaudi*, subgenital plate; 693 - *E. titschacki* male, clunial comb; 694 - *E. titschacki*, phallosome; 695 - *E. vachoni* male, clunial comb; 696 - *E. vachoni*, phallosome; 697 - *E. vachoni*, subgenital plate.

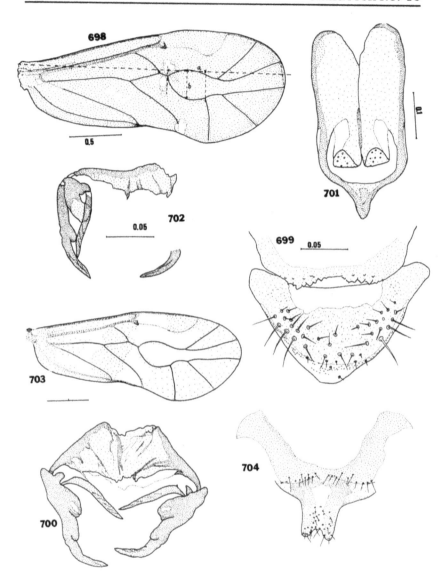

Figs. 698-704. *Kaestneriella* and *Peripsocus* spp. 698 - *K. fumosa* female, forewing showing measurements for R5 index; 699 - *K. fumosa* male, clunial comb and epiproct; 700 - *K. fumosa*, endophallic sclerites; 701 - *K. tenebrosa*, phallosome; 702 - *K. tenebrosa*, endophallic sclerites; 703 - *Peripsocus madidus* male, forewing; 704 - *P. madidus*, subgenital plate.

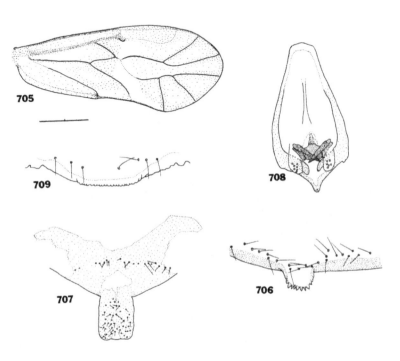

Figs. 705-709. *Peripsocus* spp. 705 - *P. reductus* male, forewing; 706 - *P. reductus* male, clunial comb; 707 - *P. reductus*, subgenital plate; 708 - *P. subfasciatus*, phallosome; 709 - *P. subfasciatus* male, clunial comb.

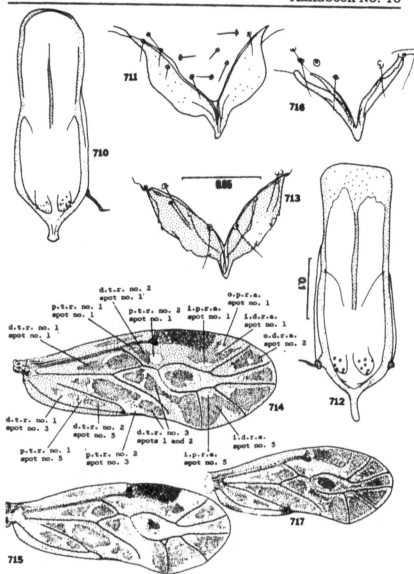

Figs. 710-717. *Peripsocus* spp. 710 - *P. alachuae*, phallosome; 711 - *P. alachuae* male, clunial process; 712 - *P. alboguttatus*, phallosome; 713 - *P. alboguttatus* male, clunial process; 714 - *P. maculosus* female, forewing (d. t. r and p. t. r. - dark and pale transverse rows; i. p. r. a. and i. d. r. a. - inner pale and dark radial arcs); 715 - *P. madescens* female, forewing; 716 - *P. madescens* male, clunial process; 717 - *P. minimus* female, forewing.

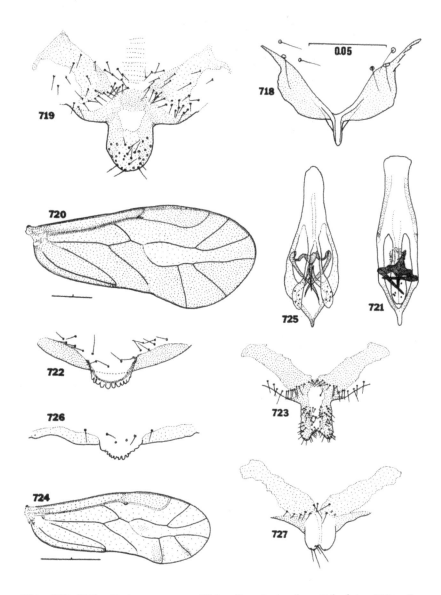

Figs. 718-727. *Peripsocus* spp. 719 - *P. potosi* subgenital plate; 720 - *P. phaeopterus*, forewing; 721 - *P. phaeopterus*, phallosome; 722 - *P. phaeopterus* male, clunial comb; 723 - *P. phaeopterus*, subgenital plate; 724 - *P. stagnivagus* female, forewing; 725 - *P. stagnivagus*, phallosome; 726 - *P. stagnivagus* male, clunial comb; 727 - *P. stagnivagus*, subgenital plate.

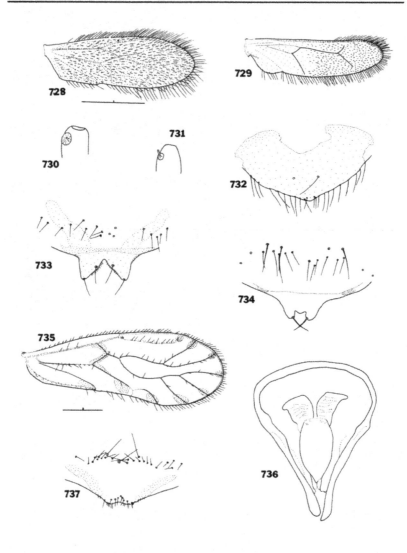

Figs. 728-737. 728 - *Archipsocus floridanus* macropterous female, forewing; 729 - *Archipsocus gurneyi* macropterous female, hindwing; 730 - *Archipsocopsis* sp., coeloconic sensillum of f6; 731 - *Archipsocus nomas*, trichoid sensillum of f6; 732 - *A. gurneyi*, subgenital plate; 733 - *Heterocaecilius* sp. , subgenital plate; 734 - *Pseudocaecilius citricola*, subgenital plate; 735 - *Trichopsocus acuminatus* female, forewing; 736 - *T. acuminatus*, phallosome; 737 - *T. acuminatus*, subgenital plate.

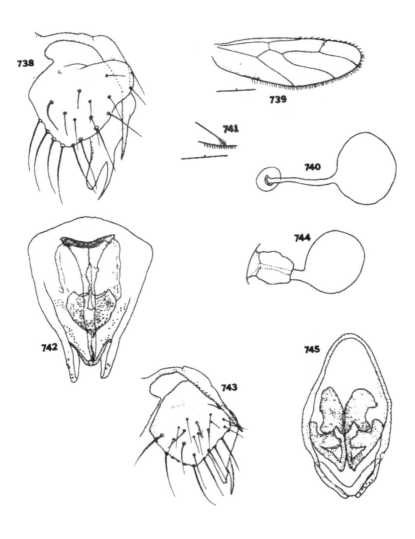

Figs. 738-745. *Trichopsocus* and *Aaroniella* spp. 738 - *T. acuminatus*, ovipositor valvulae; 739 - *T. acuminatus*, hindwing; 740 - *T. acuminatus*, spermatheca; 741 - *T. dalii*, distal end of vein Cu1 on margin in hindwing showing brown spot on both sides of vein; 742 - *T. dalii*, phallosome; 743 - *T. dalii*, ovipositor valvulae; 744 - *T. dalii*, spermatheca; 745 - *A. maculosa*, phallosome.

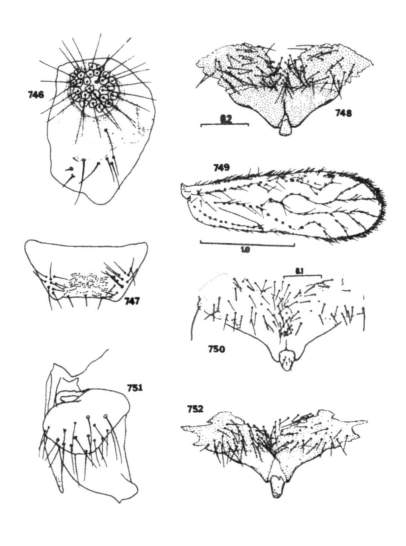

Figs. 746-752. *Aaroniella* spp. 746 - *A. maculosa* male, paraproct; 747 - *A. maculosa* male, epiproct; 748 - *A. maculosa*, subgenital plate; 749 - *A. achrysa* female, forewing; 750 - *A. achrysa*, subgenital plate; 751 - *A. achrysa*, ovipositor valvulae; 752 - *A. badonneli*, subgenital plate.

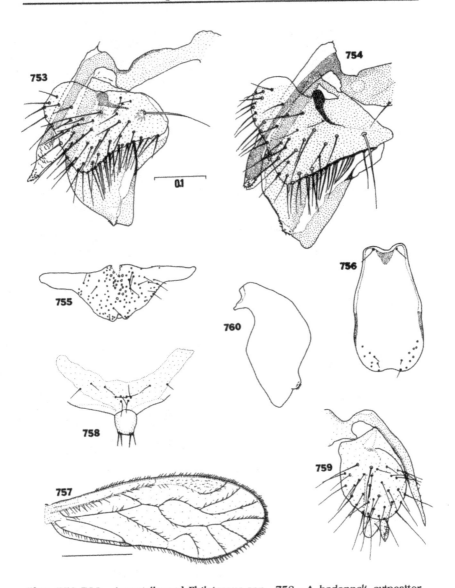

Figs. 753-760. *Aaroniella* and *Philotarsus* spp. 753 - *A. badonnelt*, ovipositor valvulae; 754 - *A. maculosa*, ovipositor valvulae; 755 - *P. kwakiutl*, hypandrium; 756 - *P. kwakiutl* male, epiproct; 757 - *P. kwakiutl* female, forewing; 758 - *P. kwakiutl*, subgenital plate; 759 - *P. kwakiutl*, ovipositor valvulae; 760 - *P. picicornis*, v2.

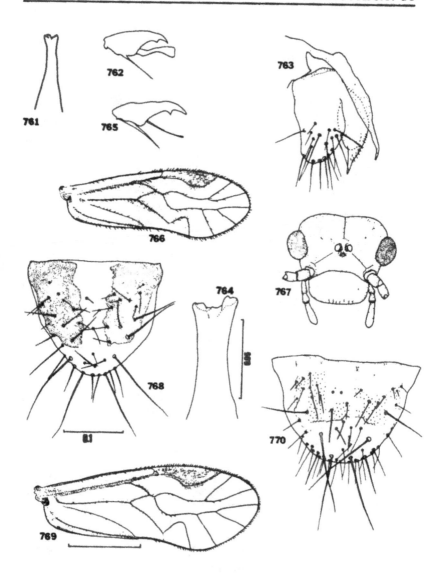

Figs. 761-770. Elipsocids. 761 - *Cuneopalpus cyanops*, lacinial tip; 762 - *C. cyanops*, pretarsal claw; 763 - *C. cyanops*, ovipositor valvulae; 764 - *Elipsocus guentheri*, lacinial tip; 765 - *E. guentheri*, pretarsal claw; 766 - *Elipsocus abdominalis* female, forewing; 767 - *E. guentheri* female, epiproctal color pattern; 768 - *Elipsocus hyalinus* female, forewing; 770 - *E. hyalinus* female, epiproctal color pattern.

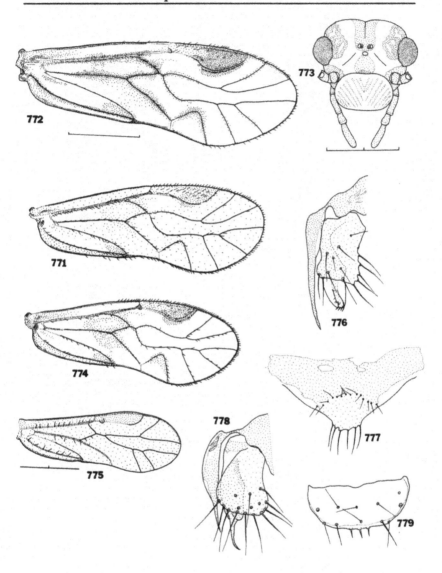

Figs. 771-779. Elipsocids. 771 - *Elipsocus moebiusi* female, forewing; 772 - *Elipsocus obscurus* female, forewing; 773 - *Elipsocus pumilis* female, head color pattern; 774 - *E. pumilis* female, forewing; 775 - *Nepiomorpha peripsocoides*, macropterous female, forewing; 776 - *N. peripsocoides*, ovipositor valvulae; 777 - *N. peripsocoides*, subgenital plate; 778 - *Palmicola aphrodite*, ovipositor valvulae; 779 - *P. aphrodite* female, epiproct.

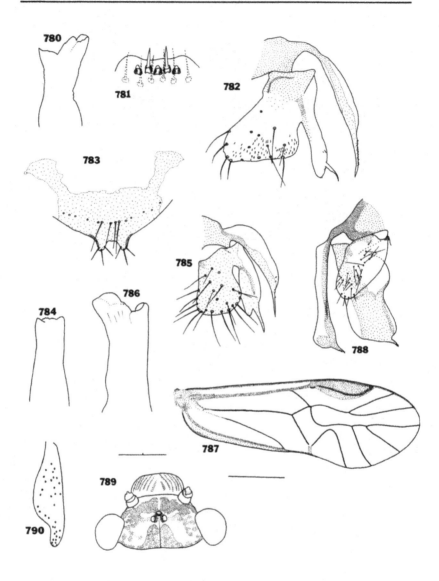

Figs. 780-790. 780 - *Propsocus pulchripennis*, lacinial tip; 781 - *P. pulchripennis*, distal inner labral sensilla; 782 - *P. pulchripennis*, ovipositor valvulae; 783 - *P. pulchripennis*, subgenital plate; 784 - *Reuterella helvimacula*, lacinial tip; 785 - *R. helvimacula*, ovipositor valvulae; 786 - *Mesopsocus unipunctatus*, lacinial tip; 787 - *M. unipunctatus* male, forewing; 788 - *M. unipunctatus*, ovipositor valvulae; 789 - *Mesopsocus laticeps*, color pattern of head; 790 - *M. laticeps*, external paramere.

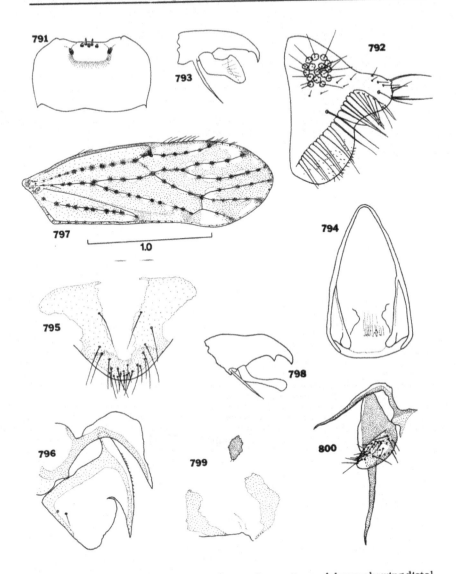

Figs. 791-800. Psocetae. 791 - *Amphigerontia montivaga*, labrum showing distal inner sensilla and sclerotic arc; 792 - *Trichadenotecnum alexanderae* female, paraproct; 793 - *Hemipsocus chloroticus*, pretarsal claw; 794 - *H. chloroticus*, phallosome; 795 - *H. chloroticus*, subgenital plate; 796 - *H. chloroticus*, ovipositor valvulae; 797 - *Hemipsocus pretiosus*, forewing; 798 - *Lichenomima sparsa*, pretarsal claw; 799 - *Lichenomima coloradensis* female, sclerotizations of ninth sternum; 800 - *L. coloradensis*, ovipositor valvulae.

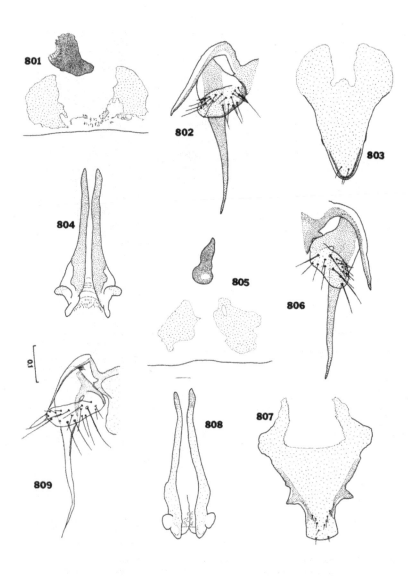

Figs. 801-809. *Lichenomima* and *Myopsocus* spp. 801 - *L. lugens* female, sclerotizations of ninth sternum; 802 - *L. lugens*, ovipositor valvulae; 803 - *L. lugens*, hypandrium; 804 - *L. lugens*, phallosome; 805 - *L. sparsa*, sclerotizations of ninth sternum.

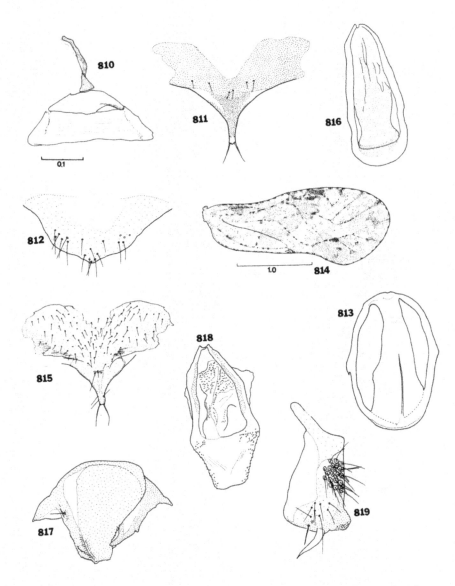

Figs. 810-819. *Myopsocus* and *Atropsocus* spp. 810 - *M. antillanus* female sclerotizations of ninth sternum; 811 - *M. eatoni*, subgenital plate; 812 - *M. eatoni*, hypandrium; 813 - *M. eatoni*, phallosome; 814 - *M. minutus* female, forewing; 815 - *M. minutus*, subgenital plate; 816 - *M. minutus*, phallosome; 817 - *A. atratus*, hypandrium; 818 - *A. atratus*, phallosome; 819 - *A. atratus* male, paraproct.

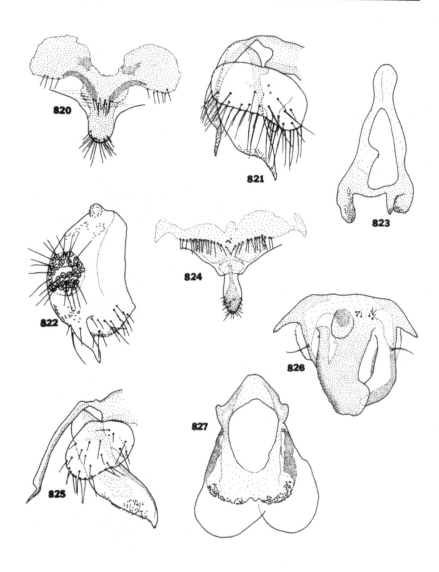

Figs. 820-827. Psocini. 820 - *Atropsocus atratus*, subgenital plate; 821 - *A. atratus*, ovipositor valvulae; 822 - *Hyalopsocus striatus* male, paraproct; 823 - *Hyalopsocus floridanus*, phallosome; 824 - *H. floridanus*, subgenital plate; 825 - *H. floridonus*, ovipositor valvulae; 826 - *Psocus crosbyi*, hypandrium; 827 - *P. crosbyi*, phallosome.

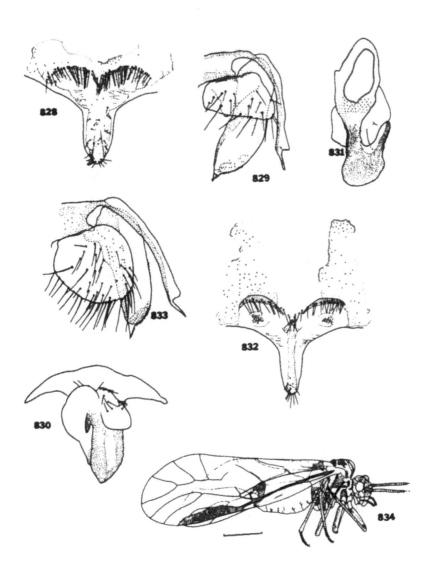

Figs. 828-834. *Psocus* and *Camelopsocus* spp. 828 - *P. crosbyi*, subgenital plate; 829 - *P. crosbyi*, ovipositor valvulae; 830 - *P. leidyi*, hypandrium; 831 - *P. leidyi*, phallosome; 832 - *P. leidyi*, subgenital plate; 833 - *P. leidyi*, ovipositor valvulae; 834 - *C. bactrianus* male, habitus.

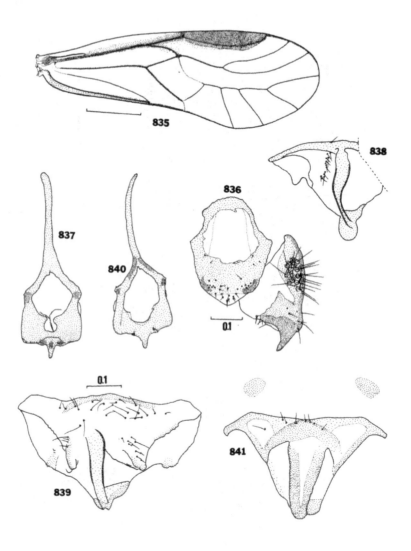

Figs. 835-841. *Camelopsocus* and *Indiopsocus* spp. 835 - *C. bactrianus* male, forewing; 836 - *C. bactrianus* male, epiproct and paraproct; 837 - *C. hiemalis*, phallosome; 838 - *C. monticolus*, hypandrium; 839 - *C. similis*, hypandrium; 840 - *C. tucsonensis*, phallosome; 841 - *I. bisignatus*, hypandrium.

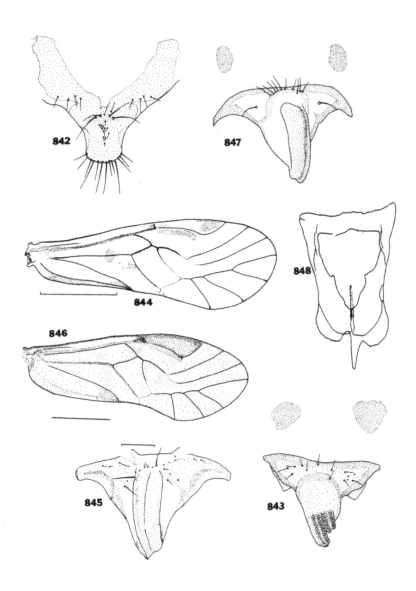

Figs. 842-848. *Indiopsocus* spp. 842 - *I. bisignatus*, subgenital plate; 843 - *I. ceterus*, hypandrium; 844 - *I. coquilletti* male, forewing; 845 - *I. coquilletti*, hypandrium; 846 - *I. infumatus* female, forewing; 847 - *I. infumatus*, hypandrium; 848 - *I. infumatus*, phallosome.

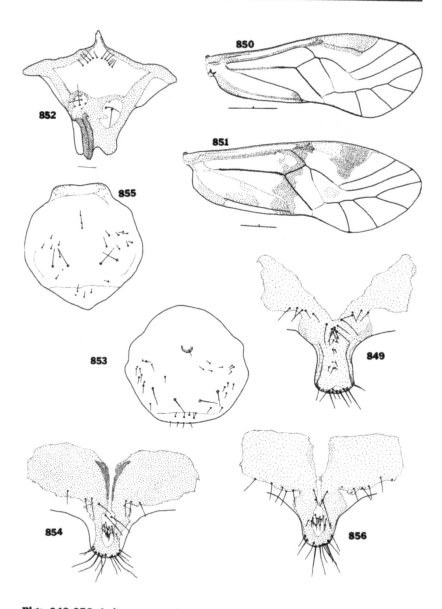

Figs. 849-856. *Indiopsocus* and *Loensia* spp. 849 - *I. infumatus*, subgenital plate; 850 - *I. campestris* male, forewing; 851 - *I. texanus* female, forewing; 852 - *L. conspersa*, hypandrium; 853 - *L. maculosa* male, epiproct; 854 - *L. maculosa*, subgenital plate; 855 - *L. moesta* male, epiproct; 856 - *L. moesta*, subgenital plate.

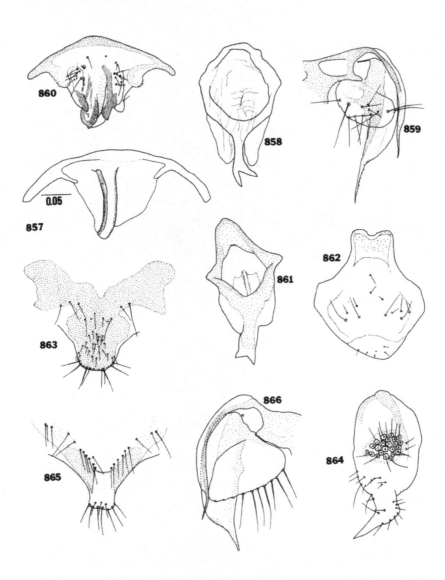

Figs. 857-866. *Ptycta* and *Steleops* spp. 857 - *P. lineata*, hypandrium; 858 - *P. lineata*, phallosome; 859 - *P. lineata*, ovipositor valvulae; 860 - *P. polluta*, hypandrium; 861 - *P. polluta*, phallosome; 862 - *P. polluta*, epiproct; 863 - *P. polluta*, subgenital plate; 864 - *S. lichenatus* male, paraproct; 865 - *S. elegans*, subgenital plate; 866 - *S. elegans*, ovipositor valvulae.

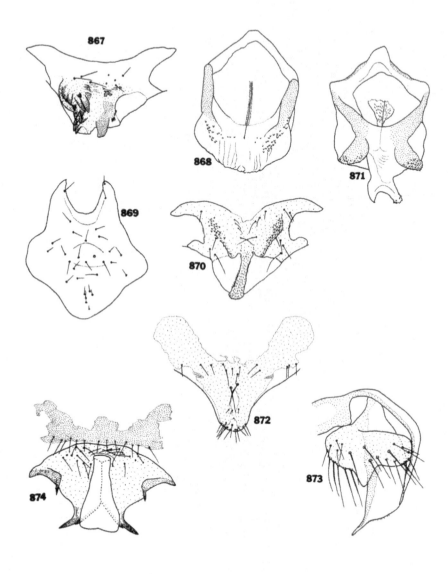

Figs. 867-874. *Steleops* and *Trichadenotecnum* spp. 867 - *S. elegans*, hypandrium; 868 - *S. elegans*, phallosome; 869 - *S. elegans* male, epiproct; 870 - *S. lichenatus*, hypandrium; 871 - *S. lichenatus*, phallosome; 872 - *S. lichenatus*, subgenital plate; 873 - *S. lichenatus*, ovipositor valvulae.

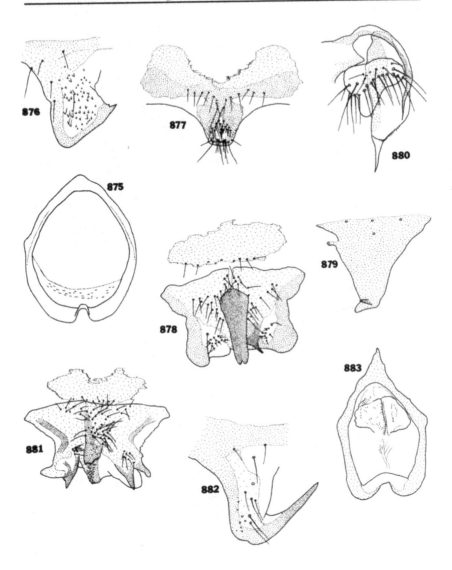

Figs. 875-883. *Trichadenotecnum* spp. 875 - *T. alexanderae*, phallosome; 876 - *T. alexanderae* male, clunial process; 877 - *T. castum*, subgenital plate; 878 - *T. desolatum*, hypandrium; 879 - *T. desolatum* male, clunial process; 880 - *T. desolatum*, ovipositor valvulae; 881 - *T. majus*, hypandrium; 882 - *T. majus* male, clunial process; 883 - *T. majus*, phallosome.

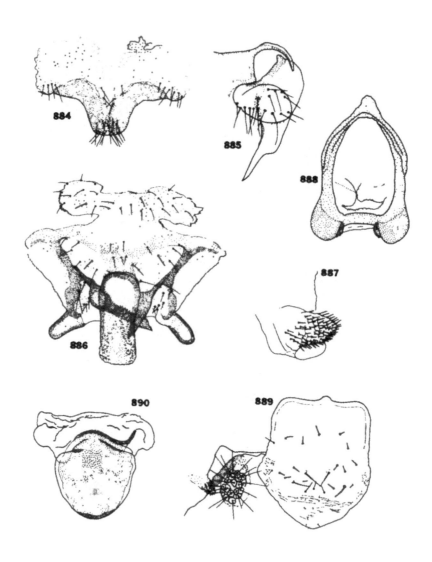

Figs. 884-890. *Trichadenotecnum* spp. 884 - *T. majus*, subgenital plate; 885 -
T. pardus, ovipositor valvulae; 886 - *T. quaesitum*, hypandrium; 887 - *T. quaesitum*
male, clunial process; 888 - *T. quaesitum*, phallosome; 889 - *T. quaesitum* male,
epiproct and base of paraproct; 890 - *T. quaesitum* female, sclerotizations of ninth
sternum.

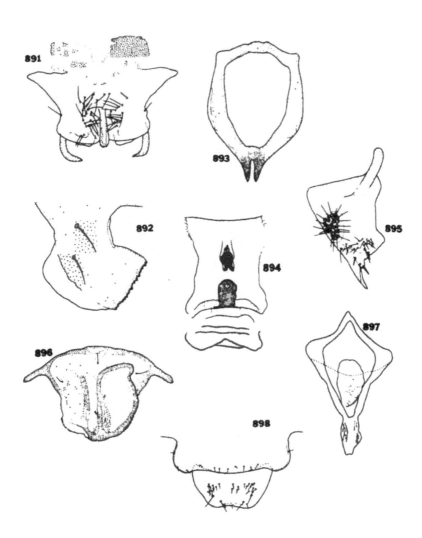

Figs. 891-898. *Trichadenotecnum* and *Metylophorus* spp. 891 - *T. slossonae*, hypandrium; 892 - *T. slossonae* male, clunial process; 893 - *T. slossonae*, phallosome; 894 - *M. novaescotiae* female, sclerotizations of ninth sternum; 895 - *M. novaescotiae* male, paraproct; 896 - *M. barretti*, hypandrium; 897 - *M. barretti*, phallosome; 898 - *M. barretti* male, epiproct and clunial shelf.

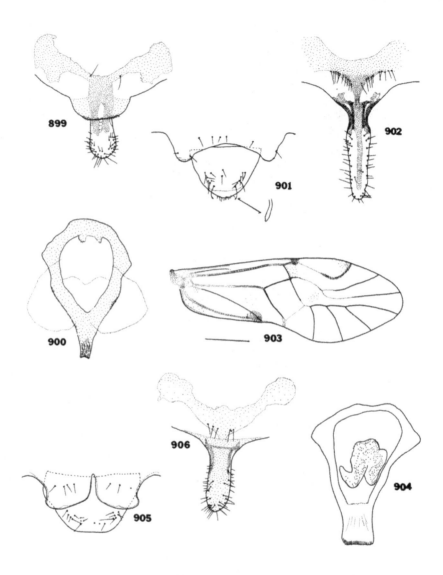

Figs. 899-906. *Metylophorus* spp. 899 - *M. barretti*, subgenital plate; 900 - *M. novaescotiae*, phallosome; 901 - *M. novaescotiae* male, epiproct and clunial shelf; 902 - *M. novaescotiae*, subgenital plate; 903 - *M. purus* female, forewing; 904 - *M. purus*, phallosome; 905 - *M. purus* male, epiproct and clunial shelf; 906 - *M. purus*, subgenital plate.

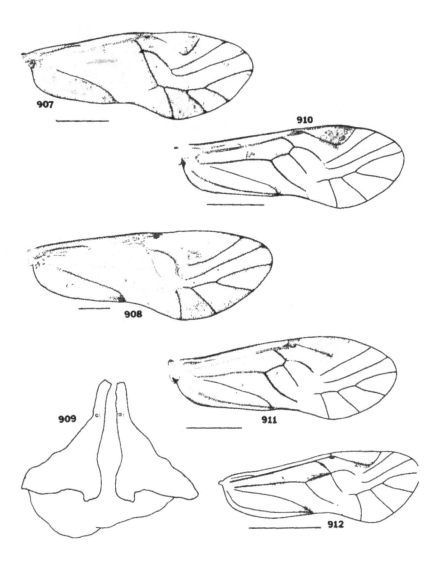

Figs. 907-912. Psocidae. 907 - *Cerastipsocus trifasciatus*, forewing; 908 - *Cerastipsocus venosus*, forewing; 909 - *Amphigerontia bifasciata*, phallosome; 910 - *Amphigerontia contaminata* female, forewing; 911 - *Amphigerontia montivaga* female, forewing; 912 - *Blaste cockerelli* female, forewing.

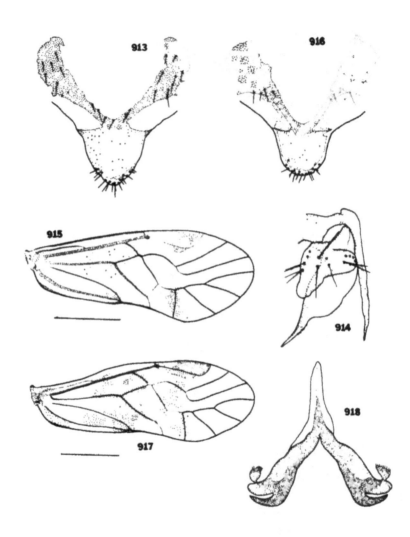

Figs. 913-918. *Blaste* spp. 913 - *B. cockerelli*, subgenital plate; 914 - *B. cockerelli*, ovipositor valvulae; 915 - *B. garciorum* female, forewing; 916 - *B. garciorum*, subgenital plate; 917 - *B. longipennis* female, forewing; 918 - *B. longipennis*, phallosome.

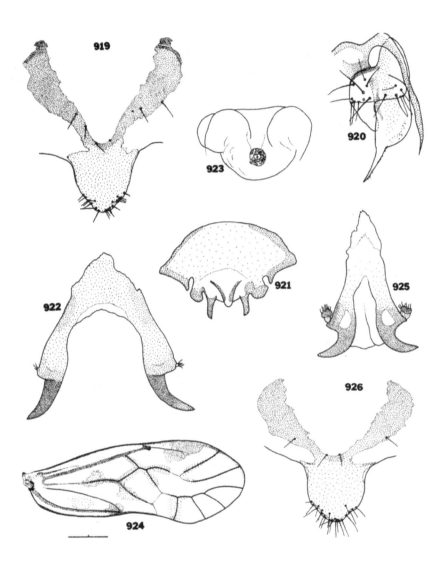

Figs. 919-926. *Blaste* spp. 919 - *B. longipennis*, subgenital plate; 920 - *B. longipennis*, ovipositor valvulae; 921 - *B. opposita*, distal segment of hypandrium; 922 - *B. opposita*, phallosome; 923 - *B. opposita* female, ninth sternum; 924 - *B. oregona* female, forewing; 925 - *B. oregona*, phallosome; 926 - *B. oregona*, subgenital plate.

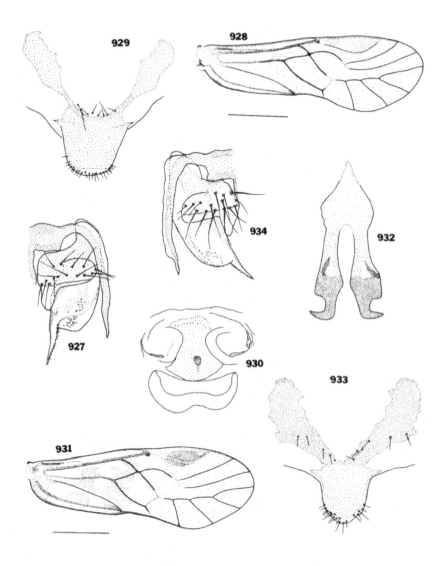

Figs. 927-934. *Blaste* spp. 927 - *B. oregona*, ovipositor valvulae; 928 - *B. persimilis* male, forewing; 929 - *B. posticata*, subgenital plate; 930 - *B. posticata* female, ninth sternum; 931 - *B. quieta* female, forewing; 932 - *B. quieta*, phallosome; 933 - *B. quieta*, subgenital plate; 934 - *B. quieta*, ovipositor valvulae.

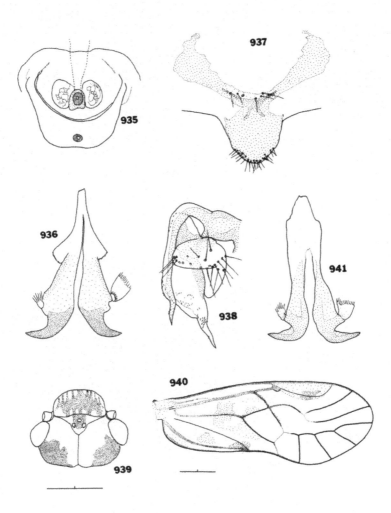

Figs. 935-941. *Blaste* spp. 935 - *B. quieta* female, ninth sternum; 936 - *B. subapterous*, phallosome; 937 - *B. subapterous*, subgenital plate; 938 - *B. subapterous*, ovipositor valvulae; 939 - *B. subquieta* female, color pattern of head; 940 - *B. subquieta* female, forewing; 941 - *B. subquieta*, phallosome.

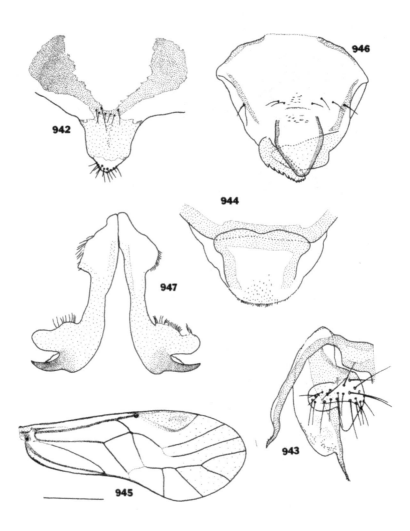

Figs. 942-947. *Blaste* and *Blastopsocus* spp. 942 - *B. subquieta*, subgenital plate; 943 - *B. subquieta*, ovipositor valvulae; 944 - *Blastopsocus lithinus* male, epiproct and margin of clunium; 945 - *B. lithinus* female, forewing; 946 - *B. lithinus*, distal segment of hypandrium; 947 - *B. lithinus*, phallosome.

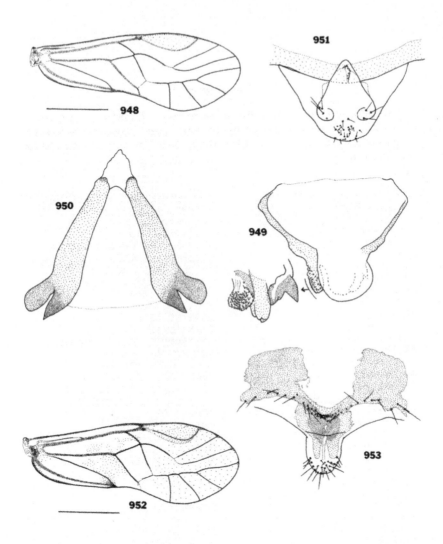

Figs. 948-953. *Blastopsocus* spp. 948 - *B. variabilis* female, forewing; 949 - *B. variabilis*, distal segment of hypandrium; 950 - *B. variabilis*, phallosome; 951 - *B. variabilis* male, epiproct and clunial margin; 952 - *B. semistriatus* female, forewing; 953 - *B. semistriatus*, subgenital plate.

Index

Note. Junior synonyms are excluded except for generic names and new and very recent species-name synonyms. Also excluded are names of species not found in the study area but only referred to for comparison.

Aaroniella (genus). 225, 226-228.
    achrysa. 225, 226.
    badonneli. 225, 226, 227
    (eertmoedi). 227.
    maculosa. 225, 226, 227.
Agave. 183.
Alaptus. 7.
(Albardia). 38.
Alder. 211, 228, 233.
Alfalfa. 160.
Algae. 6, 9.
(Amapsocus). 153.
Amber. 2, 64, 77, 315.
Amphientometae. 57, 98.
Amphientomidae. 5, 57, 58, 98.
Amphientominae. 101.
Amphientomini. 101.
Amphientomum (genus). 102.
Amphigerontia (genus). 294, 298-301.
    bifasciata. 294, 298.
    contaminata. 208, 299.
    infernicola. 300.
    montivaga. 295, 300.
    petiolata. 295, 301.
Amphigerontiinae. 251, 252, 294.
Amphipsocidae. 5, 107, 113, 115-117.
Andropogon. 126, 137, 210.
Anomopsocus (genus). 145, 153.
    amabilis. 145, 153.
Anoplura. 1.
(Antipsocus). 153.
Ants. 44, 86, 90, 95, 97, 98.
Arceuthobium. 229, 304.
Archipsocidae. 2, 5, 108, 215-219.

Archipsocopsis (genus). 3, 215, 216-218.
    frater. 215, 217.
    mendax. 217
    parvula. 215, 217.
Archipsocus (genus). 216, 218, 219.
    floridanus. 216, 218.
    gurneyi. 216, 218, 219.
    nomas. 216, 218, 219.
    panama. 218.
    puber. 218.
Archipsyllidae. 2.
Arctostaphylos. 205.
Artemisia. 93, 162, 235, 266, 305.
Asiopsocidae. 105, 113.
Asiopsocoidea. 105, 113.
Asiopsocus (genus). 113, 114.
    mongolicus. 113.
    sonorensis. 113, 314.
Aspens. 234.
Aster. 302.
Atropetae. 15, 16.
(Atropos). 44.
Atropsocus (genus). 253, 254.
    atratus. 254.
(Axinopsocus). 50.

Balliella (genus). 30, 31.
    calensis. 30, 31.
Bamboos. 60.
Banana (foliage). 26, 63, 164, 168, 173, 176, 197, 198.
Bee (nests). 42, 45, 90.
Belaphotroctes (genus). 66, 75-78.
    alleni. 68, 75.

badonnelî. 66, 67, 76.
ghesquierei. 66, 76.
hermosus. 68, 77.
simberloffi. 68, 77.
traegardhi. 75
Bertkauia (genus). 110.
crosbyana. 110, 111, 315.
lepicidinaria. 110, 111
lucifuga. 110.
Birches. 136, 226.
Birds. 5, 6, 7.
Blackberries. 158.
Blaste (genus) 294, 301-309, 315.
(californica). 304, 305.
cockerelli. 296, 302.
garciorum. 296, 302, 303.
(inornata). 307.
longipennis. 297, 302, 303.
opposita. 297, 302, 303.
oregona. 297, 302, 304, 315.
osceola. 296, 302, 305.
persimilis. 296, 302, 305.
posticata. 296, 302, 306.
quieta. 297, 302, 307.
subapterous. 297, 302, 308, 315.
subquieta. 297, 302, 308.
Blastopsocus (genus). 294, 309-312.
lithinus. 295, 309, 310.
semistriatus. 295, 309, 310.
variabilis. 295, 309, 311.
Booklice. 1.
Braconids. 7.
Bromeliads. 22, 53, 174, 249.
Broom (Scotch). 129, 211, 214, 229, 235, 236.
Buxus. 196.
(Cabarer). 232.
Cabbage (Palm). 128, 188, 199, 207, 218, 219, 226.
Cactus. 183.
Caecilietae. 105, 112.
Caeciliidae. 107, 113, 120.
Caecilioidea. 105, 115.
Caecilius (genus). 120, 125-143.
africanus. 121, 125.
africanus (group). 120, 125.
antillanus. 122, 126.
atricornis. 122, 133.
boreus. 124, 133, 314.
burmeisteri. 123, 134, 314.
caligonus (group). 121.
caloclypeus. 122, 134.
casarum. 122, 126.
confluens. 125, 129.

confluens (group). 121, 128.
croesus. 124, 135.
distinctus. 124, 131.
fasciatus (group). 121, 130.
flavidus. 124, 135, 314, 315.
flavidus (group). 121, 132.
fuscopterus. 123.
gonostigma. 124, 129.
graminis. 125, 130, 314.
hyperboreus. 123, 136, 314.
incoloratus. 124.
indicator. 121, 128.
insularum. 122, 127.
juniperorum. 124, 142.
lochloosae. 122, 137.
manteri. 122, 137.
maritimus. 123, 138, 315.
micanopi. 122, 138.
nadleri. 124, 131.
perplexus. 123, 139.
pinicola. 123, 139.
posticus. 121, 140.
posticus (group). 121, 140.
subflavus. 124, 142.
subflavus (group). 121, 141.
tamiami. 122, 140.
totonacus. 124, 132.

Caesalpinia. 167.
Camelopsocus (genus). 259, 265-267.
bactrianus. 259, 265.
hiemalis. 260, 265, 266.
monticolus. 260, 265, 266.
similis. 260, 265, 266.
tucsonensis. 260, 265, 267.
Capsicum. 183.
Carpinus. 228.
Carrizo. 168, 172.
Casuarina. 82, 127, 251.
Cattails. 137, 162, 167.
Cedar (western red). 189, 233, 236, 241.
Cerastipsocinae. 251.
Cerastipsocini. 251, 292.
Cerastipsocus (genus). 252, 292-294.
fuscipennis. 292.
trifasciatus. 292, 314.
venosus. 293.
(Cerastis). 292, 293.
Cerobasis (genus). 37, 38, 40-41.
annulata. 38, 40.
guestfalica. 38, 41, 315.
world species. 40
(Chaetopsocus). 193, 197.
Chamaedorea. 183.

Chaparral. 266, 271.
Cherry. 255.
Chrysanthemum. 197.
Citrus (trees). 127, 163, 213, 217, 222, 224.
Coccolobis. 127, 142.
Coccothrinax. 138.
Cocklebur. 160.
Coconut. 127.
Copeognatha. 8.
Cork. 250.
Corn (Indian). 137.
Cornus. 258.
Corrodentia. 8.
Cottonwood. 227.
Crataegus. 162.
Cronartium. 96.
Cucurbita. 183.
(Cuba). 41.
Cuneopalpus (genus). 230, 232.
    cyanops. 230, 232.
Cupressus. 135.
Cycads. 126.
Cymbidium. 213.
Cypress (bald). 287.
Cytisus. See broom, Scotch.

Dasydemella (genus). 115, 118.
Dasydemellidae. 105, 117.
(Deipnopsocus). 33.
Dendropogon. 36, 128.
Dianthus. 197.
Dichentomum (genus). 2.
Dieffenbachia. 131.
(Dolopteryx). 47.
Dorypteryx (genus). 46, 47, 49.
    domestica. 46, 47.
    pallida. 46, 47, 49.
Douglas (fir). 89, 139, 140, 186, 189, 205,
    233, 234, 235, 267.
Dwarf (mistletoe). See Arceuthobium.

Echinopsocinae. 16, 19, 26.
Echinopsocus (genus). 26.
Echmepteryx (genus). 19, 23-26.
    Echmepteryx (subgenus). 19, 23-25.
        hageni. 19, 23.
        intermedia. 19, 24.
        youngi. 19, 24.
    Thylacopsis (subgenus). 19, 25-26.
        falco. 19, 25.
        madagascariensis. 19, 26, 312.
        mihira. 25.
Ectopsocidae. 5, 109, 190-201.
Ectopsocopsis (genus). 190, 191, 193.

balli. 193
    cryptomeriae. 191, 193.
Ectopsocus (genus). 190, 191, 193-201.
    briggsi. 191, 193, 194, 315.
    californicus. 191, 194, 315.
    maindroni. 192, 195.
    meridionalis. 192, 196.
    petersi. 192, 196.
    pumilis. 192, 197.
    richardsi. 192, 197.
    salpinx. 192, 198.
    strauchi. 191, 199.
    thibaudi. 191, 199.
    titschacki. 192, 200.
    vachoni. 192, 200, 315.
Elipsocidae. 108, 229.
Elipsocus (genus). 230, 232-236.
    abdominalis. 231, 233, 312.
    guentheri. 230, 233, 234.
    hyalinus. 231, 234, 234, 312.
    moebiusi. 231, 233, 235.
    obscurus. 231, 233, 235.
    pumilis. 231, 233, 235.
    (westwoodi). 235.
Embidopsocinae. 65, 75.
Embidopsocus (genus). 65, 78-82.
    bousemani. 70, 78, 79.
    citrensis. 70, 78, 80.
    femoralis. 67, 79, 82, 312.
    laticeps. 70, 78, 79.
    luteus. 79.
    mexicanus. 70, 78, 80.
    needhami. 67, 79, 81.
    thorntoni. 70, 78, 81.
Eolachesillinae. 145, 152.
Epipsocetae. 103, 109.
Epipsocidae. 105, 109.
Epipsocus (genus). 109, 110.
    ciliatus. 110.
Euphoriella. 7.
Evergreens (broad-leaf). 143.

(Fabrella). 51.
Ferns. 25, 29, 224.
Ficus. 286.
(Fita). 51.
Formica. 90, 95, 97.
Fur (mammal). 5.

Galls (plant). 64.
(Gambrella). 50.
Grape. 114.
Graphocaecilini. 152.
Graphopsocus (genus). 119.

cruciatus. 119.

Hackberry. 227, 255.
Heliconia. 63.
Hemipsocidae. 5, 107, 243-245.
Hemipsocus (genus). 244-245.
	chloroticus. 244.
	pretiosus. 244.
Hemipteroids. 1, 2.
Hemlock (trees). 144, 189, 207, 210, 287.
Heterocaecilius (genus). 220, 221.
	minutus. 221.
(Heterolepinotus). 41.
(Heteropsocus). 33.
Hickory. 154.
Hives (bee). 42, 45.
Holly. 143.
(Holoneura). 241.
Homilopsocidea. 105, 144.
Hyalopsocus (genus). 253, 255-257.
	contrarius. 255.
	floridanus. 253, 255.
	striatus. 253, 255, 256.
(Hyperetes). 38.

Indiopsocus (genus). 260, 267-272.
	bisignatus. 261, 268.
	campestris. 261, 268.
	ceterus. 261, 268, 269.
	coquilletti. 261, 268, 270.
	infumatus. 260, 268, 271.
	(insulanus). 268.
	texanus. 261, 268, 271.

Jade (plant). 194.
Jeffrey (pine). 190.
Jojoba. 224, 266, 267.
Juniper. 76, 85, 88, 114, 126, 135, 139,
	142, 169, 171, 174, 186, 205, 212,
	266, 273, 301, 305, 306, 308.

Kaestneriella (genus). 202, 204-205.
	fumosa. 202, 204, 205.
	pilosa. 204.
	tenebrosa. 202, 204, 205.
Kodamatus (genus). 120.

(Labocoria). 241.
Lachesilla (genus). 145, 155-190.
	aethiopica. 151, 176.
	albertina. 147, 165, 314.
	alpejia. 150.
	andra. 148, 156.
	andra (group). 148, 156.

anna. 151, 166, 167.
arida. 148, 185.
arnae. 149, 157, 314.
bottimeri. 150, 166, 168.
centralis. 147, 163.
centralis (group). 146, 163.
chapmani. 151, 166, 168.
chiricahua. 185, 186.
cintalapa. 146, 164.
contraforcepeta. 151, 166, 169.
corona. 147, 165.
corona (group). 147, 164.
denticulata. 150, 170.
dona. 149, 158, 315.
forcepeta. 150, 166.
forcepeta (group). 150, 166, 170.
fuscipalpis. 147, 174.
fuscipalpis (group). 147, 174.
gracilis. 151, 166, 171.
greeni. 152, 176, 177.
jeanae. 148, 185, 186.
kathrynae. 151, 166, 172.
kola. 148, 159.
major. 150, 166, 172.
nita. 147, 185, 187.
nubilis. 149, 160.
nubiloides. 149, 161, 314.
nuptialis. 176.
pacifica. 152, 176, 177.
pallida. 152, 176, 178.
patzunensis (group). 146, 174.
pedicularia. 152, 156, 176, 179, 316.
pedicularia (group). 150, 175.
penta. 151, 166, 173.
perezi. 147, 164.
punctata. 149, 162.
quercus. 151, 176, 181.
rena. 152, 176, 182.
riegeli. 149, 183.
riegeli (group). 149, 183.
rufa. 148, 185, 188.
rufa (group). 147, 185.
sulcata. 146, 175.
tectorum. 151, 176, 182.
texcocana. 146, 189.
texcocana (group). 146, 189.
tropica. 149, 183, 184.
ultima. 149, 183, 185.
yakima. 148, 185, 189.
Lachesillidae. 5, 107, 145.
Lachesillinae. 145, 155.
(Lapithes). 110.
Larch. 134, 181, 211, 212, 242.
Larix. See Larch.

Larrea. 174.
Laurel (oak). 217, 218.
Laurocerasus. 102.
(Lenkoella). 114.
(Lepidilla). 28.
Lepidopsocidae. 15, 16.
Lepidopsocinae. 19, 23.
Lepinotus (genus). 38, 41-44.
    inquilinus. 38, 42.
    patruelis. 38, 42, 43.
    reticulatus. 38, 42, 43, 316.
Lepolepis (genus). 26, 27.
(Leptella). 239.
(Leptopsocus). 156.
Lichenomima (genus). 245, 246-249.
    coloradensis. 246, 247.
    conspersa. 246.
    lugens. 246, 247.
    sparsa. 246, 248.
    (virginiana). 247.
Lichens. 6, 9.
Liposcelididae (=Liposcelidae). 2, 5, 6, 57,
    64.
Liposcelidinae. 65, 82.
Liposcelis (genus). 2, 5, 65, 82-98.
    bicolor. 73, 89.
    bostrychophila. 75, 96, 316.
    brunnea. 70, 83.
    corrodens. 75, 96.
    decolor. 73, 90.
    deltachi. 72, 84.
    entomophila. 72, 85.
    formicaria. 73, 95, 315.
    fusciceps. 72, 85.
    hirsutoides. 72, 86.
    (kidderi). 92.
    (knullei). 91.
    lacinia. 74, 91.
    (liparus). 83.
    mendax. 74, 95.
    nasa. 74, 86.
    nigra. 74, 91.
    ornata. 72, 87.
    paeta. 74, 97.
    pallens. 74, 87.
    pallida. 72, 88.
    pearmani. 74, 92.
    prenolepidis. 75, 97.
    rufa. 74, 92.
    silvarum. 74, 93, 315.
    (stimulans). 92.
    triocellata. 74, 94.
    villosa. 72, 88.
Liquidambar. 238.

Lithoseopsis (genus). 98, 101-102.
    hellmani. 99, 101.
    hystrix. 99, 101, 102.
Locust (honey). 228.
Loensia (genus). 264, 273-276.
    conspersa. 264, 273.
    fasciata. 264, 273.
    maculosa. 264, 273, 274.
    moesta. 264, 273, 274.
Loneura (genus). 112.
    crenata. 112.
Lophioneuridae. 1.

Madrone. 201.
Mallophaga. 1.
Mamoncillo. 102.
Mangroves. 78, 85, 96, 142, 167, 199, 207,
    277.
Maoripsocus. 120.
Maple (trees). 89, 211, 227, 228, 233, 237,
    241.
Matsumuraiella (genus). 117, 118.
Mesopsocidae. 108, 240.
Mesopsocus (genus). 240-243.
    immunis. 241.
    laticeps. 240, 241, 242.
    unipunctatus. 241, 242.
Mesquite. 85.
Metylophorini. 251, 252, 288.
Metylophorus (genus). 289-292.
    barretti. 289.
    (hoodi). 290.
    nebulosus. 289.
    novaescotiae. 289, 290.
    purus. 289, 291.
(Micropsocus). 193.
Moss (Spanish). See Dendropogon.
Musaceous (plants). 22.
Mymarids. 7.
Myopsocidae. 5, 107, 245.
(Myopsocnema). 38.
Myopsocus (genus). 245, 249-251.
    antillanus. 246, 249.
    eatoni. 246, 249, 250.
    minutus. 246, 249, 250.
    undulosus. 249.
Myrica. 143, 217.
Myrmicodipnella (genus). 37, 44.
    aptera. 37, 44.
Myrtle (wax). See Myrica.

Nanolachesilla (genus). 146, 153-155.
    hirundo. 146, 154.
    chelata. 146, 154.

Nanopsocetae. 58, 59.
Nanopsocus (genus). 59.
    oceanicus. 59.
Neolepolepis (genus). 19.
    caribensis. 19, 27.
    occidentalis. 20, 27.
    xerica. 28.
Neotoma. 19, 35.
Neptomorpha (genus). 231, 236-237.
    crucifera. 237.
    peripsocoides. 231, 237.
Nepticulomima (genus). 18, 21.
    sakantala. 21.
Nothophagus. 64.
Notiopsocus (genus). 114.
    simplex. 114.
(Nymphopsocus). 51.

Oak (trees, foliage, etc.). Page numbers are
    omitted to save space. See the "distri-
    bution and habitat" sections for Lipos-
    celididae and all Psocomorph families.
Oats. 97.
(Ocellataria). 51.
(Ocelloria). 51.
(Onychotroctes). 59.
Orange (trees). 173, 251.
Orchid (plants, etc.). 131, 162, 170, 174,
    183, 198, 199, 200, 213, 244.

Pachytroctes (genus). 58, 60-62.
    aegyptius. 59, 61.
    neoleonensis. 59, 61.
Pachytroctidae. 57, 59.
(Paleotroctes). 64.
Palm (trees, foliage). Page numbers are
    omitted to save space. See the "distri-
    bution and habitat" sections for all
    families of Troglomorpha and
    Psocomorpha plus Pachytroctidae and
    Liposcelididae.
Palmicola (genus). 231, 237-238.
    aphrodite. 231, 237.
    solitaria. 232, 237, 238.
(Paradoxenus). 82.
(Paradoxides). 82.
Parapsocus (genus). 2.
Pea (black-eyed). 184.
Peach. 160.
Pepper (Capsicum). 183.
Peridermium. 96.
Perientominae. 18, 21.
Peripsocidae. 109, 201.
Peripsocus (genus). 202, 205-215.

alachuae. 204, 209, 210.
alboguttatus. 203, 209, 210.
maculosus. 204, 209, 211.
madescens. 204, 209, 211.
madidus. 202, 206.
minimus. 203, 209, 212.
pauliani. 203, 206, 207.
phaeopterus. 204, 206, 213.
potosi. 203, 209, 212.
(quadrifasciatus). 208.
reductus. 203, 206, 207.
stagnivagus. 204, 214.
subfasciatus. 202, 206, 208.
Philotarsidae. 5, 108, 224.
Philotarsus (genus). 225, 228-229.
    kwakiutl. 225, 228.
    picicornis. 225, 228, 229.
(Phlotodes). 249.
Phragmites. 94, 129, 217.
Picea. 211.
(Pictopsocus). 255.
Pine (trees, foliage). Page numbers are
    omitted to save space. See the "dis-
    tribution and habitat" sections for
    Lepidopsocidae, Liposcelididae and the
    Psocomorph families.
Pineapple. 200, 307.
Platanus. 91.
Pleurococcine (algae). 6.
Podocarpus. 142.
Polyporus. 97, 315.
Prionoglaridae. 16, 54.
Proentomum (genus). 18, 21-22, 312.
    personatum. 18, 21, 22, 312.
Prolachesilla (genus). 146, 155.
    mexicana. 155.
    terricola. 146, 155.
Pronotiopsocus (genus). 113, 114-115.
    amazonicus. 115.
Propsocus (genus). 230, 238-239.
    pulchripennis. 230, 238.
Prosopis. 174.
Pseudocaecilidae. 3, 5, 108, 220.
Pseudocaecilius (genus). 220, 221-222.
    citricola. 220, 221.
    tahitiensis. 220, 221, 222.
(Pseudopsocus Hong). 2.
(Pseudopsocus Chapman). 153.
Pseudopsocus Kolbe (genus). 239.
Pseudorypteryx (genus). 47, 50-51.
    mexicana. 47, 50, 51.
Pseudotsuga. See Douglas (fir).
(Psocathropos). 51.
Psocatropetae. 15, 45.

Psocatropos (genus). 47, 50.
  (floridanus). 51.
  microps. 47, 51.
Psocetae. 105, 243.
(Psochus). 257.
Psocidae. 3, 5, 6, 107, 251.
Psocina. 8.
Psocinae. 251, 252.
(Psocinella). 50.
Psocini. 251, 252, 253.
Psocomorpha. 3, 6, 11, 103.
Psocus (genus). 253, 257-259.
  bipunctatus. 257.
  crosbyi. 257.
  leidyi. 257, 258.
Psoquilla (genus). 30, 33.
  infuscata. 33.
  marginepunctata. 30, 33.
Psoquillidae. 5, 16, 29.
Psyllipsocus (genus). 7, 47, 51.
  oculatus. 47, 52.
  ramburi. 47, 53, 316.
Psyllotroctes. 60.
Ptenopsila. 115.
(Pterodela). 156.
Pteroxanium (genus). 19, 28-29.
  kelloggi. 19, 28, 312, 315.
Ptiloneuridae. 105, 112.
(Ptilopsocus). 116.
Ptycta (genus). 259, 264, 276-278.
  haleakalae. 276.
  lineata. 264, 276.
  polluta. 265, 276, 277.
Ptyctini. 251, 252, 259.

Quercus. See oak (trees).

Rapanea. 276.
Rat (pack). 27, 35, 89, 94. See also
  Neotoma.
Rattan. 198.
Rhododendron. 224.
(Rhyopsocopsis). 33.
Rhyopsocus (genus). 30, 33-37.
  bentonae. 30, 34.
  disparilis. 31, 34.
  eclipticus. 30, 34, 35.
  micropterus. 30, 34, 35.
  (pescadori). 36.
  (phillipsae). 35.
  (squamosus). 36.
  texanus. 31, 34, 36.
Rubus. 158.
Rust (fungi). 96.

Sagebrush. See Artemisia.
Salvia. 235, 266.
Sambucus. 258.
Scolopama (genus). 26.
Sedges. 126.
Soa (genus). 18, 22, 312.
  flaviterminata. 18, 22, 312.
Sorghum. 183.
Sparrow (house). See Passer.
Speleketor (genus). 54-56.
  flocki. 54, 55.
  irwini. 54, 55.
  pictus. 54, 55.
Sphaeropsocidae. 57, 63.
Sphaeropsocopsis (genus). 63-64.
  argentinus. 63.
  chilensis. 63.
Sphaeropsocus (genus). 63, 64.
  kunowii. 64.
Spruce (trees, foliage). See also Picea. 134,
  136, 139, 140, 177, 187, 189, 207,
  211, 234.
Steleops (genus). 263, 278-279.
  elegans. 264, 278.
  lichenatus. 263, 278, 279.
  punctipennis. 278.
Stenopsocidae. 105, 113, 118.
(Stenotroctes). 78.
Stimulopalpus (genus). 98, 99.
  japonicus. 99.
Strangler (fig). 226.
Sumac. 211, 235.
Sycamore. See also Platanus. 91, 227.

Tamarack. See Larch.
Tapinella (genus). 59, 62-63.
  formosana. 62.
  maculata. 59, 62.
(Tasmanopsocus). 28.
Teliapsocus (genus). 105, 117-118.
  conterminus. 105, 118, 315.
(Teratopsocus). 119.
(Terracaecilius). 156.
Thalia. 217.
Thatch (palm). 142.
Thrinax. See Thatch (palm).
Thuja. 140, 142, 214, 299.
Thylacella (genus). 18, 20-21.
  cubana. 18, 20.
  eversiana. 20.
Thylacellinae. 18, 20.
Thylacopsis (subgenus). See Echmepteryx.
(Tichobia). 38.
(Tiliapsocus). 255.

(Titella). 292.
(Trichadenopsocus). 280.
Trichadenotecnum (genus). 2, 4, 261, 280-
288.
    alexanderae. 4, 263, 280, 281.
    castum. 263, 280, 281.
    circularoides. 262, 280, 283, 312.
    desolatum. 262, 280, 283, 284.
    innuptum. 263, 280, 282.
    majus. 262, 280, 283, 285.
    merum. 263, 280, 282.
    pardus. 262, 280, 283, 286, 312.
    quaesitum. 263, 280, 286.
    sexpunctatum. 280.
    slossonae. 262, 280, 287.
Trichopsocidae. 108, 223.
Trichopsocus (genus). 223-224.
    acuminatus. 223, 312.
    (clarus). 223.
    dalii. 223, 224.
(Tricladellus). 238.
(Tricladus). 383.
(Trigonoscellscus). 78.
(Troctes). Junior synonym of Liposcelis,
    q.v.
(Trocticus). 241.

Troctomorpha. 11, 57.
Trogiidae. 5, 6, 15, 16, 37.
Trogiomorpha. 11, 15.
Trogium (genus). 37, 38, 44-45.
    pulsatorium. 38, 44.
Typha. See also cattails. 34, 63, 126, 137,
    161, 167.

(Udamolepis). 20.

Vaccinium. 211.
Vitis. 257.
(Vulturops). 50.

Washingtonia. 55.

Xanthocaecilius (genus). 120, 143-144.
    quillayute. 120, 143, 314.
    sommermanae. 120, 143, 144.
Yams. 61.
Yucca. 34, 53, 62, 85, 86, 88, 114. 135, 169,
    174.
Zea. 137, 183, 193.

(Zimia). 38.

Milton Keynes UK
Ingram Content Group UK Ltd.
UKHW021903071024
449327UK00021B/1609